DRUG STUDIES IN THE ELDERLY

Methodological Concerns

DRUG STUDIES IN THE ELDERLY

Methodological Concerns

Edited by

Neal R. Cutler, M.D.

Director, Department of Geriatrics
Cedars-Sinai Medical Center
University of California–Los Angeles School of Medicine
Los Angeles, California

and

Prem K. Narang, Ph.D.

Supervisor, Clinical Pharmacokinetics Research Laboratory
Pharmacy Department
Clinical Center
National Institutes of Health
Bethesda, Maryland

PLENUM MEDICAL BOOK COMPANY
New York and London

Library of Congress Cataloging in Publication Data

Drug Studies in the elderly.

Includes bibliographies and index.
1. Geriatric pharmacology — Research — Methodology. 2. Drugs — Testing. 3. Drugs — Metabolism — Age factors. I. Cutler, Neal R. II. Narang, Prem K. [DNLM: 1. Clinical Trials — in old age. 2. Pharmacology, Clinical — in old age. QV 38 D7945]
RC953.7.D767   1986   615.5′8′088056   86-18738
ISBN-13: 978-1-4684-1255-0   e-ISBN-13: 978-1-4684-1253-6
DOI: 10.1007/978-1-4684-1253-6

© 1986 Plenum Publishing Corporation
Softcover reprint of the hardcover 1st edition 1986
233 Spring Street, New York, N.Y. 10013

Plenum Medical Book Company is an imprint of Plenum Publishing Corporation

To Gerald Joy Dunitz, Alexander, Samantha,

&

Vishan Dutt and Krishna Narang

CONTRIBUTORS

DARRELL R. ABERNETHY, M.D., PH.D., Division of Clinical Pharmacology, Brown University; and Department of Medicine, Roger Williams General Hospital, Providence, Rhode Island 02908

WILLIAM B. ABRAMS, M.D., Scientific Development, Merck Sharp & Dohme Research Laboratories, West Point, Pennsylvania 19485; and Jefferson Medical College of Thomas Jefferson University, Philadelphia, Pennsylvania 19107

DONALD B. CALNE, M.D., Division of Neurology, Department of Medicine, Health Sciences Centre Hospital, University of British Columbia, Vancouver, British Columbia, Canada V6T 1W5

DAVID G. COVELL, PH.D., Laboratory of Mathematical Biology, National Cancer Institute, National Institutes of Health, Bethesda, Maryland 20892

NEAL R. CUTLER, M.D., Department of Geriatrics, Cedars–Sinai Medical Center, University of California–Los Angeles School of Medicine, Los Angeles, California 90048

MARINOS C. DALAKAS, M.D., Infectious Diseases Branch, National Institute of Neurological and Communicative Disorders and Stroke, National Institutes of Health, Bethesda, Maryland 20892

WILLIAM B. ERSHLER, M.D., Division of Hematology and Geriatrics, Department of Medicine, University of Wisconsin, Madison, Wisconsin 53706

ALAN FORREST, PHARM. D., Department of Clinical Pharmacy, School of Pharmacy, University of Maryland at Baltimore, Baltimore, Maryland 21201

BARBARA R. HELLER, R.N., ED.D., F.A.A.N., Department of Nursing Education, Administration and Health Policy, University of Maryland School of Nursing, Baltimore, Maryland 21201

JANICE G. HITZHUSEN, M.D., Divisions of Clinical Pharmacology and Geriatrics, University of Massachusetts Medical Center, Worcester, Massachusetts 01605

R. GARY HOLLENBECK, PH.D., School of Pharmacy, University of Maryland, The Center for the Study of Pharmacy and Therapeutics for the Elderly, Baltimore, Maryland 21201

BRIAN F. JOHNSON, M.D., Divisions of Clinical Pharmacology and Geriatrics, University of Massachusetts Medical Center, Worcester, Massachusetts 01605

K.-C. KHOO, PH.D., Department of Drug Metabolism, Hoffman-La Roche, Inc., Nutley, New Jersey 07110

HARVEY G. KLEIN, M.D., Department of Transfusion Medicine, Clinical Center, National Institutes of Health, Bethesda, Maryland 20892

ARTO LAIHINEN, M.D., Division of Neurology, Department of Medicine, Health Sciences Centre Hospital, University of British Columbia, Vancouver, British Columbia, Canada V6T 1W5

PETER P. LAMY, PH.D., School of Pharmacy, University of Maryland, The Center for the Study of Pharmacy and Therapeutics for the Elderly, Baltimore, Maryland 21201

PAUL LEBER, M.D., Division of Neuropharmacological Drug Products, Food and Drug Administration, Rockville, Maryland 20857

LAWRENCE J. LESKO, PH.D., School of Pharmacy, Clinical Pharmacokinetics Laboratory, University of Maryland at Baltimore, Baltimore, Maryland 21201

CHO-MING LOI, PH.D., Idaho State University College of Pharmacy, Pocatello, Idaho 83209; and Clinical Pharmacology and Gerontology Unit, Veterans Administration Medical Center, Boise, Idaho 83702

J. W. MASSARELLA, PH.D., Department of Drug Metabolism, Hoffman-La Roche, Inc., Nutley, New Jersey 07110

PREM K. NARANG, PH.D., Clinical Pharmacokinetics Research Laboratory, Pharmacy Department, Clinical Center, National Institutes of Health, Bethesda, Maryland 20892

WILLIAM Z. POTTER, M.D., PH.D., Section on Clinical Pharmacology, Laboratory of Clinical Science, National Institute of Mental Health, Bethesda, Maryland 20892

MAUREEN E. POWER, R.N., M.P.H., Aging Research Nursing Service, Clinical Center, National Institutes of Health, Bethesda, Maryland 20205

ELIZABETH J. READ, M.D., Department of Transfusion Medicine, Clinical Center, National Institutes of Health, Bethesda, Maryland 20892

BRUCE E. RODDA, Clinical Biostatistics and Research Data Systems, Merck Sharp & Dohme Research Laboratories, Rahway, New Jersey 07065

DAVID S. ROFFMAN, PHARM.D., Department of Clinical Pharmacy, School of Pharmacy, University of Maryland at Baltimore, Baltimore, Maryland 21201

MATTHEW V. RUDORFER, M.D., Section on Clinical Pharmacology, Laboratory of Clinical Science, National Institute of Mental Health, Bethesda, Maryland 20892

GEORGEANNE COX SANTOLLA, R.N., M.S., American Healthcare Institute, Silver Spring, Maryland 20910

ALEXANDER M. M. SHEPHERD, M.D., PH.D., Departments of Pharmacology and Medicine, Division of Clinical Pharmacology, University of Texas Health Science Center at San Antonio, San Antonio, Texas 78284

WILLIAM H. THEODORE, M.D., Clinical Epilepsy Section, National Institute of Neurological and Communicative Disorders and Stroke, National Institutes of Health, Bethesda, Maryland 20205

ROBERT E. VESTAL, M.D., Clinical Pharmacology and Gerontology Unit, Veterans Administration Medical Center, Boise, Idaho, 83702; and Division of Gerontology and Geriatric Medicine, Department of Medicine, University of Washington School of Medicine, Seattle, Washington 98195

THOMAS M. VOGT, M.D., M.P.H., Center for Health Research, Kaiser Permanente, Northwest Region, Portland, Oregon 97215

MARC E. WEKSLER, M.D., Division of Geriatrics, Department of Medicine, Cornell University Medical College, New York, New York 10021

PREFACE

Clinical trials are the most definitive tool for evaluation of the applicability of clinical investigations. The main objective of clinical investigations is to assess the potential value of a therapeutic entity in the treatment or prophylaxis of a disease or a condition. It is also deemed necessary at this stage to obtain information regarding the undesirable side effects, associated risks, and their interrelationship with clinical assessments. Most of these clinical investigations conform, in some form or fashion, to the guidelines adopted by the Food and Drug Administration (FDA) for a given class of compounds. Clinical investigations in the past have not included specific studies in special or subpopulations, e.g., the elderly. Because of an ever-increasing elderly population, newer policies for clinical investigations are now being debated with the recognition of enhanced drug sensitivity in this special population. This key research activity can lead not only to improved health care in the elderly but also to control of its costs.

For an appropriate and beneficial clinical use of a therapeutic entity, it is important that drug-dosing decisions be based not only on the information gained from the appropriately designed and executed clinical trials, but also on the wealth of knowledge available from the ongoing basic research investigations in the fields of immunology, physiology, endocrinology, hematology, and gerontology. The confounding role of environment, genetics, social activity, and physiological changes in the disease and aging process needs to be investigated so as to define variables that play a role in altering dispositions of therapeutic agents.

In order to increase effectiveness and reduce the risk of drug-related toxicity through overdosage, a two-pronged research effort has been shown to be very fruitful: one that focuses on the altered handling of drugs in the elderly and the other that emphasizes the influence of these alterations on the dynamics of response. Improved designs of clinical investigations that explore changes in both the pharmacokinetics and pharmacodynamics must keep knowledge obtained via basic research in mind so as to be able to sharpen the focus of questions being asked and the rigor of experimental designs.

The impetus for this book evolved from a meeting with an investigator from National Institutes of Health, regarding optimal sampling schedules to evaluate the kinetics and dynamics of a drug of interest. In order to be able to estimate drug disposition characteristics and estimate its parameters with minimum variance, a certain blood-sampling scheme was deemed appropriate. However, it was soon learned that frequent sampling will lead to patient stress, which in turn has been shown to result in artificially elevated levels of endogenous hormones and biogenic amines, the very measures being sought for as the dynamic attributes of end-organ response. An appropriate compromise was achieved by altering the sampling design, and the problem was avoided. We feel that if such basic information was not integrated into the design of the clinical investigation of that drug, it could have led to erroneous results and conclusions, and would have perhaps altered the future direction of research, at least in that one area.

In this book we have attempted to present information, not a solution, that we hope will prove useful when considering drug studies in the elderly. The book is divided into three parts: (I) Research Determinants in Aging; (II) Pharmacokinetics and Pharmacodynamics of some classes of drugs commonly used in the elderly, and (III) General Perspectives. Part I contains the basic and clinical information on aging of organ and system function in several different areas of biomedical research. Part II provides an overview of research pertaining to kinetics and dynamics of classes of drugs commonly used in treatment of the elderly. As suggested, each contributor has attempted to raise pertinent questions and to suggest any concerns he may have regarding methodological issues from previous literature. Role of active metabolites, type of elderly subject's population, longitudinal versus cross-sectional trials, and effect of disease and coadministered drugs have been some frequently raised concerns from past studies in the elderly. Part III pertains to some general topics that play a significant role in the design and outcome of a clinical trial.

Although some of these topics may not have a direct relationship to drug studies in the elderly, some others certainly do. Design and manufacture of dosage forms appropriate for use in the geriatric population is an important and very critical area. Application and development of sensitive and specific analytical methodologies in clinical investigation of drugs and their metabolites is of vital importance. Both regulatory and industrial perspectives on clinical trial design document the need for special attention to clinical trials in geriatrics. All of us associated with research, either directly or indirectly, seemingly lead a life looking for statistically significant differences among different treatments, and therefore the appropriate use of statistical estimations, as it pertains to drug disposition data obtained from clinical trials, is discussed along with statistical concerns. Use of replicate analysis as a means of decreasing intersubject variability and increasing power of trials is suggested along with appropriate modeling techniques for the purposes of predicting future research design performance. This sequence is intended to provide the reader with most of the pertinent issues as they may relate to design of drug studies in the elderly. The content of some contributions has taken on a different perspective and that has primarily reflected on a lack of pertinent data related to a class of compounds and/or disease states poorly understood in the elderly.

One of the main purposes of a preface is to give the authors an opportunity to thank those without whose help the work could not have been completed as efficiently. We are indebted to all the contributors for their help in allowing us to compile this book over a short period of time. Our thanks also to all our secretarial staff, whose efficiency reached its peak during this period. Finally, we would like to express our gratitude to the staff of Plenum Press for their help in meticulous editing of the manuscripts and the book. Special thanks go to Ms. Janice Stern, Senior Medical Editor, for her timely editing, critique, and suggestions in putting this book together.

Finally, the views expressed in this book are those of authors and editors alone, and do not necessarily reflect the views of the National Institutes of Health.

Neal R. Cutler
Prem K. Narang

CONTENTS

CHAPTER 2

Age: A Complex Variable

Prem K. Narang

CHAPTER 3

Physiological Changes with Aging: Relevance to Drug Study Design

Alexander M. M. Shepherd

CHAPTER 4

Immunity and Aging

William B. Ershler and Marc E. Weksler

CHAPTER 5

Cardiovascular Changes with Aging

Brian F. Johnson and Janice C. Hitzhusen

CHAPTER 6

The Effects of Age on Hepatic Drug Metabolism

Cho-Ming Loi and Robert E. Vestal

CHAPTER 7

Hematological Effects of Aging: Considerations for Clinical Trials

Elizabeth J. Read and Harvey G. Klein

CHAPTER 11

Effect of Age on the Clinical Pharmacokinetics
of Antiarrhythmic Drugs

J. W. Massarella and K.-C. Khoo

CHAPTER 12

Beta Blockers in the Elderly

David S. Roffman and Alan Forrest

CHAPTER 13

Antiepileptic Drugs in the Elderly

William H. Theodore

CHAPTER 14

Pharmacokinetics and Bioavailability of Corticosteroids in the Treatment of Neurological Diseases of the Elderly

Marinos C. Dalakas

CHAPTER 15

Pharmacological Treatment of Parkinson's Disease

Donald B. Calne and Arto Laihinen

CHAPTER 16

Cognitive Enhancers in Alzheimer's Disease

Neal R. Cutler and Prem K. Narang

PART III

General Perspectives

CHAPTER 17

Dosage Form Considerations in Clinical Trials Involving
Elderly Patients

R. Gary Hollenbeck and Peter P. Lamy

CHAPTER 18

Clinical Trial Design—Industry Perspective

William B. Abrams and Bruce E. Rodda

CHAPTER 19

Methodological Issues: A Regulatory Perspective

Paul Leber

CHAPTER 20

Statistical Analysis of Drug Disposition Data

David G. Covell and Prem K. Narang

CHAPTER 21

Analytical Methods

Lawrence J. Lesko

Chapter 22

Nursing Perspectives on Clinical Trials in Geriatrics

Barbara R. Heller, Maureen E. Power,
and Georgeanne Cox Santolla

PART I

RESEARCH DETERMINANTS IN AGING

CHAPTER 1

EPIDEMIOLOGY AND DEMOGRAPHY OF AGING
SOME LESSONS IN SHORTSIGHTEDNESS

THOMAS M. VOGT

> ... in this century a major change has taken place in both the absolute number and relative proportion of older people. This change should be regarded as a triumph.... Unfortunately, our society was unprepared for the "demographic revolution"; we did not have the social institutions, the medical care system, the employment policies, and the nursing homes to properly respond to the change.
>
> Robert N. Butler[1]

1. INTRODUCTION

The phenomenon known as the "demographic revolution" has been with us for more than 200 years. Falling birth rates, rising incomes, and declining death rates have followed industrialization. Between 1880 and 1920 the impact of industrialization on demographic profiles was dramatically enhanced by a public health revolution, during which acute infectious diseases and tuberculosis became minor causes of death for the first time in history.

By 1920, the population of the United States was not only growing,

---

THOMAS M. VOGT • Center for Health Research, Kaiser Permanente, Northwest Region, Portland, Oregon 97215.

it was aging, particularly if the effects of immigration were removed. In the 1940s when antibiotics disposed of pneumonia as a major cause of death and the baby boom was in full swing, the handwriting was on the wall—tomorrow's world would see a startling increase in the number of persons over 65 years of age.

The implications of the growing aged population on our way of life have been only slowly recognized, and efforts to deal with this phenomenon have been woefully inadequate.

2. Demographic Trends in Aging

> It's a worldwide demographic revolution. Clearly we're headed for a profoundly different society. The changes cross all boundaries of race, sex, and nationality because aging is a lifelong process that affects all of us.
>
> Robert N. Butler[1]

History can be a marvelous teacher, but it has little to tell us about the changing age structure of our population. There are no precedents,

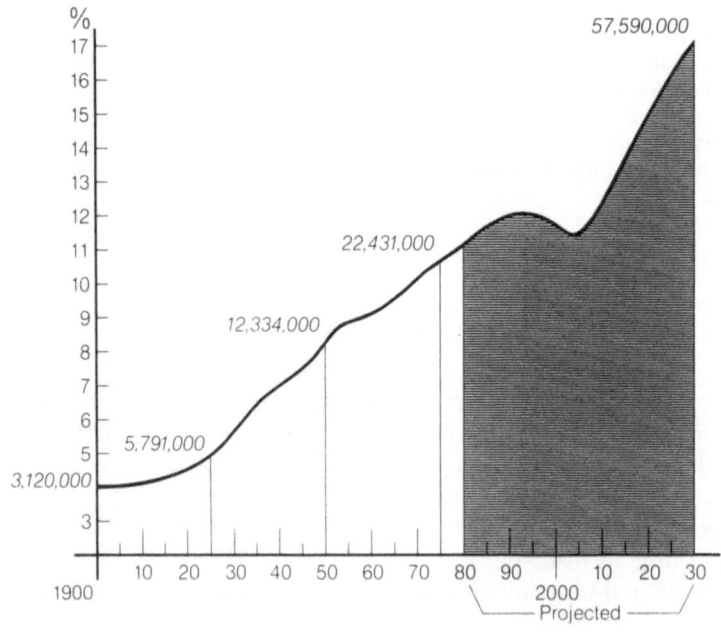

FIGURE 1   Percent of U.S. population aged 65 and older, 1900–2030. From Butler.[1]

TABLE I PERCENT DISTRIBUTION OF POPULATION 65 YEARS AND OLDER
BY AGE FOR THE YEARS 1950–2020[a]

| Age (years) | 1950 | 1960 | 1970 | 1976 | Projections | | | | |
					1980	1990	2000	2010	2020
65+	100	100	100	100	100	100	100	100	100
65–69	41	38	35	36	35	34	29	33	35
70–74	28	29	27	26	27	26	26	23	27
75–79	17	19	19	18	17	18	20	17	17
80–84	9	10	12	12	12	12	13	13	10
85+	5	6	7	9	9	10	12	13	11

[a] Source: U.S. Bureau of the Census.[84]

no historical footnotes to guide us in decision making. We are on our own, just as we are in dealing with issues such as toxic wastes, nuclear proliferation, and disposal of radioactive materials. The increase in the population of aged persons is a serious challenge to our society's ability to provide a high quality of life to its citizens. The number of persons in the United States aged 65 and above will double between 1976 and 2020 to a total population of about 45 million. Figure 1 illustrates the growth in numbers of aged from 1900 to 2030 according to U.S. Census estimates.[1] In that 130-year period the number of aged will have increased by a factor of 18. Table I indicates how, even among the over-65 group, the mean age is increasing dramatically. The proportion of persons in that group who are over 85, for example, will rise by more than twofold in the 70 years from 1950 to 2020.

These changes will have repercussions. For example, Table II presents sex ratios in various age groups from the years 1950 through 2020. The increasing differential in male/female survival coupled with the increasing length of survival results in increasing sex imbalances with advancing age as well as with advancing calendar time. Thus, in 1950, there were 89.5 males over 65 for every 100 females over 65 years of age, whereas in 2020, there will be only 69.3 males per 100 females. Why is that so important? Because women and men have different needs and require different types and amounts of services, medical care, and consumer goods. They exhibit different voting patterns, get different illnesses, and use different drugs at different rates. Unless society prepares for population changes, the facilities, services, and goods available will not be adequate to the needs of the older population.

TABLE II   MALES PER 100 FEMALES IN U.S. POPULATION BY AGE: 1950–2020[a]

Age (years)	1950	1960	1970	1976	Projections				
					1980	1990	2000	2010	2020
All ages	99	97	95	95	95	95	95	96	96
Under 15	104	103	104	104	105	105	105	105	105
15–29	99	98	98	100	101	101	102	102	102
30–44	97	97	95	96	97	97	97	98	97
45–59	100	97	93	95	95	95	96	95	95
60–64	100	91	88	88	88	99	90	91	91
65–69	94	88	81	79	80	81	82	82	83
70–74	91	85	73	74	72	73	75	75	76
75–84	85	77	66	61	60	60	60	61	61
85+	70	64	53	47	45	41	39	39	39
65+	90	83	72	69	68	67	67	67	69
75+	83	75	63	58	56	55	54	53	54

[a] Source: U.S. Bureau of the Census.[84]

2.1. MIGRATION OF THE ELDERLY

Migration is a common phenomenon of the retired. Freed of the necessity to remain in a given location in order to maintain economic self-sufficiency, many older persons move to more temperate climates or to be near children and other family members. Figure 2 indicates U.S. Census estimates for the proportion of persons over 65 in various states. Florida and the midwestern states have a much higher proportion of elderly than do Colorado, Utah, New Mexico, South Carolina, and Nevada.[2] Central cities contain 34% of persons 65 and over, and 27% of them are in rural areas.[3]

National patterns of migration have changed with time.[4] The sunbelt migration is a recent trend, and counties with high levels of recreation and vacation amenities are increasing their proportion of elderly. Arkansas, Florida, Arizona, Oregon, California, and Texas account for more than 50% of all elderly migration.[5,6] Ethnic and cultural factors must be considered as well: elderly blacks are more urbanized than elderly whites.

Each region must evaluate its particular situation—areas with a high proportion or those that can expect to attract a high proportion of elderly persons should be planning medical care facilities, housing and zoning laws, access for handicapped persons, and countless other aspects of life.

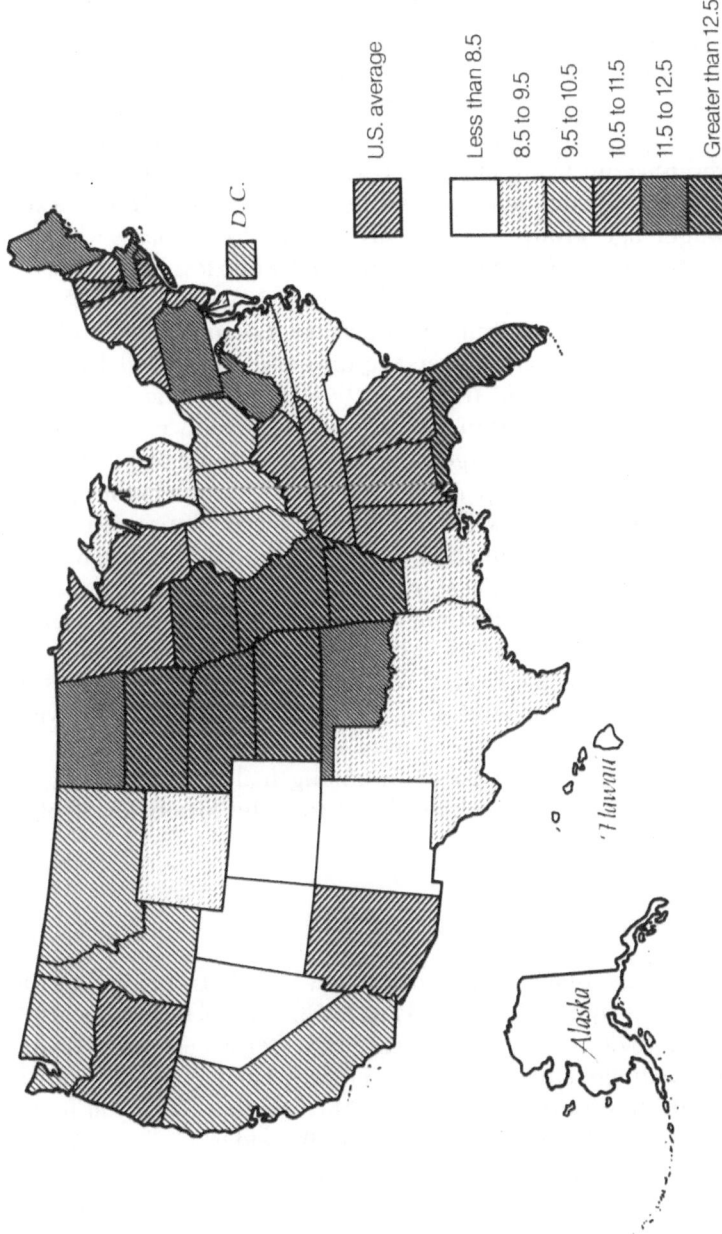

FIGURE 2   Percent of persons 65 years and older by state, 1976. From Siegel.[2]

2.2. Patterns of Life Changes

Patterns of life change considerably as people move into the geriatric age group. For example, the proportion of persons over 65 without a spouse (68%), the proportion who live alone (56%) and who are not employed (90%), the fraction of those who do not own their own homes (63%), who have no living children (27.5%), or who find public transportation wholly inadequate to their needs (50%) all differ markedly from younger age groups.[7] There are five times as many widows as widowers, and 77% of older men are married while less than half (48%) of older women are still married.[8] The aged have had less schooling than younger persons as well. In the Duke Longitudinal Studies on Normal Aging 63% of those over age 65 never graduated from high school compared to 26% in the 18- to 64-year age range.[9] Of course, those percentages will continue to change as current, better-educated cohorts move into the older age ranges. Mean income of persons over 65 is about one-half that of persons aged 55–64,[9] and a higher proportion of the elderly are below federal poverty levels. Retirement is associated with a substantial reduction in income as well as a switch from a variable income, adjusted to cost-of-living changes, to one that is fixed, and largely unaffected by inflationary changes. The growth in numbers of aged persons also means an increase in numbers of persons with poverty level incomes that do not respond to economic swings, as well as a steady rise in the number of dependents (nonworkers) per individual in the work force. In the Duke studies, change in financial status was the best predictor of life satisfaction among males, with those whose living standard had most declined indicating the lowest levels of life satisfaction.[9]

2.3. Life Expectancy

Death rates for older persons decreased by 19% from 1940 to 1954, then actually rose by 4.8% from 1954 to 1968, and fell by another 14.3% from 1968 to 1977.[2] Table III presents U.S. Census estimates of these changes by each age subgroup of the elderly. Changes and trends are similar for all subgroups. Many persons believe that there will be little further increase in life expectancy among the elderly[10–12] and that research and public attention should turn more to improving the quality of life rather than increasing the lifespan. A 1979 HHS report described the federal government's goals more in terms of maintaining function and vigor among the elderly rather than in terms of lengthening life.[13] Federal programs will only go so far; ultimately individuals must become

TABLE III DEATH RATES PER 1000 POPULATION BY AGE AND PERCENT
CHANGE FOR POPULATION 55+ YEARS OLD: 1940–1977[a]

Year	Age (years)				
	55–64	65–74	75–84	85+	65+
1940	22.2	48.4	112.0	235.7	72.2
1954	17.4	37.9	86.0	181.6	58.6
1968	17.0	37.2	82.9	195.8	61.4
1976	14.8	31.3	73.3	155.7	54.3
1977	14.3	30.5	71.5	145.9	52.6
	Percent change (years)				
1940–54	−21.6	−21.7	−23.2	−23.0	−18.8
1954–68	− 2.3	− 1.8	− 3.6	+ 7.8	+ 4.8
1968–77	−15.9	−18.0	−13.8	−25.5	−14.3

[a] Source: National Center for Health Statistics data as presented in Siegel.[2]

more involved in their own health to achieve goals of shortening the
dying process and living freer of disability and serious discomfort.[14]

Another important issue that must be addressed when one discusses
life expectancy is the fact that life expectancy in the United States at
birth is about 8 years less for a male than for a female.[15] The deterio-
rating position of males compared to females in terms of life-span is
illustrated in Table IV. Since 1900, the relative advantage of females to
males has increased from around 10% to 70–75%.[2] Major causes of this
difference are male mortality rates over twice as high as for females for
the following conditions: accidents, suicides, cirrhosis, lung cancer, emphysema, and coronary heart disease.[16] Each of these conditions has a
strong behavioral component.

TABLE IV RATIOS OF MALE TO FEMALE DEATH RATES BY AGE
FOR THE ELDERLY: 1900–1976[a]

Year	Age (years)				
	55–64	65–74	75–84	85+	65+
1900	1.14	1.11	1.08	1.05	1.06
1940	1.45	1.29	1.17	1.08	1.17
1954	1.82	1.57	1.29	1.06	1.30
1968	2.07	1.88	1.46	1.18	1.44
1976	1.99	1.97	1.58	1.26	1.46

[a] Source: U.S. Bureau of the Census data as presented by Siegel.[2]

3. Epidemiology of Health and Illness among the Elderly

What we recognize as "aging" or "oldness" is the emerging tendency to over-adapt to one's own routines and expectations rather than to adapt flexibly and resourcefully to the world at large.

R. Kastenbaum[17]

There is a great deal of misunderstanding about illness in the elderly. Typical stereotypes include both the hale and hearty nonagenarians and the useless and feeble elderly living in lonely rest homes. Like all stereotypes, each of these examples exists, but is not the norm. The truth is that older people have fewer acute conditions than the young, but they are more debilitated by such illnesses when they do occur.[18] Women have more frequent acute illness than men and spend more days in bed with each episode of illness. Men tend to reduce activity when sick, whereas women are more likely to go to bed.

3.1. Prevalence of Various Medical Conditions in the Elderly

Table V indicates the proportion of elderly men and women with various chronic conditions as determined by the National Health Survey. The top five conditions are similar in both sexes, although the order differs somewhat. Arthritis is the most common current condition in both sexes, but is considerably more prevalent among elderly women than among men. Thirteen percent of women have limited activity from arthritis. Men have less pain, swelling, and stiffening than women.[19] Knees are the most common site for arthritic impairment; arthritis among women involves the knee more often than it does in men.

The major causes of activity restriction among the aged are heart disease, orthopedic impairments, and cerebrovascular disease. Major causes of death, on the other hand, are heart disease, cancer, and stroke. Self-report and medical examination reveal similar complaints, although dermatological and dental problems are more often noted on examination.[18]

Table VI presents complaints that lead to medical care use by the elderly along with the principal diagnoses made at those visits and at discharge from the hospital.[18] For both men and women patients, muscle and bone chief complaints are by far the most common symptoms that patients asked to be evaluated for by physicians. Circulatory diseases, however, are the principal diagnoses made by physicians. Respiratory diseases are next most common for men, and musculoskeletal conditions are next for women. This difference is notable among hospital discharges as well. Fractures are the third most common cause of hospitalization for women (20.6%), but only the seventh most common for men (9.7%).

TABLE V   PERCENT OF MEN AND WOMEN OVER 65
WHO REPORT VARIOUS CHRONIC CONDITIONS IN THE
UNITED STATES HEALTH INTERVIEW SURVEYS[a]

	Women	Men
Had in past 12 months		
Arthritis	45.0	28.7
Hypertension	24.1	14.1
Hearing problem	26.3	33.8
Vision problem	22.0	18.3
Heart condition	19.8	19.9
Diabetes	9.1	6.0
Orthopedic hip problem	9.0	6.5
Condition that limits activity		
Arthritis	12.7	7.7
Heart condition	9.6	12.5
Vision	4.6	4.3
Hypertension	4.7	3.0
Diabetes	3.3	2.9
Leg or hip problem	3.0	2.5
Cerebrovascular disease	1.9	2.9
Ever had		
Heart attack	8.4	14.3
Cancer	6.1	3.4
Stroke	3.7	5.4
Wrist fracture	7.9	5.4
Hip fracture	3.2	1.5
Spine fracture	1.5	2.1

[a] Source: National Center for Health Statistics data and adaptations from Verbrugge.[18]

Falls and fractures are, of course, a common liability of the elderly, but this liability is disproportionately burdensome for women. Greater longevity and progressive osteoporosis probably account for that differential. Malignant neoplasms are the second most common cause of hospitalization among the elderly.

Cardiovascular problems are common in both sexes, but there is a shift with age from hypertension to heart failure, another indication of the preventive potential from adequate treatment of hypertension. Women seem to be less bothered in daily life by the problems that ultimately kill them than are men. They report more current cancer and more past cancer than men (Table V), but older men are more often hospitalized for cancer and have higher cancer death rates. Women have higher rates of benign cancers[20] and a better prognosis than men. They have more

TABLE VI LEADING COMPLAINTS AND DIAGNOSES FOR AMBULATORY CARE VISITS AND LEADING DIAGNOSES FOR HOSPITAL DISCHARGES AMONG OLDER WOMEN AND OLDER MEN, UNITED STATES[a]

Principal complaints for office visits (general types) (percent of symptomatic visits)	Principal diagnoses for office visits (general types) (percent of "sick" visits)	Principal specific diagnoses for office visits (percent of all visits)	Principal specific diagnoses for hospital discharges (discharges per 100,000 pop.)
Women 65+			
Muscle/bone, 22.6%	Circulatory, 25.0%	Hypertension, 11.0%	Ischemic heart disease, 41.3
Respiratory, 15.7	Musculoskeletal, 12.4	Chronic ischemic heart disease, 6.1	Malignant neoplasms, 27.4
Digestive, 12.7	Nervous system/sense organs, 10.2	Diabetes mellitus, 4.0	Fractures, all sites, 20.6
Cardiovascular, 8.9	Respiratory, 8.5	Osteoarthritis and allied conditions, 3.3	Cerebrovascular disease, 19.6
Eyes/ears, 8.0	Endocrine/nutritional/metabolic, 6.8	Cataract, 3.0	Cataract, 11.7
General, 7.8	Digestive, 5.9	Arthritis, unspecified, 2.1	Diabetes mellitus, 10.6
Genitourinary, 7.0	Genitourinary, 5.7	Symptomatic heart disease, 1.8	Arthritis/rheumatism, 10.1
Nervous, 7.0	Accidents/poisoning/violence, 5.1	Other diseases of eye, 1.7	Pneumonia, all forms, 9.9
Men 65+			
Muscle/bone, 21.5%	Circulatory, 28.2%	Not available	Ischemic heart disease, 51.9
Respiratory, 20.3	Respiratory, 12.0		Malignant neoplasms, 42.8
Digestive, 11.0	Nervous system/sense organs, 9.4		Cerebrovascular disease, 21.1
Skin/hair, 8.6	Muscle/bone, 7.9		Hyperplasia of prostate, 17.8
Genitourinary, 8.3	Endocrine/nutritional/metabolic, 6.9		Pneumonia, all forms, 13.1
General, 7.8	Genitourinary, 6.5		Congestive heart failure, 9.7
Eyes/ears, 7.4	Neoplasms, 5.6		Fractures, all sites, 9.7
Cardiovascular, 6.8	Accidents/poisoning/violence, 5.4		Cataract, 9.4

[a] Source: Verbrugge.[18] Used by permission.

TABLE VII   DEATH RATES FOR THE 10 LEADING CAUSES OF DEATH FOR AGES 65 AND OVER, BY AGE: 1976 (DEATHS PER 100,000 POPULATION)[a]

Cause of death by rank	65 years and over	65 to 74 years	75 to 84 years	85 years and over
All causes	5,428.9	3,127.6	7,331.6	15,486.9
1. Diseases of the heart	2,393.5	1,286.9	3,263.7	7,343.3
2. Malignant neoplasms	979.0	786.3	1,248.6	1,441.5
3. Cerebrovascular diseases	694.6	280.1	1,014.0	2,586.8
4. Influenza and pneumonia	211.1	70.1	289.3	959.2
5. Arteriosclerosis	122.2	25.8	152.5	714.3
6. Diabetes mellitus	108.1	70.0	155.8	219.2
7. Accidents	104.5	62.2	134.5	306.7
Motor vehicle	25.2	21.7	32.3	26.0
All other	79.3	40.4	102.2	280.7
8. Bronchitis, emphysema and asthma	76.8	60.7	101.4	108.5
9. Cirrhosis of liver	36.5	42.6	29.3	18.0
10. Nephritis and nephrosis	25.0	15.2	34.1	64.6
11. All other causes	677.5	427.8	908.6	1,683.8

[a] Source: Siegel,[2] based on National Center for Health Statistics data.

chronic, but less serious, health problems in general than do men—more hypertension, diabetes, arthritis, varicose veins, anemia, migraine, sciatica, digestive and urinary problems (except ulcer and hernia), allergies, and orthopedic problems. Such conditions are often symptomatic, but seldom fatal.[18] Males have more heart disease, stroke, arteriosclerosis, pneumonia, emphysema, and asthma, all conditions with a reasonable chance of being fatal. Between 80 and 86% of the aged have one or more chronic diseases compared to 40% of those under 65.

Table VII presents the 10 leading causes of death for ages 65 and above by age subgroup. As age increases, stroke and nonvehicular accidents become more common causes of death, although heart disease and cancer remain as the number one and two causes of death, respectively.[2]

3.2. CARDIOVASCULAR DISEASE AMONG THE ELDERLY

Serum lipids, blood pressure, glucose tolerance, smoking, diet, and physical inactivity are all powerful predictors of cardiovascular disease. In the elderly, however, the strength of the associations between these risk factors and vascular disease declines gradually. This is especially true for cholesterol and cigarette smoking. Although smoking remains a strong predictor of pulmonary disease and lung cancer, its relation to total mortality diminishes owing to the decreasing association with heart disease. Thus, some risk factors among middle-aged individuals may, to some extent, lack validity when applied to the elderly. Both cholesterol and smoking decrease with age beginning at about age 65.

Blood pressure, on the other hand, remains a very powerful predictor of stroke, congestive heart failure, and myocardial infarction in the elderly. Mean systolic blood pressure rises steadily with age to about 150 mm Hg at age 70. Mean diastolic pressure forms a curve, with a peak at about age 55–60 (83–85 mm Hg), declining thereafter to around 78 mm Hg at age 70.[22] Women have consistently lower mean diastolic pressures than men after age 40, but there is little difference in systolic pressure by sex. These relationships are summarized in Fig. 3.

The incidence of various manifestations of cardiovascular disease changes with age, although cardiovascular conditions remain the principal cause of death from early middle age onward. Coronary heart disease increases with age up to age 65[22] and remains constant in men thereafter. The incidence continues to rise into old age among women, however. Cerebrovascular accidents, congestive heart failure, and peripheral vascular disease all rise steadily as age advances. The relative incidence rates of these conditions for men and women by age is shown in Table VIII. The well-known decline in myocardial infarction rates has been evident in all age groups, but particularly pronounced at ages 35–44 and 65–74.[22] This suggests that the potential for prevention is at

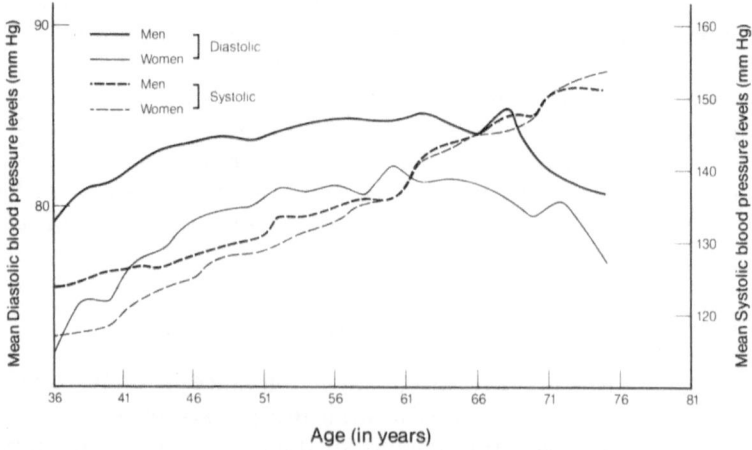

FIGURE 3   Mean systolic and diastolic blood pressure levels by age for the Framingham study cohort. Adapted from Siegel.[2]

TABLE VIII INCIDENCE PER 1000 OF VARIOUS MANIFESTATIONS
OF CARDIOVASCULAR DISEASE BY AGE AND SEX IN THE FRAMINGHAM STUDY,
MEN AND WOMEN, AGED 45–74, 20-YEAR FOLLOW-UP[a]

Age (years)	Coronary heart disease		Cerebrovascular accident		Peripheral arterial disease		Congestive heart failure	
	M	F	M	F	M	F	M	F
45–54	9.9	3.1	2.0	0.9	1.8	0.6	1.8	0.8
55–64	20.8	9.5	3.2	2.9	5.1	1.9	4.3	2.7
65–74	20.4	14.5	8.4	8.6	6.3	3.8	8.2	6.8

[a] Source: Kannel and Gordon.[22]

least as great in the geriatric population as it is in younger age groups. Among the aged, the key correctable risk factor for cardiovascular illness is hypertension, yet in the Framingham Study, the elderly were less likely to be treated than younger persons with hypertension.[22]

The prevalence of selected risk factors by age in the Framingham cohort is shown in Table IX. These data do not include persons over 65 years of age (except for cigarette smoking), but show clear evidence that advancing age is associated with special problems of hypertension and inactivity. The importance of paying attention to these risk factors is illustrated not only by the well-known statistical relationships between hypertension and total mortality, but also by studies which indicate that hypertensive persons, whether middle-aged or geriatric, perform less well than normotensives on tasks of intellectual function and psychomotor response.[23–26] Administration of vasodilators may reverse this pattern.[23]

3.3. AGING AND DEMENTIA

Senile dementia is a mental disorder of gradual onset and continual progression, characterized by a certain type of memory impairment—namely, an amnestic syndrome, the essential features of which are severe impairment of recent and remote memories and of recall, shortened retention span, disorientation, and, sometimes, confabulations.[27] There is strong evidence that the average duration of senile dementia increased between 1947 and 1957 in Sweden.[28] This increased duration is probably the result of extension of life due to treatment of intercurrent infections and other forms of medical care. The age-specific incidence of senile dementia increases steadily up to the early eighties at which point it declines somewhat.[29]

TABLE IX PERCENT PREVALENCE OF SELECTED CARDIOVASCULAR RISK FACTORS IN THE U.S. POPULATION[a]

Age (years)	Inactivity	Obesity	Hypertension	Smoking	Diabetes	High cholesterol	ECG-LVH
Men							
35–44	12.1	12.5	13.5	48.6	1.1	20.2	2.9
45–54	16.9	14.7	18.3	43.1	1.1	15.7	4.8
55–64	21.0	12.5	22.3	37.4	3.3	23.5	10.1
65–74	27.1	12.7	27.1	22.8	3.2	21.6	7.1
Women							
35–44	13.3	20.1	8.5	38.8	0.8	12.9	0.9
45–54	19.3	24.2	18.2	36.1	2.9	28.0	3.6
55–64	30.8	30.9	31.2	24.2	3.2	49.7	4.1
65–74	39.0	27.2	47.6	10.2	6.1	51.0	9.6

[a] Source: Kannel and Gordon.[22]

"Cognitive changes vary by cohort, change as a function of intervention, and differ vastly among individuals. More than at any time in the life span there is diversity in old age. In different individuals, cognition in old age ranges from senility to wisdom, and in a given individual some abilities fade while others are maximized."[30] Remarkably little is known of the epidemiology of the vast and varied changes that occur with advancing age.[29] The relation of cognitive function to blood pressure has been noted elsewhere in this chapter. That relation is clearly complex, and little is known about underlying mechanisms. Although the data suggest that reduction in blood pressure might improve cognitive functioning, there has been little direct investigation of this issue. When I was in medical school in the late 1960s, it was taught as axiomatic that antihypertensive drugs would reduce cerebral perfusion in elderly brains with already compromised circulation and thereby increase the risk of both stroke and cognitive impairment. That "truth" was not based on data. Nevertheless, the relative value of antihypertensive therapy in the elderly remains highly controversial. The Systolic Hypertension in the Elderly Program (SHEP) is a nationwide, collaborative clinical trial currently addressing that question. The pilot phase of that project carefully examined the relation of short-term (1–2 years) cognitive changes in relation to drug therapy and found no evidence of harm from antihypertensive treatment.[31] It is not unlikely, though, that the various classes of antihypertensive agents might have a differential impact on cognitive function since they act by very diverse mechanisms.

Much more work must be done in this area before epidemiologic data can suggest causal and preventive hypotheses that relate to changes in cognitive function.

3.4. Injuries and the Elderly

Although much has been written about the risk of injuries among the elderly, in general, injury rates among that group are lower than among younger populations.[32] In 1974, injury rates per 100 persons were 32, 36, 32, 19, and 17 for ages 6 and under, 6–16, 17–44, 45–64, and 65 and more, respectively.[33] However, the elderly have much higher case fatality rates, disability, and days of hospitalization and restricted activity after an injury than do younger persons. Three types of accidents account for 75% of injuries: (1) falls (70% of deaths due to falls are in persons 65 years and older, a fact that is related to osteoporosis and increasing bone fragility[34,35]), which might be preventable to some degree by increasing fluoride concentrations in the water[36]; (2) fire and heat injuries (7% of injury deaths among the elderly); and (3) vehicular crashes. Nine percent of all licensed drivers are 65 and older, and 12.3% of all vehicular deaths involve that age group, although one-third of those are pedestrian deaths.[32] Visual field narrows progressively beginning in the late thirties,[37] and the rate of dark adaptation decreases with age as well.

Each year 20% of elderly women suffer an injury requiring medical care or restricted activity,[38] and among the elderly, female injury rates are higher than male, a distinct change from the pattern in younger persons. Like other aspects of health and illness in the elderly, the patterns of injury in that age group are so different from younger age groups that the problem may look worse or better depending on the statistic selected and how it is viewed. The elderly have different patterns, which demand different approaches. Effective response requires an understanding of this fact.

3.5. Well-Being and the Elderly

Most of us assume that our general sense of well-being will decline with age, and that negative situational factors such as illness and poverty will be increasingly devastating because of our increased vulnerability. Most research, however, contradicts these "obvious" assumptions.[39] Although depression does increase with age, worrying declines and life satisfaction increases.[40,41] Although the meaning of such apparently contradictory findings is not completely clear, it is abundantly evident that aging does not require unambiguous decline in subjective well-being.

Age is differentially related to performance in different areas. Scientists, mathematicians, philosophers, and creative writers usually are most creative and productive in their thirties.[42–47] This is probably due to a variety of factors such as closeness to advanced education, mental flexibility, and fewer administrative burdens.[42] The importance of the latter is emphasized by the fact that administrators and executives make their maximal contributions in their fifties.[48] In a sense, the normal progression of persons in academic professions is through two careers. The first is a professional one, which requires 10 or 15 years until success and seniority move the individual into a more powerful (i.e., more administrative) position. The second is maximal success in an administrative career some 10 or 15 years after that. Many persons who engage in mental work show little evidence of decline before age 70, although the variability among different individuals does increase with age.[49–51] This variation in decline of mental function is mirrored in the interindividual differences that occur with age-related physiological changes.[52]

The factors that contribute to subjective well-being do change with age, but not greatly. The principal factors related to well-being are economic status, health, and residential environment. Health becomes more important with age, as does the presence of children. Hobbies and spare-time activities are less related to well-being as age increases. In general, subjective well-being in older individuals is associated with higher levels of religiosity, social desirability and familiarity, decreased obligations, fewer stressful events, and increased freedom.[39]

3.6. SUICIDE

Suicide is the most dramatic manifestation of depression. Since depression increases with age, it is not surprising that suicide does as well. This increase is more pronounced among men than among women.[53] Interestingly, the number of attempted suicides declines with age, although the success rate of those attempts rises dramatically.[54] Only 1 attempt in 20 is successful in persons under age 40, while one in four is successful in persons over age 60. Figure 4 indicates the suicide rates by age in the United States during 1975. The rate is fairly stable through adult years until age 65. Subsequent to that age, there is a fairly steady increase in suicide rate.[54]

3.7. DISABILITY

The most frightening specter of old age for most persons is the possibility that one will be unable to care for oneself, and the fear of

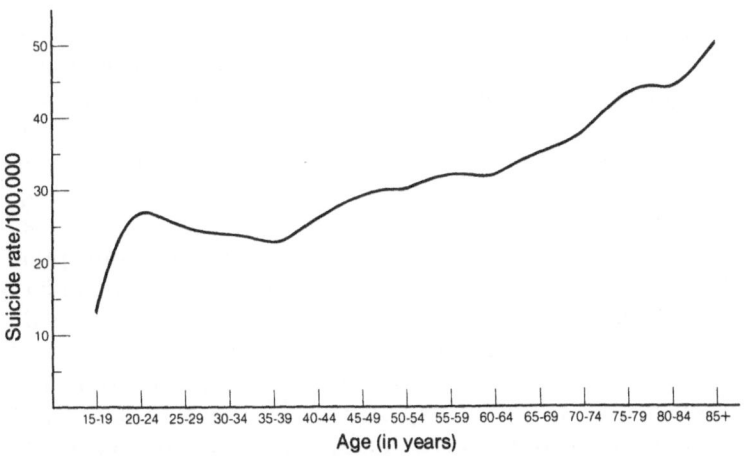

FIGURE 4   Annual suicide rates per 100,000 by age in the United States. Source: *Vital Statistics of the United States*.[85]

becoming so dependent on others that one's self-respect is lost. This scenario occurs less commonly than assumed. Of the noninstitutionalized elderly, only 5–7% are confined to the house.[3,55] Fourteen percent of that group has some significant impairment of physical mobility.[3] One of ten elderly whites, and twice as many elderly blacks, have troubles with chores, washing and bathing, and dressing. The most common complaint of the elderly is difficulty in cutting toenails and in navigating stairs. These problems are reported by 40% of whites and 50–60% of blacks over the age of 80.[3]

Only 5% of the elderly, mostly very old and female, are in institutions.[8] The oldest 10% of the population uses 30% of the medical care,[56] and yet, despite this, Medicare does not cover the greatest medical need of the elderly: chronic care for long-term illness. The percent of persons aged 65 and above who have no medical coverage other than Medicare is higher among nonwhites, the very old, those with poorer health, and those with lower levels of income and education.[57] In other words, medical insurance for the elderly is most available to those who least need it.

4. Drugs and the Elderly

> Drug treatment based on ignorance—ignorance of the course of symptomatology, ignorance of the drug, the interaction of drug and drug, as well as drug and nutrition, and/or ignorance of the emotional and social environment of the patient—often leaves the older person with an exacerbation of symptoms, in the decrement of function and in the persistence of disease.
> R. B. Weg[58]

Drug therapy is a major weapon against disability and death among the elderly. It can permit an independent life-style where one would otherwise have been impossible.[59] On the other hand, the old are highly susceptible to drug problems. They have multiple conditions, which can increase the numbers of drugs taken, alter the way in which drugs are metabolized, and raise the probability of minor or serious drug side effects. The complex drug regimens often followed by older persons also increase the probability of noncompliance. Delayed absorption, decreased plasma volume, increased body fat, decreased extracellular fluid, smaller lean body mass, reduced cerebral blood flow, and delayed renal excretion all may contribute to adverse drug effects.[60–67] Adverse effects of drugs increase with age and are directly proportional to the number and frequency of drug-dose exposures.[68–71] About 30,000 deaths and 1.5 million hospital admissions annually result from drug effects.[72] One in five geriatric patients in general hospitals exhibits a disorder directly attributable to the effects of prescribed drugs, and about 1 of 19 hospital admissions in persons over age 50 is the result of a reaction to a prescribed drug.[73] In Seattle, one-fourth of older county outreach program participants showed drug-related problems of cognition.[74] On the other hand, serious drug reactions may be less common in the elderly than most anticipate. Inciardi *et al.* found that only 2.6% of severe, acute drug reactions in a Dade County hospital involved elderly persons.[75] This may be due to the low levels of illicit drug use among the geriatric population.

Compliance with a prescribed medication schedule, so essential to achieving optimal benefit at minimum risk, is related to a variety of social and demographic factors including number of medications taken,[76] age and household composition,[77] cost, and degree of understanding or inadequate information concerning the drug's action.[59,78] The safe prescription of drugs requires that the prescriber be aware of factors in the patient's background or environment that are likely to result in noncompliance.

Table X lists the 10 most common classes of prescription drugs as determined in two studies.[21,79] Cardiovascular, analgesic, psychoactive, and gastrointestinal drugs constitute the four most frequently prescribed types. Psychoactive drugs are widely given to elderly persons as a means of coping with changing life situations. Analgesic drugs are common because of arthritis and other common causes of aches and pains in the elderly. Drugs in three of the four leading categories are primarily for symptomatic improvement rather than for specific therapy of an illness. Among the cardiovascular drugs, antihypertensive medications are the most common.

TABLE X   MOST FREQUENTLY PRESCRIBED CLASSES OF DRUGS FOR THE ELDERLY ACCORDING TO GUTTMAN[21] AND HURD[79]

Rank	Guttman	Hurd
1	Cardiovascular	Cardiac/blood
2	Analgesic/antiarthritic	Analgesics
3	Sedative/tranquilizer/antidepressant	Gastrointestinal
4	Gastrointestinal	Sedative/tranquilizer/antidepressant
5	Hormonal and diabetic	Salts, diuretics
6	Other drugs	Serums, toxoids
7	Respiratory	Antiinfectives
8	Topical	Hormones
9	Nervous system	Antihistamines
10	Antibacterial	Autonomic

The most common nonprescription drugs are internal analgesics, laxatives and other digestive drugs, vitamins, antihistamines, nasal and liquid decongestants, and sleeping aids, in that order.[21] Persons over 65 consume 25% of all drugs,[80] and they spend three times as much on nonprescription drugs as do other persons.[63]

The great variety of drugs and dose forms available and the rapid increase in new varieties combine to make rational prescribing and use extremely difficult. Even in 1968 Koch-Wesser[81] told an HEW Task Force on Prescription drugs that ". . . lack of knowledge and sophistication in the proper therapeutic use of drugs is perhaps the greatest deficiency of the average American physician today." Since that time, of course, the pharmacological armamentarium has increased greatly, and the competition for time in medical curricula has tended to reduce rather than expand physicians' drug educations. In summary, the number and complexity of drug regimens in the elderly is increasing. These drugs are, in many instances, lifesaving, and in others, they contribute substantially to the comfort and well-being of the individual. The risk of adverse consequences, however, grows with age and with numbers of prescribed drugs. Although alcohol consumption declines after age 60,[82] alcohol in combination with medications is responsible for a large number of adverse drug effects in the aged.[83] Rational use of medications in the elderly requires attention to education about new drugs and also attention to the potential interactions of all pharmaceutically active preparations being taken and to the factors that influence compliance.

5. OVERVIEW

Information provided by the sciences of epidemiology and demography is critical for social and economic planning but is often ignored because it concerns the future and not the present. Many characteristics of our society are changing, but the changes in age structure are among the most drastic we face. We must come to terms with the fact that more of us than ever before will live to be rather old, and that we would like to continue to be useful contributors, and be regarded as such. Now is the time to address the difficult issues raised by epidemiological and demographical studies such as the following:

1. How will society pay for acute medical care for the aged when they constitute 17% of the population, as they will in 35 years, instead of 10%, as they do now? Will society pay for such care?

2. How will society pay for long-term care when such care is hardly affordable now except through family bankruptcy?

3. How will we decide who receives very highly technological care such as organ transplants and artificial implants, which are never going to be generally available owing to high cost?

4. How do we decide who lives (or dies) and when to stop artificial life support systems?

5. How will the small group of post-baby boom adults economically support the larger group of baby boomers when their time to retire comes along? The latter are now in their peak productive years and still we are unable to adequately support the costs of the present elderly medical care system.

6. How can physicians keep pace with pharmaceutical changes and continue to prescribe rational, helpful drugs for their patients? Improved use of computers and computerized medical record information systems are likely. So are an increase in numbers of well-designed clinical trials to determine therapeutic efficiencies and removal of ineffective drugs from the market.

7. How can we learn to effectively use the data we collect to plan for a better future?

The list could be considerably longer. A humane old age for all of us depends on serious attempts to learn from the data that are now available to tell us how to prepare for the future.

REFERENCES

1. Butler RN: Introduction, in Haynes SG, Feinleib M (eds): *Second Conference on the Epidemiology of Aging*. USDHHS, Public Health Service, NIH Publication No 80–969, Washington, D.C., 1980, p 4.
2. Siegel JS: Recent and prospective demographic trends for the elderly population and some implications for health care, in Haynes SG, Feinleib M (eds): *Second Conference on the Epidemiology of Aging*. USDHHS, Public Health Service, NIH Publication No 80–969, Washington, D.C., 1980, pp 289–314.
3. Shanas E: Self-assessment of physical function: White and black elderly in the United States, in Haynes SG, Feinleib M (eds): *Second Conference on the Epidemiology of Aging*. USDHHS, Public Health Service, NIH Publication NO 80–969, Washington, D.C., 1980, pp 269–281.
4. Graff TO, Wiseman RF: Changing concentration of older Americans. *Geographical Rev* 1978; 68:379–393.
5. Wiseman RF: Regional patterns of elderly concentration and migration, in Golant SM (ed): *Location and Environment of Elderly Population*. Washington, DC, VH Winston and Sons, 1979.

6. Wiseman RF: *Spatial Aspects of Aging*. Washington, DC, Association of American Geographers, 1979.

7. Chien CP, Townsend EJ, Ross-Townsend A, et al: Substance use and abuse among the community elderly: The medical aspect, *Addictive Dis* 1978; 3:357–372.

8. Steffl BM: Facts about old people and health manpower, in Steffl M (ed): *Handbook of Gerontological Nursing*. New York, Van Nostrand Reinhold Company, 1984.

9. Palmore E: *Social Patterns in Normal Aging: Findings from the Duke Longitudinal Study*. Durham, NC, Duke University Press, 1981.

10. Keyfitz N: What difference would it make if cancer were eradicated? An examination of the Taeuber paradox. *Demography* 1977; 14:411–418.

11. Keyfitz N: Improving life expectancy: An uphill road ahead. *Amer J Public Health* 1978; 68:954–956.

12. Tsai SP, Lee ES, Hardy RJ: The effect of a reduction in leading causes of death: Potential gains in life expectancy. *Am J Public Health* 1978; 68:966–971.

13. Office of the Assistant Secretary for Health, USDHEW: *Healthy People: The Surgeon General's Report on Health Promotion and Disease Prevention*. Washington, DC, DHEW Publication No (PHS) 79-55071, 1979.

14. Knowles JH: The responsibility of the individual. *Daedalus* 1977; (Winter):57–80.

15. Waldron I: Why do women live longer than men? *Soc Sci Med* 1976; 10:349–362.

16. Waldron I: Sex differences in longevity, in Haynes SG, Feinleib M (eds): *Second Conference on the Epidemiology of Aging*. USDHHS, NIH Publication No 80–969, Washington, D.C., 1980, pp 163–186.

17. Kastenbaum R: When aging begins: A lifespan developmental approach. *Res Aging* 1984; 6:105–118.

18. Verbrugge L: A health profile of older women with comparisons to older men. *Res Aging* 1984; 6:291–322.

19. Maurer K: Basic data on arthritis (knee, hip, and sacroiliac joints) in adults ages 25–74 years, United States, 1971–1975. *Vital and Health Statistics*, Series 11, No 213. Hyattsville, MD, National Center for Health Statistics, 1979.

20. National Cancer Institute: *Surveillance, Epidemiology, and End Results: Incidence and Mortality Data, 1973–1977*, National Cancer Institute Monograph No 57, NIH Publication No 81-2330. Bethesda, MD, National Institutes of Health, 1981.

21. Guttman D: Patterns of legal drug use by older Americans. *Addictive Dis* 1978; 3:337–356.

22. Kannel WB, Gordon T: Cardiovascular risk factors in the aged: The Framingham Study, in Haynes SG, Feinleib M (eds): *The Second Conference on the Epidemiology of Aging*. USDHHS NIH Publication 80–969, Washington, D.C., 1980.

23. Spieth, W: Slowness of task performance and cardiovascular diseases, in Weford AT, Birren JE (eds): *Behavior, Aging, and the Nervous System*. Springfield, IL, Charles C Thomas, 1965, pp 366–400.

24. Abrahams JP: Physiological correlates of cardiovascular diseases, in Elias MF, Eleftheriou BE, Elias PK (eds): *Experimental Aging: Special Review of Progress in Biology*. Bar Harbor, ME, EAR Inc, 1976, pp 330–350.

25. Eisdorfer C, Wilkie F: Stress, disease, aging and behavior, in Birren JE, Schaie KW (eds): *Handbook of the Psychology of Aging*. New York, Van Nostrand Reinhold, 1977, pp 251–275.

26. Veterans Administration Cooperative Study Group on Antihypertensive Agents: Effect of treatment on morbidity in hypertension. Results in patients with diastolic blood pressure averaging 115 through 129 mm Hg. *JAMA* 1967; 202:1028–1034.

27. Kral VA: Senile dementia and normal aging. *Can Psychiatr Assoc J* 1972; 17(suppl 2):SS25–SS30.

28. Gruenberg EM, Hagnell O: The rising prevalence of chronic brain syndrome in the elderly, in Levi L, Kagan AR (eds): *Society, Stress and Disease: Aging and Old Age.* London, Oxford University Press, 1985, vol 5.

29. Gruenberg EM: Epidemiology of senile dementia, in Haynes SG, Feinleib M (eds): *Second Conference on the Epidemiology of Aging.* USDHHS, NIH Publication No 80–969, Washington, D.C., 1980, pp 91–97.

30. Woodruff DS: A review of aging and cognitive processes. *Res Aging* 1983; 5:139–154.

31. Gurland B, Greenlick MR, Luhr JC, et al.: Changes in systolic hypertension in relation to changes in cognition and expression over selected bimonthly intervals. 1986 (submitted for publication).

32. Hogue CC: Epidemiology of injury in older age, in Haynes SG, Feinleib M (eds): *Second Conference on the Epidemiology of Aging.* USDHHS, NIH Publication No 80–969, Washington, D.C., 1980, pp 127–135.

33. National Center for Health Statistics: *Current Estimates from the Health Interview Survey.* US 1974, Series 10, No. 100, USDHEW Publication No (HRA) 76–1527, Washington, D.C., 1975.

34. Iskrant AP, Smith RW: Osteoporosis in women 45 years and over related to subsequent fracture. *Public Health Rep* 1969; 84:33–38.

35. Smith D, Khairi MRA, Johnston CC: The loss of bone mineral with aging and its relationship to risk of fracture. *J Clin Invest* 1975; 56:311–318.

36. Hegsted DM: Fluoride and mineral metabolism. *Ann Dent* 1968; 27:134–143.

37. Planek TW: The aging driver in today's traffic: A critical review, in Waller PF (ed): *Aging and Highway Safety: The Elderly in a Mobile Society.* North Carolina Symposium on Highway Safety 7. Chapel Hill, University of North Carolina Safety Research Center, 1974; pp 3–38.

38. Metropolitan Life Foundation: Health of the elderly. *Stat Bull* 1982; 63:2–5.

39. Herzog AR, Rodgers WL, Woodworth J: *Subjective Well-Being among Different Age Groups.* Research Report Series, Survey Research Center, Institute for Social Research. Ann Arbor, University of Michigan, 1982.

40. Campbell A, Converse PE, Rodgers WL: *The Quality of American Life: Perceptions, Evaluations, and Satisfactions.* NeW York, Russell Sage, 1976.

41. Gurin G, Veroff J, Feld S: *Americans View Their Mental Health.* Ann Arbor, University of Michigan Press, 1960.

42. Lehman HC: *Age and Achievement.* Princeton, NJ, Princeton University Press, 1953.

43. Lehman HC: The creative production rates of present versus past generations of scientists. *J Gerontol* 1962; 17:409–417.

44. Bromley DB: Some experimental tests of the effect of age on creative intelligence output. *J Gerontol* 1956; 2:74–82.

45. Sward K: Age and mental ability in superior men. *Am J Psychol* 1945; 58:443–479.

46. Dennis W: Creative productivity between the ages of 20 and 80 years. *J Gerontol* 1966; 21:1–8.

47. Bayer AE: *College and University Faculty: A Statistical Description.* Washington, DC, American Council on Education, June 1970.

48. Lester RA: Age, performance, and retirement legislation, in Somers AR, Fabian D (eds): *The Geriatric Imperative.* New York, Appleton-Century-Crofts, 1981.

49. Schaie KW: Age changes in adult intelligence, in Woodruff DA, Birren JE (eds): *Aging: Scientific Perspectives and Social Issues.* New York, Van Nostrand, 1975, pp 111–124.

50. Botwinick J: Intellectual abilities, in Birren JE, Schaie KW (eds): *Handbook of the Psychology of Aging.* New York, Van Nostrand Reinhold, 1977, pp 580–605.

51. Palmore E: Intelligence, summary and the future, in Palmore E (ed): *Normal aging:*

Reports from the Duke Longitudinal Study, 1955–1969. Durham, NC, Duke University Press, 1970, pp 418–420.

52. Rowe JW: Research in geriatrics and gerontology, in Somers AR, Fabian DR (eds): *The Geriatric Imperative.* New York, Appleton-Century-Crofts, 1981.

53. Atchley RA: Aging and suicide; Reflection of the quality of life? in Haynes SG, Feinleib M (eds): *Second Conference on the Epidemiology of Aging.* USDHHS, NIH Publication No 80–969, Washington, D.C., 1980, pp 141–158.

54. Blazer DG: *Depression in Late Life.* St. Louis, CV Mosby Co, 1982.

55. US Bureau of the Census: Social and economic characteristics of the older population 1974. *Current Population Reports,* Series P-23, No 57. Washington, DC, US Government Printing Office, 1975.

56. Harris CS: *Fact Book on Aging: A Profile of America's Older Population.* Washington, DC, National Council on Aging, 1978.

57. National Center for Health Services Research: *Private Health Insurance Coverage of the Medicare Population.* Data preview 18, DHHS Publication NO (PHS) 84-3362, Washington, D.C., 1984.

58. Weg RB: Drug interaction with the changing physiology of the Aged: Practice and potential, in Kayne RC (ed): *Drugs and the Elderly.* Los Angeles, University of Southern California Press, 1978, p 136.

59. Lundin DV: Medication taking behavior in the elderly—A pilot study. *Drug Intell Clin Pharm* 1978; 12:518–522.

60. Cadwallader DE: Drug interactions in the elderly, in Petersen DM, Whittington FJ, Payne BP (eds): *Drugs and the Elderly: Social and Pharmacological Issues.* Springfield, IL, Charles C Thomas, 1979.

61. Gotz BE, Gotz VP: Drugs and the elderly. *Am J Nursing* 1978; 78:1347–1351.

62. Lamy PP: Considerations of drug therapy in the elderly. *J Drug Issues* 1979; 9:27–45.

63. Lamy PP, Kitler ME: Drugs and the geriatric patient. *J Am Geriatr Soc* 1971; 19:23–33.

64. Lamy PP, Vestal RE: Drug prescribing for the elderly. *Hosp Prac* 1976; 11:111–118.

65. Raffoul PR, Cooper JK, Love DW: Drug misuse in older people. *Gerontologist* 1981; 21:146–150.

66. Vestal RE: Drug use in the elderly: A review of problems and special considerations. *Drugs* 1978; 16:358–382.

67. Ziance RJ: Side effects of drugs in the elderly, in Petersen DM, Whittington FJ, Payne BP (eds): *Drugs and the Elderly: Social and Pharmacological Issues.* Springfield, IL, Charles C Thomas, 1979, pp 53–79.

68. Caranosos GJ, Stewart RB, Cluff LE: Drug-induced illness leading to hospitalization. *JAMA* 1974; 228:713–717.

69. Hurwitz N: Predisposing factors in adverse reactions to drugs. *Br Med J* 1969; 1:536–539.

70. O'Malley K, Judge TG, Crooks J: Geriatric clinical pharmacology and therapeutics, in Avery GS (ed): *Drug Treatment.* Philadelphia, Lea & Febiger, 1976, pp 123–142.

71. Seidl LS, Thornton GF, Smith JW, et al: Studies on the epidemiology of adverse drug reactions. II: Reactions in patients on a general medical service. *Bull Johns Hopkins Hosp* 1966; 119:299–315.

72. Basen MM: The elderly and drugs—Problem overview and program strategy. *Public Health Rep* 1978; 92:43–48.

73. Petersen DM, Thomas CW: Acute drug reactions among the elderly, in Petersen DM, Thomas CW (eds): *Drugs and the Elderly: Social and Pharmaceutical Issues.* Springfield, IL, Charles C Thomas, 1979, pp 41–50.

74. Eisdorfer C, Basen M: Drug misuse by the elderly, in Dupont RL, Goldstein A,

O'Donnell J (eds): *Handbook of Drug Abuse*. Washington, DC, National Institute on Drug Abuse, 1979.

75. Inciardi JA, McBride DC, Russe BR, et al: Acute drug reactions among the aged: A research note. *Addictive Dis* 1978; 3:383–388.
76. Hemminki E, Heikkila J: Elderly people's compliance with prescriptions, and quality of medication. *Scand J Soc Med* 1975; 3:87–92.
77. Schwartz D, Wang M, Zeitz L, et al: Medication errors made by elderly, chronically ill patients. *Am J Public Health* 1962; 52:2018–2029.
78. Parkin DM, Henney CR, Quirk J, et al: Deviation from prescribed drug treatment after discharge from hospital. *Br Med J* 1976; 2:686–688.
79. Hurd PD: The role of self-supporting attribution bias in health-related behavior. Doctoral thesis, University of Minnesota, 1979.
80. Butler RN: *Why Survive? Being Old in America*. New York, Harper and Row, 1975.
81. Koch-Wesser J: *US Task Force on Prescription Drugs: The Drug Prescribers*. Washington, DC, US Government Printing Office, 1968.
82. Gallup G: The rising number of drinkers. *Washington Post,* June 10, 1974.
83. Pascarelli EF, Fischer W: Drug dependence in the elderly. *Int J Aging Hum Dev* 1974; 5:347–356.
84. US Bureau of the Census: *Current Population Reports*, Series P-25, No 311, 519, 614, 643, 704. Washington, DC, US Government Printing Office, 1965, 1974, 1975, 1977.
85. *Vital Statistics of the United States, 1975*. USDHEW, Washington, DC, 1978.

CHAPTER 2

AGE: A COMPLEX VARIABLE

PREM K. NARANG

1. INTRODUCTION

Interest in geriatrics has grown exponentially over the past decade. It is not merely a function of sentiment but rather a response to the realities dictated by demographic changes. As can be seen from Fig. 1, based on current projections, the present U.S. population of older people (those 65 years of age and over) will have more than doubled by the year 2030.[1] It is expected that over 50 million people will belong to this group and will constitute 15–17% of the total population. Given achievements and improvements in health care, the figure of 17% could climb as high as 25–30%. These projections are not influenced by changes in fertility rates. The older generation in 2030 will by and large be the baby boom population of the 1940s and 1950s grown older.

Looking back over this century, the growth in the absolute number and relative proportion of older people is obvious. This has been considered a major achievement for the society that has always measured success in health care by the standard of longevity. But there is also a tragic side to growing older, which, particularly in the case of women, makes people more prone to crime, impoverishment, and a variety of diseases. We also recognize the considerable impact of senile dementias, especially Alzheimer's type, which are a major contributing factor to the

PREM K. NARANG • Clinical Pharmacokinetics Research Laboratory, Pharmacy Department, Clinical Center, National Institutes of Health, Bethesda, Maryland 20892.

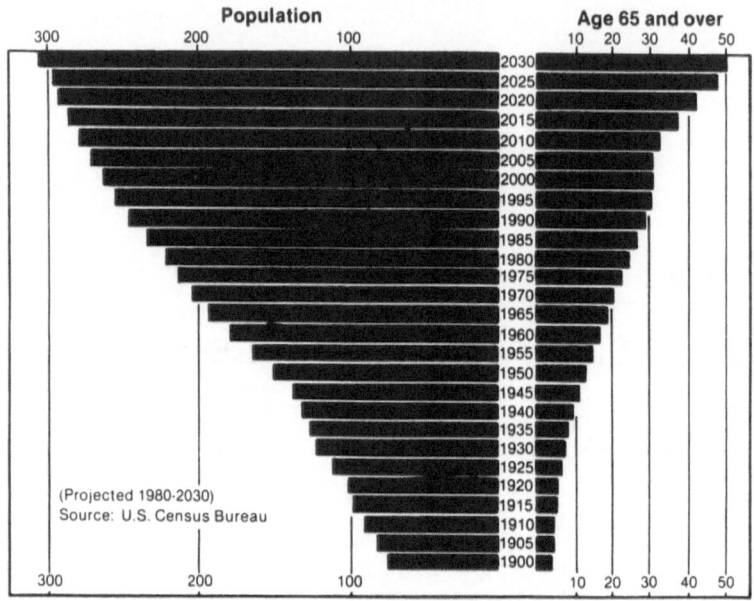

FIGURE 1   The elderly (those 65 years and over) as a function of total population of the
United States for years 1900–2030.

ever-increasing nursing-home admissions. This demographic revolution
is slowly and increasingly influencing social institutions, employment
policies, the medical care system, and our overall social fabric.

Before investigating the problem of aging, it is of utmost importance
that new knowledge be gained to enhance our understanding of the
biology and/or pathology of the aging process. It is also vital that dis-
tinctions be made between the effects of disease and the process of aging
before a researcher can understand the confounding nature of disease
and the aging process.

In this chapter, I shall first examine the problem that revolves around
the age variable. Understanding this variable is of vital importance to
the researcher in the field of epidemiology of aging and to the clinician
designing clinical trials with pharmacotherapeutic modalities. Second, I
shall present various investigative efforts that have explored alternatives
to the variable "chronological age," e.g., functional and biological age,
and finally a brief discussion on how this information may be helpful,
if not in a broader sense of understanding the age variable, at least in
the design of clinical drug trials in the geriatric population.

2. DEFINING THE AGE VARIABLE: DOES IT DESCRIBE AGING?

In order to define the variable "age," it is important that one realize some of the difficulties associated with it. The aging process can be and often is described as intrinsic, universally progressive, deleterious, and irreversible. Separation of the aging process from the effect of disease and other environmental factors must be made in research in an attempt to define and evaluate the process. The problem of defining the aging process is complicated by interindividual variation in how people age, both physiologically and psychologically. One must also bear in mind that body tissues, organs, and their functional capacities age differently, with the extent varying in different individuals. Therefore, the term "old age" (given a chronological age, e.g., 65 years) cannot be defined by some constant number of years. Application of epidemiological methods to the study of the aging process therefore presents more difficulties than their application to a disease process. A clear, valid, and reliable definition of aging still needs to be formulated. For example, classification of old age varies in the entitlements and eligibilities sponsored by the federal government, e.g., the lower limit for the "aged" is 60 years in the nutrition program (under Title VIII of the Older American Act), 62 years in the HUD housing program (No. 202), and 65 years in the Medicare program.[2] The definition of the age variable is further complicated by our inability to completely understand the impact of genetic makeup in reflecting the individual differences in the process of aging.

2.1. BIOLOGICAL AGE

Presupposition of a "biological age," the measurement of which is based on parameters other than time, is the fundamental thesis behind gerontological research. Inductive evidence for a biological age is afforded by the observation that individuals may be young or old in relation to their number of years or chronological age. Therefore, one can pose a fundamental question about aging: what is meant by the term "biological aging"? Does it mean that specific populations are characterized by finite maximal life-spans that differ greatly from one population to another? Do all members of a given population eventually show a progressive decline with time in their physiological performance? Methods or experimental approaches to objectively assess such ideas are either uncertain or their significance varies under different biological conditions. For instance, the amount of lipofuscin increases not only as a function of age but also as a result of malnutrition.[3] Is it, then, accurate to deduce from this observation that malnutrition accelerates the aging process or that lipofuscin levels below those normally encountered at a given age reflect a biologically younger individual?

It would, therefore, be extremely valuable to be able to predict, reproducibly, a verifiable discrepancy between the biological and the chronological age of individuals. To achieve this, a researcher must be able to assess or estimate biological age. If it can first be achieved in animals, such a model could then be scaled up for humans. Such estimates can be of considerable interest, not only in preventive medicine but also in identifying the basic aging process, and thus provide definition to the elusive variable "age."

The approach adopted for measuring biological age may depend on the researcher's concept of aging. If one accepts a distinct aging process, biologically occurring in all species, one is inclined to overlook and disregard the underlying, often fatal, disease in an attempt to assess age effects. On the other hand, if biological age is postulated as an expression of disease, also referred to as polypathy[4] or multiple pathology,[5] its assessment would demand a complete documentation of all lesions detectable by *in vivo* testing and/or postmortem studies. The choice between these two approaches is further complicated by a lack of precise definition of an "underlying disease process."

The question still remains: what might be considered as reliable and reproducible parameters of aging? Several biological changes correlate with passage of chronological time as individuals in a given population approach their apparent maximal life-span. For example, it is widely accepted that older individuals are characterized by increased susceptibility to most diseases, diminished organ function, reduced metabolism, and decreased organ perfusion.[6] Some parameters, such as the capability for repair of a specific type of DNA damage[7] and the susceptibility of DNA to chemically induced mutagenesis,[8] may even correlate in some species with maximal life-span. One feature that probably characterizes all aging populations is the progressive, modified ability to adapt to changes in the surrounding environment. One biochemical expression of this manifestation of aging is the altered capacity for the production of key enzyme molecules in response to some environmental challenge. Adelman[9] in 1971 reported one such biochemical expression of the aging process as the increase in the duration of lag period, or the time that elapses between the administration of inducing stimulus, glucose in this case, and the observed initiation of increased glucokinase (a liver enzyme) activity in Sprague-Dawley rats. As can be seen from Fig. 2, the lag period increases progressively with increasing age and is apparently directly proportional to the chronological age. Adelman and co-workers[10] have

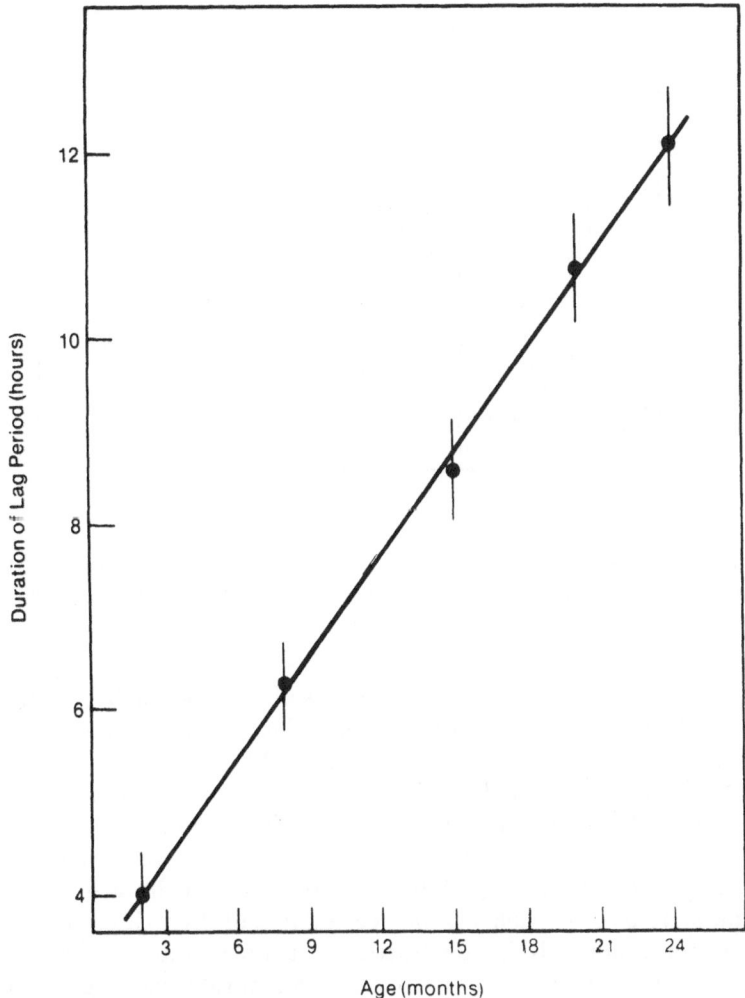

FIGURE 2   Regression of chronological age of the rat and duration of the lag period of
the glucokinase induction.

also shown other such enzymatic adaptations of tyrosine aminotransferase in response to ACTH and of cytochrome reductase in response
to phenobarbital in young and old animals (Fig. 3). Several other parameters have been evaluated to assess biological age and its discrepancy
from chronological age. A list of previously investigated parameters that
detect changes in rates of aging between a cohort aging at a normal rate
and one in which the rate is expected to be different is given in Table

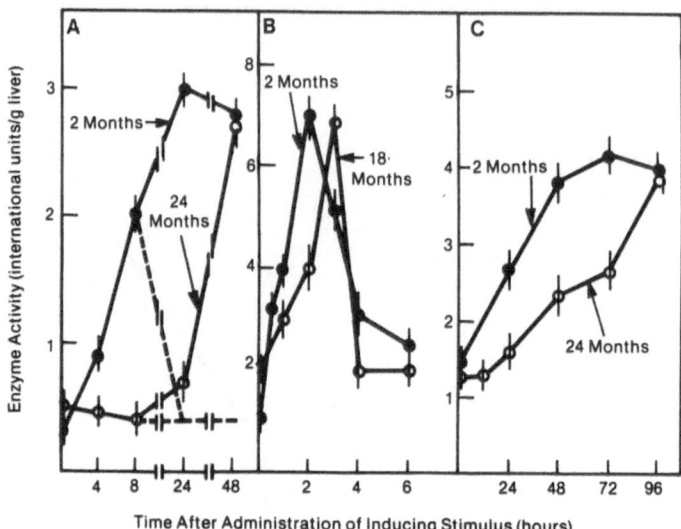

FIGURE 3   Age-dependent hepatic enzyme regulation in rats ($n = 6$). (A) Glucokinase-glucose. (B) Tyrosine aminotransferase ACTH. (C) NADPH: Cytochrome C reductase phenobarbital.

I. An excellent review of these parameters of biological aging within the framework of this discussion has been published by Ludwig and Smoke.[11]

Of the 12 parameters listed (Table I), thyroxine degradation rate and cell-doubling potential of skin fibroblasts, from the two human studies, need further comment. Similar to what has been observed by Adelman and co-workers for age-dependent hepatic enzyme regulation,[10] there appears to be an age-dependent decrease in the activity of the enzyme system responsible for the degradation of thyroxine, which leads to a decrease not only in the fractional turnover rate but also in the apparent thyroxine distribution space,[12] a measurable parameter to evaluate biological aging and perhaps in the formulation of a definition for the variable "age." Hayflick[13] in 1980 reported that the longevity of the skin fibroblast cultures is negatively correlated with the age of the donor. This observation suggests the phenomenon of senescence, as it reflects on the limited potential of somatic cells to double. Aging is associated with a loss in skin elasticity. Is it possible that these estimations of the doubling potential of skin somatic cells may provide an insight into changes in skin structure with age? However, we find ourselves in a dilemma again, as these involutional changes, e.g., loss of skin elasticity, bone brittleness, neuronal cell loss, and decreased immunological competence, are themselves genetically controlled. These changes increase

TABLE I   COHORT SIZE NEEDED TO REVEAL A 10% OR A 20% DEPARTURE FROM ASSUMED LIFE EXPECTANCY, FOR EACH OF 12 PARAMETERS OF BIOLOGICAL AGE[a,b]

Nature of measurement	Species	Assumed life-span in years	Slope	S.D.	n for given % of life-span	
					10%	20%
1. Thickness of glomerular basement membrane, 10^3 Å	Rat	3	0.090	.23	13	5
2. Neuronal cells, 10^6/g brain	Rat	3	−0.450	1.50	20	7
3. Chromosomal aberrations, % of abnormal mitoses at anaphase and telophase	Guinea Pig	6	0.310	2.00	19	7
4. Thymus weight, mg	Mouse	2	−0.740	2.20	35	10
5. Lipofuscin accumulation in the Purkinje cell layer of cerebellum, % tissue volume	Rat	3	0.110	0.48	34	10
6. Reactive astrocytes in hippocampus, square root of %	Rat	3	0.140	0.85	64	18
7. Thermic contractility of tail tendon, mm	Rat	3	0.063	0.42	76	20
8. Physiological capacitance in the stressed organism (gram caloric output)	Mouse	2	−5.300	25.00	86	23
9. Spleen cell response to T-cell mitogens, ratio PHA to Con A tritiated thymidine incorporation. Mean count/min	Mouse	2	−0.013	0.10	222	60
10. Thyroxin degradation rate, mg thyroxin iodine per day	Man	70	−0.055	15.00	220	60
11. Chromosomal aberrations. % of abnormal mitoses at anaphase and telophase	Hamster	3	0.500	6.40	270	71
12. Cell doubling potential of skin fibroblasts, number of divisions of cultured cells	Man	70	−0.017	8.00	660	170

[a] From Ludwig and Smoke[11] with permission from Beech Hill Publishing Co. © 1980.
[b] Rate of change per month

geometrically[14] beginning with the fourth decade of life and indicate a time-dependent, irreversible process predisposing to, but not identical to, disease. These changes could then be termed biological aging, or simply aging. Therefore, the aging process should be regarded not as an observed but as an inferred process, unless, of course, other approaches can be formulated to quantitate aging and define age of a given individual in a population.

Brown and Forbes[15] have suggested that an index of biological age must fulfill two basic requirements: (1) it must reflect the probability of

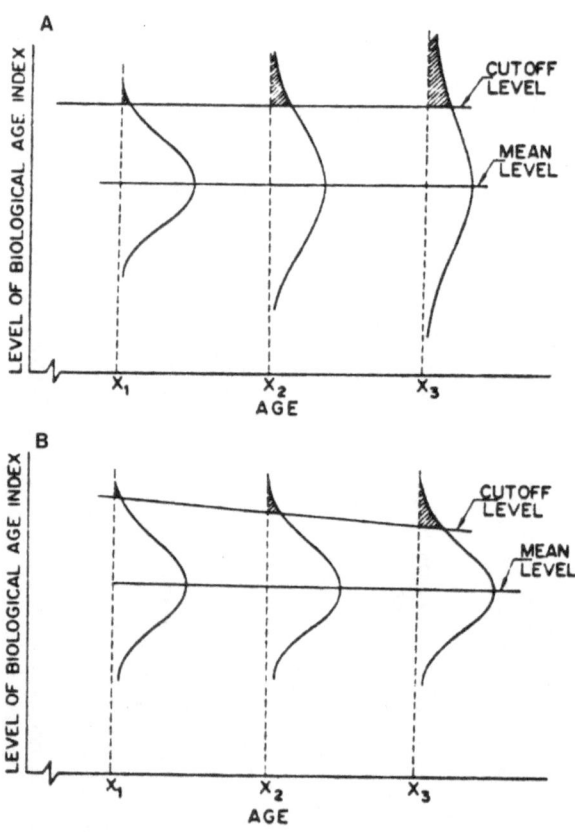

FIGURE 4 Hypothetical biological age indices whose means are uncorrelated with age, but which predict increased mortality with age: (A) increase in variance with age; (B) cutoff level decreases with age. Biological age indices at different ages are shown by the distributions, and the shaded areas represent the probability of death for a randomly selected individual of that age.

death at any chronological age, and (2) the individuals who die must have an abnormal level of such an index relative to living individuals of the same chronological age. Several mathematical models have been proposed by these authors[16,17] that relate the decline in physiological function(s) to the increased risk of death with age. As can be seen from Fig. 4A, the variance of the parameter (uncorrelated with chronological age) increases with age, and thus individuals in higher age groups with elevated levels of the biological parameter have a greater probability of death; in Fig. 4B, the cutoff level decreases with age, and again older individuals show abnormal levels and a greater probability. Analysis of data on biological age estimators, which are suspect owing to their susceptibility to significant variation from the differences in sex, strain, species, environmental interactions in experimental population, and genetic makeup, may need to be modeled stochastically to provide better understanding of the process of aging.

2.2. FUNCTIONAL AGE

Scientific evidence and common sense concur in seeing "age" as a very powerful variable for ordering information about the status and functioning of organisms. Chronological age, which roughly indexes the aging process, be it a controlled or random process, certainly affects a wide variety of biological, psychological, and social processes. It is also apparent that chronological age is not an explanatory variable, but more likely a kind of index that is used to stand for the progressive and deleterious changes we call aging.

The concept of functional age is not new. Researchers in the past have developed psychological profiles for psychological functioning, relating test score means to various age levels. Developmental psychologists have derived the intelligence quotient on the basis of the concept of mental age, which is the functional age in the areas of intellectual functioning. Age norms developed in medicine with respect to various organ systems also embody the idea of functional age. Similarly, functional age can be conceptualized in all sorts of sectors and be different from the chronological age.

The motivation for the functional-age concept lies in the fact that certain individuals can be observed to move more rapidly than others in directions associated with aging on a variety of selected age-related variables. Therefore, individuals can be classified as functionally older or younger than their true age in years. Functional age is one of a family of concepts that have been proposed in an attempt to take deviations in the chronological-age model into account and produce an alternative of equal generality but significantly higher precision. Functional age is thus a composite of many age-related measures, which in the present state of the art must be approached through rigorous statistical analysis.

The term "functional age" was first introduced in 1958 by Mc-Farland and Philbrook.[18] The concept of functional age rests primarily on the notion of differential rates of aging in individuals. Therefore, at any given time, those who age rapidly will be closer to death than those who age slowly, and they will, on the average, perform less well on age-related measures of functioning. Heron and Chown[19] identified eight separate variables that showed a linear relation to chronological age. These functional age measures were forced expiratory volume, grip strength, sitting height, low nonverbal intelligence, poor perceptual maze performance, poor digit coding, systolic blood pressure, and hearing loss. These authors, however, did not combine these variables to form functional-age scores because of significant variability and statistically poor nonsignificant correlations. Comfort[20] provided a set of such vari-

TABLE II   VARIABLES USED IN FUNCTIONAL AGE PREDICTION EQUATIONS
IN THREE STUDIES

Dirken[21]	Furukawa et al.[22]	Normative aging study[23]
Resting EKG	Glomerular filtration rate	Grayness of hair
Exercise EKG	Renal plasma flow	Disassemble (GATB)
Maximum energetic load	Systolic blood pressure	Speech reception loss
Maximum pulse	Diastolic blood pressure	FEV_1
frequency	Cholesterol	Length of ear
Maximum system blood	Red blood cells	% hemoglobin
pressure	White blood cells	Chances for advancement
VC_1	Hemoglobin	Perceived position in
FEV_1	concentration	company
Maximum breathing rate	Icterus Index	Surgency
Urinary 17-oxogenic	Total protein	Expected age of
17-Oxosteroids	Height	retirement
Creatinine	Weight	Plans to remain with same
Psychomotor positioning	Vital capacity	company
Tapping rate	Ocular accommodation	Hearing loss for 8-kHz
Hand dynamometry	Vibratory sensation	tone
Speech audiometry	Grasping power	
Pitch ceiling	Tapping rate	
Presbycusis	Body flexibility	
Visual acuity	Heart rate recovery	
Picture completion		
Reaction time		
Concentration		

ables or measures, which included graying of hair, height, skin thickness and elasticity, thorax size, systolic and diastolic blood pressure, total vital capacity, tidal volume, and forced expiratory volume. Table II shows the variables that were employed in three major studies[21-23] in assessing the functional age from prediction equations using a multiple linear regression model.

Assessment of functional age is a persistent practical problem in job assignment, retirement policies, and driver and pilot licensing. Early detection of aging, as an index to functional decrement, would be extremely useful in the clinical management of a patient, especially if the physician knew the extent to which decrements in certain organ systems reliably presage decrements in others. Therefore, for all practical purposes, functional age can be operationally defined as the age predicted by the tests used. Such estimates are usually based on research done cross-sectionally, rather than longitudinally. Results from these studies, however, need validation from longitudinal trials. Damon[23] used a step-wise multiple-regression program to evaluate 18 independent variables. Fifteen of them were body measurements along with grip strength, grayness of hair, and baldness. The contribution of each independent variable entered in the regression equation toward improvements in R was assessed. Table III shows the results of their analysis. They found grayness of hair to be the best readily available indicator of age as it con-

TABLE III   ORDERING OF PHYSICAL VARIABLES,[a] BY CONTRIBUTION TO PREDICTION OF AGE, 600 HEALTHY VETERANS, AGED 25–75[b]

Variable	Multiple R	R^2	R^2 change	Simple r, with age
Grayness, hair	0.64	0.41	0.41	0.64
Grip, right hand	0.67	0.45	0.04	0.31
Ear breadth	0.69	0.48[c]	0.04[c]	0.26
Sitting height	0.71	0.50	0.02	0.23
Ear length	0.72	0.52	0.02	0.30
Nose breadth	0.73	0.54	0.01	0.22
Bideltoid breadth	0.74	0.55	0.01	0.25
Abdominal depth	0.76	0.57	0.02	0.21
Triceps skinfold	0.76	0.58	0.01	0.11
Baldness	0.77	0.59	0.01	0.35

[a] Best of eighteen; the eight noncontributory variables were height, weight, chest depth and expansion, bi-iliac breadth, upper arm and calf circumferences, and nose length.
[b] From Damon[23] with permission.
[c] Discrepancy due to rounding off.

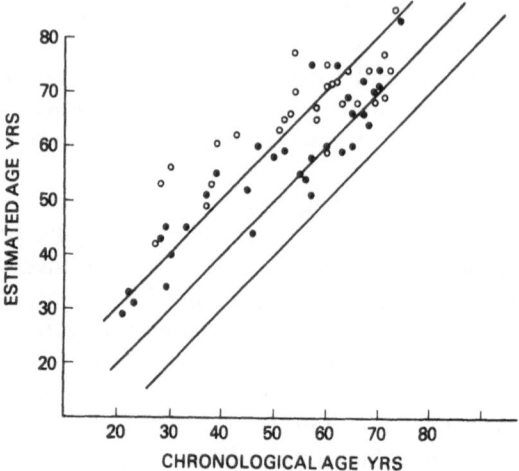

FIGURE 5   Estimated age of hypertensives as a function of their chronological age. The middle line is the regression function obtained from normal healthy subjects with upper and lower solid lines as tolerance limits ($\pm$ 1 S.D.). $\bigcirc$, Female; $\bullet$, male.

tributed almost 41% to the total R^2 of 59%. Their study concluded that physiological, functional traits may predict age much better than anatomical or structural ones.

Furukawa et al.[22] in 1975 showed that the estimated age for hypertensive subjects, computed by the regression equation obtained from healthy subjects, was significantly higher than their chronological age (Fig. 5). Therefore, hypertensives were suggested to be functionally older than normals. Webster and Logie[24] predicted the functional age from seven clinical variables in a sample of 1080 apparently healthy female subjects ranging in age from 21 to 83. The authors took additional steps by attempting to validate the prediction by selecting a subsample of 97 nonsmokers, who were shown to have a mean predicted age less than their chronological age by 1.5 years. Although problems still remain in this study in the conceptual confounding of the disease with aging, the attempt to predict external criteria makes this study a good evidence for validity of the functional-age concept.

2.3. PITFALLS OF AGE CONCEPTS

Enormous research efforts and resources have been invested not only in the concepts of functional and biological age but in other such concepts, each attempting to assess and develop an alternative variable to chronological age. However, these concepts have primarily survived on the basis of their scientific appeal. As a researcher in the field of aging and clinical pharmacology, one must be aware of the potential methodological concerns with such concepts. Predictable declines are observed in several functions with age. Measures such as the cardiac output, vital capacity, glomerular filtration rate, renal plasma flow, grip strength, and reaction time show decrements, whereas some others, like blood volume, sugar level, plasma osmotic pressure, serum electrolyte content, and verbal intelligence, do not show any age-dependent changes. These observations reflect an independence in these changes rather than a tendency to covariate. The time of onset of these changes varies, as do the curves for such decrements in individuals. Therefore, if these alternative age variables are measured at one time and shown to be predictive or somehow correlated with chronological age, there is no reason to believe that inter- and extrapolation to another date and time would maintain their predictive value.

Most studies for developing alternative age variables have used a multiple-regression model to predict chronological age by using a set of age-related variables. This model formulation does not appear to be an appropriate method, as, other than assigning weights to a set of variables that will maximize the correlation of the resulting linear combination with some criterion variable, it assumes that the relationship between the predictors and criterion variable is linear. In fact, as we have already seen, many variables show different rates of decline with age and hence may not be appropriate for use in the multiple-regression model. Separate regressions for young and old cohorts (Table IV) clearly demonstrate poorer correlations of age with "age-related" variables in the young compared to those seen in the old. Use of a "tool-matching" variable, as a test of perceptual ability, appears to be unrelated to age among young men (Table IV) and therefore cannot form a good variable to assess functional age on the basis of their scores.[25] Also, as Costa and McCrae adequately point out, there appears to be a kind of circularity in using chronological age as the criterion for developing an alternative to chronological age.[25] The logic of combining age-related measures to maximize the correlation with chronological age seems rather absurd because if one obtained a perfect regression (i.e., $R^2 = 1$) with chronological age, it would also be a perfectly useless alternative to the chronological age. Although statistically estimated ages do not correlate strongly with the chronological ages, they have allowed the researchers to explore the sources of error and variation that bear on the chosen method of analysis. A detailed critique of the alternative age concepts may be found in the excellent report of Costa and McCrae.[25]

TABLE IV   REGRESSIONS PREDICTING CHRONOLOGICAL AGE IN YOUNG
AND OLD GROUPS[a]

	Young					Old				
	Simple R	Beta	Mean	S.D.	n	Simple R	Beta	Mean	S.D.	n
T_1 age	—	—	32.63	(3.22)	762	—	—	55.42	(6.43)	504
Grayness	0.30	0.30	0.65	(0.86)	511	0.41	0.30	2.80	(1.42)	407
Speech reception threshold	0.07	0.08	1.32	(4.21)	543	−0.42	0.32	5.81	(7.52)	353
Tool matching	0.01	0.03	28.37	(5.76)	367	−0.31	−0.15	23.75	(5.39)	282
Disassemble	−0.04	0.01	30.44	(3.61)	367	−0.41	−0.23	26.11	(4.36)	282
Ear length	0.20	0.19	67.46	(3.98)	693	0.21	0.09	70.61	(4.34)	471
	Multiple R: .362					Multiple R: .643				

[a] From Costa and McCrae[25] with permission.

3. CLINICAL TRIALS IN GERIATRICS: PHARMACOLOGICAL BASIS

With the recognition of the fact that the elderly will constitute an ever-increasing proportion of our patient population and with the inherent problems associated with an accurate description of the variable age, it is easy to understand the extraordinary amount of attention that is being directed into geriatric research. The elderly are a much more heterogeneous group than are young normal subjects. In order to effectively gain information from the study in the elderly and about aging, one must gain insights not only into the area of epidemiology but also in areas of pharmacology, toxicology, toxicokinetics, and pharmacokinetics. A significant amount of information in these areas is obtained from the cross-sectional and longitudinal clinical trials in geriatrics. It should be emphasized that cross-sectional studies provide information about age-related differences as contrasted with age-related changes. In order to gain maximum information from each such trial, it is critical that study designs be balanced and statistically effective.

It is common knowledge that the incidence of various diseases and illnesses increases with increasing age, and in seeking relief, the elderly patient receives more drugs per capita than those in the younger age groups. The potential for drug interactions in this population is therefore significantly enhanced owing to multiple-drug therapy. Several surveys[26,27] document the increased propensity for toxic effects of drugs used in the elderly. It has been suggested that the likelihood of adverse drug reactions to drugs in the elderly population is three to seven times higher than in the young group.[27] This enhanced "sensitivity" of the elderly to therapeutic agents can be attributed to aspects of drug disposition, viz., pharmacokinetics and/or pharmacodynamics. For most drugs, the kinetic differences are generally consistent with age differences in renal function, vital capacity, cardiac index, basal metabolic rate, protein binding, tissue perfusion, body composition (e.g., total body water and lean body mass), and a host of other factors that decrease with age. Figure 6 summarizes the age regression for a number of physiological functions. One could easily consider the slope of these regressions to reflect the rate of aging, but this figure also reveals that there is no single index to describe the aging process; the rate of conduction of a nerve impulse falls 10–15% between the ages of 30 and 80 years, and the resting cardiac output falls by almost 50% over the same time. Therefore, in the design of clinical trials in the elderly, special attention must be directed at appropriate aging parameters to effectively assess the pharmacodynamics of that particular class of therapeutic agents.

In the past, somewhat less investigative attention has been devoted to evaluating the dynamic aspects of drug response in the elderly. Available data are scant when one seeks information regarding the influence of aging on drug sensitivity. There is, however, an ever-increasing literature on the effects of age on pharmacokinetics, e.g., the time course associated with drug absorption, distribution, metabolism, and elimi-

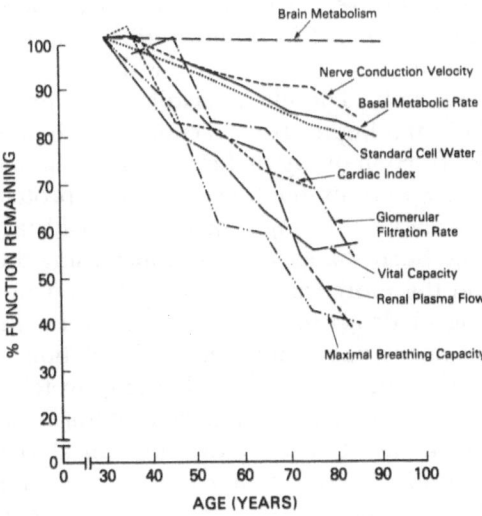

FIGURE 6   Dependence of physiological function on chronological age.

nation. Drug absorption may be altered in the elderly because of a reduction in the number of absorbing cells or because of a delay in gastric emptying and decreased motility. Distribution of drugs in the elderly can be expected to be different because of smaller stature and altered body composition. Longitudinal studies have shown that the aging process generally leads to a decrement in the actual lean body mass per unit of total body weight. It is also postulated that the aging process may result in an altered intrinsic metabolic capacity of the liver for drugs, leading to a reduced drug or metabolic clearance in the elderly. The largest amount of information on the effects of aging exists in renal physiology. Reductions in both the glomerular filtration and tubular secretory and reabsorptive functions show a significant impairment in drug elimination (Table V) in the elderly. Interested readers should review the excellent works of Vestal[28] and Goldberg and Roberts.[29] Table V shows some of the predominant physiological factors that are altered in the elderly and their possible effect on the absorption and drug disposition in the aged.[30]

The majority of clinical trials in the past have focused predominantly on evaluating the disposition of the therapeutic moiety from a measurable biological pool, e.g., the blood (or plasma) or urine, but it must be realized that the free or the unbound drug is the real determinant of both the drug pharmacokinetics, i.e., distribution and elimination, and its pharmacodynamics in most cases. Alterations in binding of the drugs to plasma proteins, red blood cells, and other tissues of the body may reflect the response differences seen in the elderly and also enhanced sensitivity. Cammarata et al.[31] have shown that albumin concentrations are reduced by up to 20% in the elderly with a simultaneous increase in the globulin fraction. Therefore, acidic drugs primarily bound to albumin will result in a higher free fraction, and some basic drugs will exhibit lower free fraction owing to their binding to globulins. Although the number of receptors or binding sites on a protein is genetically controlled, conformational changes in proteins with age can result in decreased binding. Such changes result in increasing the concentration of free drugs and therefore can potentiate a therapeutic or toxic drug response. Binding of drugs to biological ligands is, therefore, an important area that needs significantly more elucidation. Ligand concentration, number of binding sites of the binding protein, drug receptor affinity, and binding capacity are important determinants of the free drug that must be evaluated when conducting clinical trials in the elderly. Classes of compounds where binding may play an important role in altering drug effect will be discussed in individual monographs in Part II of this book.

TABLE V PHARMACOKINETIC CONSEQUENCES OF AGE-RELATED ALTERATIONS[a]

Organ	Physiologic alteration	End-organ effect	Effect on disposition of drug
Kidney	↓ Renal blood flow ↓ Permeability Thickening of the basement membrane of Bowman's capsule	↓ Glomerular filtration rate	↓ Drug clearance ↑ Half-life Altered tissue uptake
Liver	↓ Hepatic blood flow ↓ Metabolic capacity Altered hepatic tissue uptake	↓ Rate and extent of delivery	↑ Bioavailability ↓ Metabolic clearance[b] ↑ Half-life
Muscle	↓ Muscle mass ↑ Fat/lean mass ratio Altered catabolism ↓ Total body water	↑ Absorption of lipophilic agents	Altered apparent distribution volume Altered tissue uptake <—> Half-life
Gastrointestinal tract	↓ Splanchnic blood flow ↓ Gastric motility ↓ Secretory activity of gastric mucosa ↓ Gastric pH Atrophic changes in mucosal epithelium	↓ Altered drug ionization ↓ Transport efficiency	↓ Fraction absorbed ↓ Absorption rate ↑ Peak time Gastric pH
Heart and large arteries	↓ Cardiac output ↓ Elastic tissue	↓ Ejection fraction ↓ Permeability	↓ Drug clearance ↑ Half-life

[a] From Cutler and Narang[30] with permission.
[b] Depends on the extraction ratio, intrinsic clearance, and whether clearance is blood flow limited for a given drug.
↑ = increase; ↓ = decrease; <—> = no change.

4. CONCLUSIONS

Although significant progress has been made in understanding the effects of age on pharmacology, much research is still needed in the area of drug interactions and adverse drug reactions in the elderly. Very few, if any, systematic investigations have focused on the toxicology of drugs in the elderly and in evaluating the extent to which the age variable per se may predispose older subjects to drug interactions. There can be no generalizations about influence of age unless studies are performed that are designed to control for severity of disease and prevalance of other drug use as well as age. It seems appropriate that the age-related alterations in the pharmacokinetics and pharmacodynamics be first delineated before alterations due to a disease process are superimposed. This requirement demands that clinical trials evaluating the variable "age" be designed with extreme caution. Careful attention must be given to subject selection, their nutritional, environmental, psychological, and genetic status during the protocol design. Use of cohorts matched for all variables except age has been recommended. Such designs should also attempt to simulate the actual clinical use of the drug under investigation. This requirement may dictate a multidose clinical trial rather than one based on a single dose. Whatever the design, a clear understanding of the hypothesis being tested is not only desirable but critical. Studies in clinical pharmacology evaluating the age variable have been predominantly cross-sectional. Longitudinal studies are unfortunately difficult, time consuming, and expensive in design. Distinction must be made between these two design aspects: as distinct from the variable age, the cross-sectional design includes both time- and environment-related variables.

Another important facet of these study designs should be an attempt to estimate intraindividual variation in drug disposition. Diurnal rhythms in endogenous neurochemical and neuroendocrine function and oscillations in biological response modulators may significantly influence dynamic response measurements during a trial in study subjects. Drug tolerance and/or sensitivity in the elderly to prolonged drug administration are issues that need to be investigated thoroughly by experimental designs so as to achieve the goal of optimal drug dosing in the elderly. A great deal of attention must be given to the statistical analysis of data generated from clinical investigations in order to ensure the validity of estimates of not only safety and efficacy but also those describing the data. The number of subjects chosen for such studies should be based on statistical principles of power analysis by describing the desired precision.

Several approaches have been tried in the field of gerontology to define the variable age, but most of them have met with minimal success. This is partly due to the complex process of aging. Influence of social, psychological, environmental, and genetic factors on this process is a major stumbling block when designing studies that evaluate aging or those determining the influence of age on alteration of drug response. It is, therefore, important that study designs should have strict and explicit exclusion criteria based on the hypothesis under investigation.

Exploration of alternative age variables and changes in physiological functions as a function of chronological or functional age suggests that it is important that appropriate parameters be chosen when clinical trials are executed in the elderly. Estimates of cardiac index or ejection fraction when evaluating antiarrhythmics and measurements of the metabolic capacity of the liver in cases where the drug is extensively metabolized may prove to be better physiological parameters for evaluating influence of age than chronological age. Attempts should also be made to obtain a measurement of end-organ response, e.g., measurement of vital capacity and/or maximal breathing capacity when evaluating broncodila-tors. Studies that address the practical question of influence of age must do so more mechanistically by evaluating not only the pharmacokinetics but also the pharmacodynamics of therapeutic agents in the elderly. Application of information gained from basic research with the use of *in vitro* animal model systems must be kept in perspective during the design and conduct of clinical trials in geriatrics.

ACKNOWLEDGMENT. The author thanks Dr. Harvey Klein, Chief of the Department of Transfusion Medicine, for his extremely useful suggestions and Ms. Nya J. Holland for the efficient typing of this manuscript.

REFERENCES

1. *Demographic Aspects of Aging and the Older Population in the United States.* Current Population Reports, Bureau of Census, Series P-23, No 59, Bethesda, MD, 1976.
2. *Second Conference on the Epidemiology of Aging.* U.S. Department of Health and Human Services, NIH Publication No 80-969, Bethesda, MD, 1980, p 7.
3. Miquel J, Oro J, Bensch KG, Johnson JE Jr: Lipofuscin: Fine structural and bio-chemical studies, in *Free Radicals in Biology.* New York, San Francisco, London, Academic Press, Inc, 1977, vol III, pp 133–152.
4. Franke H: Aktuelle Probleme der Gerontologic und Geriatrie. *Naturwissenschaften* 1978; 61:150–156.
5. Howell TH: Multiple pathology in nonagenarians. *Geriatrics* 1963; 18:899–902.
6. Timiras PS: *Developmental Physiology and Aging.* New York, London, The Macmillan Co, 1972.
7. Hart RW, Setlow RB: Correlation between DNA excision-repair and life-span in a number of mammalian species. *Proc Nat Acad Sci USA* 1974; 71:2169–2173.

8. Schwartz AG: Correlation between species life-span and capacity to activate DMBA to a form mutagenic to mammalian cell. *Exp Cell Res* 1975; 94:445–447.

9. Adelman RC: Age-dependent effects of enzyme induction—a biochemical expression of aging. *Exp Geront* 1971; 6:75–87.

10. Adelman RC, Freeman C, Cohen BS: Enzyme adaptation as a biochemical probe for development and aging, in Weber G (ed): *Advances in Enzyme Regulation*. Oxford, New York, Pergamon Press, 1972, vol 10, pp 365–382.

11. Ludwig FC, Smoke ME: The measurement of biological age. *Exp Aging Res* 1980; 6(6):497–521.

12. Gregerman RI, Gaffney GW, Shock NW, Crowder SE: Thyroxine turnover in euthyroid man with special reference to changes with age, *J Clin Invest* 1962; 41:2065.

13. Hayflick L: The cell biology of human aging. *Sci Amer* 1980; 242:58–65.

14. Gompertz B: On the nature of the law of human mortality and new mode of determining life contingencies. *Philosophical Trans Royal Soc London (Series B)* 1825; 115:513–585.

15. Brown KS, Forbes WF: Concerning the estimation of biological age. *Gerontology* 1976: 22:428–437.

16. Brown KS, Forbes WF: A mathematical model for aging process. *J Gerontol* 1974; 29:46–51.

17. Brown KS, Forbes WF: A mathematical model for aging process II. *J Gerontol* 1974; 29:401–409.

18. McFarland RA, Philbrook FR: Job placement and adjustment for older workers: Utilization and protection of skills and physical abilities. *Geriatrics* 1958; 13:802–807.

19. Heron A, Chown S: *Age and Function*. Boston, Little, Brown Co, 1967.

20. Comfort A: Test battery to measure aging rate in men. *Lancet* 1969; 2:1411–1415.

21. Dirken JM (ed): *Functional Age of Industrial Workers*. Groningen, Netherlands, Netherland Institute for Preventive Medicine TNO, Walter Noordhoff Publishing, 1972.

22. Furukawa TL, Inoue M, Kajiya F, et al: Assessment of biological age by multiple regression analysis. *J Gerotol* 1975; 30:422–434.

23. Damon A: Predicting age from body measurements and observations. *Int J Aging Hum Dev* 1972; 3:169–174.

24. Webster IW, Logie AR: A relationship between functional age and health status in female subjects. *J Gerontol* 1973; 19:1–19.

25. Costa PT, McCrae RR: *Functional Age: A Conceptual and Empirical Critique. Second Conference on Epidemiology of Aging*. U.S. Department of Health and Human Services, NIH Publication No 80-969, Bethesda, MD, 1980, p 30.

26. Hurwitz N: Predisposing factors in adverse reactions to drugs. *Br Med J* 1969; 1:536–539.

27. Seidl LG, Thornton GF, Smith JW, et al: Studies on epidemiology of adverse drug reactions. III. Reactions in patients on a general medical service. *Bull Johns Hopkins Hosp* 1966; 119:299.

28. Vestal, RE: Drug use in the elderly: A review of problems and special considerations. *Drugs* 1978; 16:358–382.

29. Goldberg PB, Roberts JR: Pharmacological basis for developing rational drug regimens for elderly patient. *Med Clin North Am* 1983; 67(2):315–331.

30. Cutler NR, Narang PK: Implications of dosing tricyclic antidepressants and benzodiazepines in geriatrics. Symposium on clinical psychopharmacology: *Psychiatr Clin North Am* 1984; 7(4):845–861.

31. Cammarata RJ, Rodman GP, Fennell RH: Serum antigamma-globulin and anti-nuclear factors in the aged. *JAMA* 1967; 199:115–118.

CHAPTER 3

Physiological Changes with Aging
Relevance to Drug Study Design

Alexander M. M. Shepherd

1. Introduction

The purpose of this chapter is to provide an overview of those aspects of physiological change which occur with normal aging in humans and which may be of importance in influencing the design and conduct of drug studies. This will be done by briefly examining those physiological changes in humans that might be expected to influence disposition of and response to drugs. This approach is being taken for several reasons.

First, study design depends on certain assumptions about drug disposition. For example, many developmental drug studies will examine the oral bioavailability of a desired formulation of a drug. This is usually done by comparing the area under the plasma concentration versus time curve of the formulation given by mouth versus that of the same dose of drug in solution given by mouth and not by the intravenous route. Thus, relative and not absolute bioavailability is obtained. For a drug that is relatively well absorbed in the young, this information may be of clinical and predictive value. However, if, for example, gastrointestinal motility, gastric emptying time, and liver blood flow were to alter with aging, this could significantly reduce the oral bioavailability of the oral

ALEXANDER M. M. SHEPHERD • Departments of Pharmacology and Medicine, Division of Clinical Pharmacology, University of Texas Health Science Center at San Antonio, San Antonio, Texas 78284.

solution. The bioavailability of the desired formulation would then be compared with a markedly different standard. In this instance, study design to examine the absolute bioavailability of the formulation (compared with intravenous drug) would give a better indication of the probable drug effect in the elderly, compared with the young.

Another example may be seen in studies examining the magnitude of drug response. Studies of the effects of hypnotics in the elderly frequently do not take into account the different baseline in the elderly from which the drug effect is being measured. The elderly will be expected to be less well coordinated in the drug-free state than will the young, and it is vital to design studies of crossover or parallel design, where there is a control group that is age- and capability-matched with the drug groups.

Second, it is important to know which parameters of physiological function actually change consistently with age. Based on knowledge of these changes and of the interaction of these changes with the known chemical and physical properties of the drug to be studied, some predictions can be made about the expected behavior of the drug in the body. Thus, studies can be designed to specifically examine the predicted behavior of the drug. For example, if liver blood flow and liver-metabolizing capability both fell consistently with age, the blood profile of an extensively metabolized drug with high hepatic extraction, dependent on liver blood flow, would behave differently in the aged from one with lower hepatic extraction in which the rate-limiting step was the rate of hepatic metabolism of the drug.

Third, only those changes demonstrated in humans will be examined in this chapter. This is because animal studies, which are far more numerous than those in humans (it being easier and cheaper to do longitudinal and invasive studies in small animals), frequently study changes in physiology that do not pertain to the human situation. For example, studies of the effect of age on a venodilator should not use dogs, which can rapidly and effectively maintain venous return to the heart by contraction of the splanchnic and splenic circulations. An age-related change in this capability in the dog would be irrelevant to the human situation. In addition, elderly animals are usually a homogeneous population, unlike humans, most of whom have been exposed in varying degrees to the ravages of cigarettes, alcohol, obesity, and potent enzyme inducers and inhibitors. Thus, aging in animals is a rather pure process and may have little relevance to what actually happens to humans as they age. There are as many documented dissimilarities between the effects of aging in animals and in humans as there are similarities. Thus, although animal studies are of some value in predicting the effects of aging in humans, they should not be thought of as replacing well-thought-out human studies that examine the effect of age on drug disposition and response in humans.

Age-related physiological changes are best examined by describing changes in functions that could have a major impact on drug disposition and response. Changes that might affect the following parameters of drug handling will be described:

1. Drug absorption
2. Drug distribution
3. Drug metabolism
4. Drug renal excretion
5. Drug response

2. DRUG ABSORPTION

After drug administration by all except the intravascular route, there is an absorptive step before the drug enters the "central" compartment of the body. This is true whether the drug is given by the oral, the rectal, the subcutaneous, or the intramuscular route. In determining the impact of the absorptive step, it should be remembered that some drugs are administered directly into the desired site of action and that greater absorption into the central compartment may actually decrease the magnitude and duration of desired action and increase the systemic toxicity of the drug. This may be true for inhaled aerosols of bronchodilators or steroids, intrathecal antibiotics or cancer chemotherapeutic agents, or intraarticular administration of steroids.

Most drugs are given by the oral route for convenience and are absorbed from the small intestine by passive diffusion, in the absence of an active transport mechanism such as is present for some nondrug entities (for example, calcium and iron). They are best absorbed when present in the small intestine in the un-ionized form in solution, since it is in that form that they will best be transported across the lipid layers of the small intestinal endothelium. Thus, age-related factors that might alter drug absorption include

1. Reduced gastric motility
2. Increased gastric pH
3. Reduced small intestinal absorptive surface area
4. Reduced portal circulation

2.1. REDUCED GASTRIC MOTILITY

This will delay emptying of the drug into the usual absorptive compartment, the small intestine. This increases the likelihood of the drug being broken up at the highly acid pHs in the stomach. In addition, the profile of the plasma concentration of the drug will alter so that a lower, flatter area of plasma drug concentration versus time curve will be obtained. This may have significance, especially for antibiotics where the presence, even for a short time, of high blood concentrations is considered necessary to provide the bactericidal effect. Conversely, a lower peak concentration may be of therapeutic utility, since the acute adverse effects of some drugs are dependent on high peak concentrations. Examples of this are the vascular headaches seen with the arterial vasodilator drug hydralazine and the sedating and dry mouth effects of the central α_2-agonist antihypertensive drug clonidine.

What information do we have in humans that there is a trend to reduced gastric motility in the elderly? No definitive studies on the effect of age in gastric emptying have been performed. However, there is no doubt that the incidence of chronic atrophic gastritis is higher in the elderly.[1-5] The likelihood of delayed gastric emptying is higher in aged subjects with atrophic gastritis.[1,2] Thus, delayed gastric emptying into the small intestine may be anticipated in a sizable proportion of the elderly population.

2.2. INCREASED GASTRIC pH

Gastric pH may affect the degree of breakdown of a drug while it is in the stomach. In addition, the degree of ionization of weakly acidic and weakly alkaline drugs may change, thus altering the likelihood of their being completely absorbed in the small intestine. There is reasonably good evidence of reduced basal gastric acid secretion[4-9] and acid secretion stimulated either by histamine[6,10-12] or by psychological stress.[13]

2.3. REDUCED SMALL INTESTINAL ABSORPTIVE SURFACE AREA

The absorptive area in the small intestine is comparatively large because of the presence of villi and crypts that provide additional surface area. No human studies have examined this aspect of physiological aging, but there are suggestions, based on mouse studies, that increasing age is associated with atrophic, widely separated crypts, thus reducing the absorptive area available for drug absorption.

2.4. REDUCED PORTAL CIRCULATION

Reduced liver blood flow would reduce the chemical gradient across the small intestinal endothelium, thus, at least theoretically, reducing the rate of drug absorption. Several studies, starting as early as 1945, have examined this. The consensus is that there is significant and progressive reduction in liver blood flow from about 1550 ml/min to 1220 ml/min over the age range 20–75 years.[14–16] This difference is statistically significant. Whether it is clinically significant is not clear. For well-absorbed drugs, there is probably no age-related change. However, since the drug is in the small intestine for only a finite time, the absorption of a poorly absorbed drug could fall significantly if passage across the small intestinal endothelium were impaired by a reduced chemical gradient.

3. DRUG DISTRIBUTION

Drug is usually transported to its site of action in the blood. It is present in the plasma and to a varying degree in the red blood cells. In the plasma it may be free, in solution, or bound to plasma proteins, which may most conveniently be regarded as a circulating storage compartment. The proportion of drug in the blood will depend to a major extent on its lipid/water partition coefficient as well as its binding in and out of the vascular compartment. Thus, changes in the fat/lean body mass relationship as well as change in plasma proteins, principally albumin, will affect drug distribution.

Factors affecting drug distribution may usefully be divided into

1. Plasma protein levels
2. Plasma volume
3. Lean-to-fat body ratio
4. Red-blood-cell changes
5. Distribution into the urinary tract

3.1. PLASMA PROTEIN LEVELS

It has long been known that the concentration of albumin in blood falls with increasing age.[17] The average fall over the range 40–70 years is from about 4.1 to 3.0 g/100 ml. Thus, it might be expected that there would be an increase in the free, active proportion of a drug that is normally bound to albumin. In theory, the bound and free fractions of an albumin-bound drug in plasma may be calculated from the derivation by Goldstein[18] of the Langmuir isotherm[19,20] where

$$\frac{\text{Bound drug}}{\text{Total drug}} = \frac{1}{1 + \dfrac{K}{np} + \dfrac{[x]}{np}}$$

where K = dissociation constant; n = number of receptor groups per albumin molecule; p = molar concentration of albumin; and $[x]$ = concentration of unbound drug.

From this equation, it may be calculated that for a change in albumin concentration from 4.5 to 3.0 g/100 ml, the percentage of unbound drug will range from a 45% increase for a very highly bound drug (99.9%) at low concentration, to a 24% increase for a much less avidly bound drug (32%) at very high concentration. Thus, sizable changes are to be expected. However, several factors must be taken into account when trying to quantify the degree of increase in free concentration to be expected and its clinical significance.

For example, the change in protein binding may be relevant only when a drug is relatively highly bound, to an extent of 80% or more. Second, for a drug with a very large volume of distribution, only a small proportion of the total drug amount in the body is present in the blood (the rest is in the tissues). Thus, displacement of drug from albumin-binding sites in the blood would result in little change in the free concentration in the plasma as the additional free drug would then be taken up in the tissues.

Third, the rate of elimination, by hepatic metabolism, of a drug with high hepatic extraction is dependent on the free, unbound concentration of the drug. Thus the free, active concentration of such a drug with first-order elimination would remain the same but at a lower total blood concentration.

3.2. PLASMA VOLUME

There is no evidence of an age-related alteration in plasma volume in a cross-sectional study of humans aged 20–90 years[21] or in a longitudinal normal study of men studied over a 17-year period.[22] However, Edelman and Leibman[23] found that total body water decreases with increasing age.

3.3. LEAN-TO-FAT RATIO

Several measurements of lean body mass[24–29] and muscle mass[30–33] have indicated a fairly significant reduction in these parameters with increasing age. For example there is probably about a 6- to 7-lb loss of lean mass per decade of increase in age.[24] This is consistently accompanied by a gradual increase in body fat content, at least until the sixth decade of life.[25,27–30,34–41] This increase is exemplified by the gradual increase in skinfold thickness that was seen in the study by Parizkova.[42]

Novak[27] found that the proportion of fat in men and women increases from 18 to 36% and from 33 to 45%, respectively, between the ages of 20 and 75 years. There is, therefore, likely to be an increase in the lean-to-fat ratio in the body with increasing age.

This change may have relevance to the volume of distribution of drugs. Lipid-soluble drugs will distribute more to the peripheral fat compartments in the elderly, making the volume of distribution larger. Conversely, water-soluble drugs may be more concentrated in the central compartment, and the volume of distribution will thus be smaller. This inconsistent change in distribution volume will make comparison of plasma half-lives, as a measure of drug elimination in the young and elderly, less useful. Studies in the young and elderly should certainly try to determine plasma clearance as the parameter of drug elimination.

3.4. RED-BLOOD-CELL CHANGES

The information we have on changes in red blood cells is conflicting. Some studies have indicated a fall in the number of red cells per milliliter of blood with increasing age, whereas others indicate that there is no age-related change.[43-45] Hemoglobin concentration does not change consistently with age.[44-53] However, there are some indications of functional change in red cells with age. The distribution of meperidine changes such that the red cell binding decreases with increasing age.[54] This may have relevance in study design since most drug assays will measure drug levels in plasma and assume a constant red-cell-to-plasma ratio of drug concentration. Changes in this ratio may introduce error in pharmacokinetic comparisons.

3.5. DISTRIBUTION INTO URINE

Some drugs depend on their presence in urine for their effect, as is the case with antibiotics as used in the treatment of urinary tract infections. The plasma half-lives of both amoxicillin and mecillinam increase with increasing age, and urinary concentrations correspondingly fall with consequent greater likelihood of failure to clear the urinary tract infection.

4. DRUG METABOLISM

Many drugs undergo metabolic change in the body. The liver is the main site of this biotransformation. In most cases the net effect is to decrease lipid solubility, irrespective of whether the metabolic change is an initial phase I reaction (for example, oxidation, reduction, or hydrolysis) or a subsequent phase II conjugative reaction (for example, glucuronidation or sulfation). This has implications for elimination of the drug from the body since a less lipid-soluble compound will be less able to traverse the lining of the renal tubules from the urine to the plasma. Thus, the tendency will be to increase clearance of the drug. Reduction in the ability of the liver to metabolize drugs will therefore usually increase the extent and duration of their action in the body and will increase the likelihood of toxicity being seen with fixed drug doses in the elderly. However, a reduced metabolic rate will sometimes reduce drug action since some prodrugs depend on an intervening metabolic step to produce all or part of their action. For example, this is true for acetanilid, used for many years as an antipyretic and analgesic agent. Part of its action results because a major metabolite, p-hydroxyacetanilid (acetaminophen), is itself an analgesic antipyretic agent.[55,56]

In measurement of the ability of the body to eliminate drugs, the clearance concept is most useful. The clearance of a drug is that fraction of the apparent volume of distribution from which drug is removed in unit time. An extension of this is the term "intrinsic clearance," which is the clearance of a drug from the blood by the liver that could be achieved if liver blood flow were not a rate-limiting factor. Drugs may usefully be divided into those with high, intermediate, or low extraction ratios: i.e., those of which the liver removes a large, moderate, or small amount from the incoming portal venous and hepatic arterial blood. For drugs with high hepatic extraction ratios (≥ 0.70) and high intrinsic clearances, such as lidocaine, propranolol, and verapamil, the liver blood flow becomes rate limiting and hepatic clearance becomes equal to, and limited by, liver blood flow. If the hepatic extraction ratio is low (≤ 0.30), as with diazepam, diphenylhydantoin, theophylline, and warfarin, the rate of presentation of the drug to the liver is always sufficiently high to maintain the hepatic clearance at the level of the intrinsic clearance. In this instance, the hepatic clearance is dependent on intrinsic clearance and on the extent of protein binding of the drug and is independent of liver blood flow. For drugs with intermediate extraction ratios, changes in intrinsic clearance, protein binding, and liver blood flow will all affect the rate of clearance of drug by the liver. Thus, those potential age-related factors that may affect the rate and extent of elimination of drug by the liver will depend on the drug's individual propensity to be extracted by the liver.

The hepatic extraction ratio is also related to the oral bioavailability of a drug. Drugs with high hepatic extraction ratios will have low oral bioavailability, and drugs with low hepatic extraction ratios may, conversely, have high oral bioavailabilities. Reduced liver metabolizing capacity due to increased age may therefore markedly increase oral bioavailability of a drug with high first-pass hepatic extraction, due to a decrease in presystemic clearance. Conversely, there will be little effect of reduced liver metabolizing capacity on the oral bioavailability of a drug with low first-pass hepatic uptake.

Those changes which might occur with age and which could affect hepatic drug biotransformation are listed below and will be discussed in turn.

1. Liver blood flow
2. Liver size
3. Number and function of individual hepatocytes
4. Hepatic metabolizing capacity
5. Enzyme induction

4.1. LIVER BLOOD FLOW

Lack of good noninvasive methods of accurately measuring the liver blood flow to functioning hepatocytes has precluded extensive research on the effect of age on hepatic blood flow. Several studies, however, have shown decreased splanchic circulation of about 20%, with aging in humans between 20 and 70 years.[14–16] This reduction in total liver blood flow has been quantified, in another study, to be about 0.3–1.5% per year of age.[1] This is of sufficient extent that increased systemic bioavailability and reduced hepatic clearance of drugs with high extraction ratios may be seen in the elderly.

4.2. LIVER SIZE

The liver itself appears to decrease in size with increasing age, both when expressed as absolute weight (reduced by about 35% between the ages of 20 and 80 years) and as a percent of total body weight (reduced by about 30%).[57]

4.3. NUMBER AND FUNCTION OF INDIVIDUAL HEPATOCYTES

There is no doubt that there are fewer functioning liver cells[58,59] and hepatic mitochondria[59–61] per unit volume of liver tissue in the elderly. This will compound the effect anticipated from the reduced liver size in the aged.

4.4. Hepatic Metabolizing Capacity

There is a lot of information on the effect of aging on hepatic metabolizing capacity in animals. In general, there is evidence of decrease in metabolic capacity with age, but many of the studies show conflicting results indicating probable differences in techniques and in the response of different strains of animals to the aging process. Perhaps the best indicator of the effect of age in man lies in the ability of the liver to clear drugs with low intrinsic clearance, where elimination depends on liver function and not on liver blood flow. There is evidence of reduced plasma clearance of several drugs (e.g., phenazone, phenylbutazone), but there is no significant change in plasma clearance of the majority of drugs studied (for example, diazepam, isoniazid, warfarin). For drugs with high extraction ratios and thus the possibility of flow-dependent elimination, there is more consistent reduction in elimination, reinforcing the impression of reduced liver blood flow in the aged. Thus, there are some indications for reduced hepatic metabolizing capacity with age, and drug study design should take into account the probability that higher blood levels and thus greater effect and greater toxicity may be seen when the elderly are given the same doses as younger subjects.

4.5. Enzyme Induction

Chronic administration of some drugs, for example, barbiturates, increases their own metabolism as well as that of other drugs. This subject has been well reviewed.[62] The process, known as enzyme induction, may be of considerable clinical importance, particularly in the case of drugs with a narrow range of therapeutic concentration when even a small reduction in circulating level of the drug can result in loss of therapeutic effect. A good example of this is the oral anticoagulant drug warfarin. The enzymes responsible for much of the phase I hepatic biotransformation are the cytochrome P-450 isoenzymes. Several isoenzymes have already been identified, and there are some indications that there may be more than 60 of them. Different enzyme-inducing agents result in different patterns of increase in isoenzyme concentration and function. Thus, cigarette smoking (which results in polycyclic hydrocarbon exposure) and barbiturate ingestion result in stimulation of the metabolism of different drugs. Published data indicate that both cigarette smoking and subcute administration of dichloralphenazone (a mixed-function oxidase inducer) fail to cause enzyme induction responses in the elderly whereas both increase drug-metabolizing capacity in the younger age group.[63-65] This finding of reduced ability to induce metabolizing enzymes in the elderly indicates that great care should be taken to exclude prior or concurrent exposure to inducing agents when making comparisons between young and elderly in their responses to drugs. If this is not done, the responses will be measured from a different baseline, with the young showing more induction than the elderly.

5. RENAL DRUG EXCRETION

Many drugs are excreted by the renal route. Some, such as benzylpenicillin and the aminoglycoside antibiotics, almost entirely, while others, such as digoxin, are partly metabolized and partly excreted through the kidneys unchanged. Impaired renal function could result in retention of these drugs in the body and higher circulating blood levels. For drugs with a high therapeutic ratio, for example, benzylpenicillin, impaired renal function may not give rise to toxic levels when normal doses are given. However, with drugs like digoxin and the aminoglycoside antibiotics, toxicity will almost certainly occur.

Some changes occur in renal morphology which indicate that there may be reduction in renal function with increasing age. For example, the weight and volume of the kidney decreases between 20 and 30% between the ages of 30 and 90 years.[66-68] In addition, the number of glomeruli decreases significantly, by between 30 and 50% during late adult life.[67,69] Of the glomeruli that remain, there is a 10-fold increase in the number that are sclerosed in the elderly, compared with younger adults.[70] There are some indications too of blood bypassing the functioning glomeruli since angiographic studies have shown increased number of shunts in the juxtamedullary glomeruli.[71-73]

The classic early studies of Davies and Shock[74] clearly demonstrated that in normal healthy adults, glomerular filtration rate (measured by plasma inulin clearance), effective renal plasma flow (measured by plasma Diodrast clearance), and renal tubular excretory capacity (measured by maximum excretory capacity of Diodrast) fall with increasing age. Miller et al.[75] also showed that the maximal rate of tubular reabsorption of glucose fell with increasing age in subjects considered to be free of cardiovascular or renal disease. These and other data indicate that the average healthy person of 65 years of age would be expected to have lost approximately 32% of the glomerular filtration rate, 41% of the effective renal plasma flow, 20% of the renal tubular excretory capacity, and 31% of the tubular reabsorptive capacity that was present at the age of 20 years.

This uniform finding of reduced function of each of the renal excretory parameters is reinforced by the finding of impaired renal excretion of most of the drugs that have been studied in the elderly compared to the young. Many of these drugs have excretion that correlates well with creatinine clearance.[76] However, care should be taken when using serum creatinine levels to indicate the level of renal function. This is because normal serum creatinine levels often occur in the elderly despite significantly impaired creatinine clearance (down to about 30 ml/min), because of reduced endogenous creatinine production, probably related to reduced muscle mass.[77] Thus, interpretation of a normal serum creatinine value as implying normal renal function may lead to toxic doses of renally excreted drugs with low therapeutic ratios being prescribed to elderly subjects. Creatinine clearance should therefore be directly measured by collecting a 24-hr urine sample and correlating the urinary creatinine excretion with the serum creatinine concentration. Alternatively, a nomogram, designed to take account of the fact that there is reduced endogenous creatinine secretion in the elderly, may be used. The latter method is probably more satisfactory since accurate 24-hr samples of urine are hard to get.

When drug toxicity is suspected, it should also be remembered that there is an increasing incidence of proteinuria after age 65.[78] Thus, care should be taken to quantitate renal protein leak prior to drug administration when renal toxicity is anticipated. Also, it is reasonable that in order to recruit sufficient elderly patients for study, the entry criteria for proteinuria should be relaxed a little in the elderly, otherwise reduced patient recruitment will result.

The elderly are unable to conserve sodium as rapidly and as efficiently as are younger subjects when salt is restricted.[79] This is likely to result in significant dehydration when large doses of loop diuretics are given to the elderly, and care should be taken to avoid significant dehydration, resulting in prerenal uremia and orthostasis.[79] Response to some drugs, particularly antihypertensive drugs, depends on initial plasma renin activity in the blood. The trend with increasing age is to lower plasma renin activities. Whether this results in improved responsiveness to drugs such as diuretics and calcium channel antagonists is not clear, but this should be anticipated when designing studies.[80–83]

6. DRUG RESPONSE

At least two different types of age-related physiological change may alter the response seen on drug administration.

First, the starting point for measured change induced by drug may have altered. As an example, the action of cardioactive drugs will be measured in normals from a different cardiac index (which falls about 35% over the age range 20–80 years) and peripheral resistance (which rises about 85% over the same age range) in young and elderly.[84] In hypertensives, too, the resting hemodynamics tend to change with age, with the elderly having lower cardiac output and higher peripheral resistance than younger hypertensive subjects for the same blood pressure. Thus, the profile of anticipated change of a sympatholytic antihypertensive drug may differ markedly in the young and elderly, with blood pressure changes being similar. In the young, cardiac output may fall to normal, while in the elderly cardiac output may become even lower than previously, resulting in symptoms of tiredness and inability to perform normal exercise functions.

Other hemodynamic resting and stimulated parameters may also change. Resting heart rate probably does not alter much with age, but intrinsic heart rate (devoid of sympathetic and parasympathetic control) falls with age in line with the equation[85]

$$\text{Intrinsic heart rate} = 118 - 0.6 \times \text{age}$$

In line with this are the findings of lesser vagal restraint in the resting heart[86,87] but similar resting sympathetic tone[88] in the elderly. The maximal heart rate during exercise falls with age in line with the equation[88,89]

$$\text{Maximal heart rate} = 204 - 0.6 \times \text{age}$$

Thus, studies of cardiac β-sympathetic and muscarinic blockade or stimulation will not only be measuring responses from different intrinsic baselines but will have a different balance of autonomic control and an altered maximal achievable response. Thus, studies must use fairly closely aged-matched controls to obviate potential bias.

These cardiovascular changes serve as examples of the kind of baseline changes to be anticipated in the elderly. Other systems will have similar changes. For example, changes in gut motility will alter anticipated response to laxatives, and decreased mentation will increase the likelihood of clinically manifest drug-induced confusion with centrally acting drugs. The chest wall becomes less compliant[90,91] and the timed force expiratory volume falls with age.[92–97] These last two changes will alter the response anticipated with administration of bronchodilator drugs, whether inhaled or systemically administered.

The second type of age-related change is alteration in end-organ responsiveness to a fixed degree of stimulation or inhibition by a drug. In general, many of these age alterations may be thought of as a decreasing ability of the body to maintain homeostasis following a drug-induced change. A good example of this is the reduced baroreceptor sensitivity reported in the elderly.[98] This is manifest by reduced heart rate (and presumably cardiac output) response to alteration of pressure within the aorta and carotid arteries. From a functional point of view, this means that the elderly are more likely to become symptomatically orthostatic even without drugs.[99,100] Volume depletion with diuretics may exacerbate this. From a therapeutic point of view, this also has several implications. First, vasodilators such as hydralazine may often be used as second-line drugs in hypertension in the elderly, after diuretics, without sympatholytics to suppress the increase in cardiac output. Second, it makes the use of drugs that have orthostasis as an anticipated effect less desirable in the elderly.

Other systems have demonstrated altered sensitivity to drugs with increasing age. There is increased warfarin efficacy in inhibiting clotting factor synthesis, which is not related to altered pharmacokinetic parameters in the elderly.[101] The brain also appears to be more sensitive to the same levels of diazepam,[102] nitrazepam,[103] and the inhalation anesthetic agent halothane.[104]

7. Conclusion

Pharmacodynamic changes have not been as intensively investigated as pharmacokinetic changes. This is because of the greater difficulty in obtaining noninvasive quantitative measures of drug-induced change than in obtaining blood, fecal, or urine samples to measure drug disposition. Some body systems lend themselves to the measurement of drug-induced response. For example, the effects of anticoagulants and of antihypertensive drugs are relatively easily and quantitatively measured. On the other hand, the effect of, for example, digoxin on cardiac function or of antidepressant drugs on depression is much more difficult to quantitate. The development of noninvasive quantitative measures of pharmacodynamic change is an important area that is essential for investigators to explore in the next few years.

References

1. Bhanthumnavin K, Schuster M: Aging and gastrointestinal function, in Finch C, Hayflick L (eds): *Handbook of the Biology of Aging.* New York, Van Nostrand Reinhold, 1977, p 209.

2. Isokoski M, Krohn K, Varis K, et al: Parietal cell and intrinsic factor antibodies in a Finnish rural population sample. *Scand J Gastroenterol* 1969; 4:521–527.

3. Hebbel R, The topography of chronic gastritis in otherwise normal stomachs. *Am J Pathol* 1949; 25:125–138.

4. Andrews G, Haneman B, Arnold B, et al: Atrophic gastritis in the aged. *Aust Ann Med* 1967; 16:230–235.

5. Hradsky M, Groh J, Langr F, et al: Chronische Gastitis bei jungen und atten Personen Histolische and histochemische Untersuchung. *Gerontol Clin* 1966; 8:164–171.

6. Baron J: Studies of basal and peak acid output with an augmented histamine test. *Gut* 1963; 4:136–144.

7. Levin E, Kirsner J, Palmer W: A simple measure of gastric secretion in man: comparison of one hour basal secretion, histamine secretion, and 12 hour nocturnal gastric secretion. *Gastroenterology* 1951; 19:88–98.

8. Pollard W: Histamine test meals. An analysis of 988 consecutive tests. *Arch Intern Med* 1933; 51:903–919.

9. Davies D, Jones J: Investigation into gastric secretion of 100 normal persons over age of 60. *Q J Med* 1930; 23:1.

10. Grossman M, Kirsner J, Gillespie I: Basal and histalog-stimulated gastric secretion in control subjects and in patients with peptic ulcer or gastric cancer. *Gastroenterology* 1963; 45:14–26.

11. Bloomfield A, Keefer C: Gastric acidity: Relation to various factors such as age and physical fitness. *J Clin Invest* 1928; 5:285–294.

12. Bloomfield A, Keefer C: Gastric motility and the volume of gastric secretion in man. *J Clin Invest* 1928; 5:295–301.

13. Necheles H, Maskin H: Studies in constitution and peptic ulcer. I. Appetite secretion in normal persons and in ulcer patients. *Am J Dig Dis Nutr* 1936; 3:90–92.

14. Bradley SE, Ingelfinger FJ, Bradley GP, et al: The estimation of hepatic blood flow in man. *J Clin Invest* 1945; 24:890–897.

15. Myers JD: The hepatic blood flow and splanchic oxygen consumption of man—Their estimate from urea production on bromsulphalein excretion during catherization of the hepatic veins. *J Clin Invest* 1947; 26:1130–1137.

16. Sherlock S, Beam AG, Billing BH, et al: Splanchnic blood flow in man by the bromsulphalein method: The relation of peripheral plasma bromsulphalein level to the calculated flow. *J Lab Clin Med* 1950; 35:923–932.

17. Rafsky HA, Brill AA, Stern KG, et al: Electrophoretic studies on the serum of "normal" aged individuals. *Am J Med Sci* 1952; 224:522–528.

18. Goldstein A: The interaction of drugs and plasma proteins. *Pharmacol Rev* 1949, 1:102–157.

19. Langmuir I: The constitution and fundamental properties of solids and liquids. Part I. Solids. *J Am Chem Soc* 1916; 38:2221–2295.

20. Langmuir I: The constitution and fundamental properties of solids and liquids. Part II. Liquids. *J Am Chem Soc* 1917; 39:1848–1906.

21. Cohn JE, Shock NW: Blood volume studies in middle-aged and elderly males. *Am J Med Sci* 1949; 217:388–391.

22. Chien SS, Usami S, McAllister FF: Blood volume and age: Repeated measurements on normal men after 17 years. *J Appl Physiol* 1966; 21:83.

23. Edelman IS, Leibman J: Anatomy of body water and electrolytes. *Am J Med* 1959; 27:256–257.

24. Forbes GB: The adult decline in lean body mass. *Hum Biol* 1976; 48:161–173.

25. Parizkova J, Eiselt E: A further study on changes in somatic characteristics and body composition of old men followed longitudinally for 8–10 years. *Hum Biol* 1971; 43:318–326.

26. Forbes GB, Reina JC: Adult lean body mass declines with age; some longitudinal observations. *Metabolism* 1970; 19:653–663.
27. Novak L: Aging, total body potassium, fat-free mass, and cell mass in males and females between ages 18 and 55 years. *J Gerontol* 1972; 27:438–443.
28. Weinsier RL, Fuchs RJ, Kay TD, et al: Body fat; its relationship to coronary heart disease, blood pressure, lipids and other risk factors measured in a large male population. *Am J Med* 1976; 61:815–823.
29. Burmeister W, Bingert A: Quantitative changes of the human cell mass between the 8th and 90th year of life. *Klin Wochenschr* 1967; 45:409–416.
30. Malina RM: Quantification of fat, muscle and bone in man. *Clin Orthop Relat Res* 1969; 65:9–38.
31. Cohn SH, Vaswani A, Zanzi I, et al: Changes in body chemical composition with age measured by total-body neutron activation. *Metabolism* 1976; 25:85–95.
32. Bugyi B: Age dependent changes of body constitution based on adipose tissue and musculature evaluation. *Z Alternsforsch* 1967; 20:327.
33. Matsuki S, Yoda R: An evidence for the decrease of body muscle mass due to aging by means of height, weight and upper arm circumference measurement. *Endocrinol Jpn* 1972; 19:401.
34. Tran MH, Lellouch J, Richard JL: Fat body mass. II. Its relationships with some biological parameters, blood pressure, and physical training in a population of 8660 men aged 20 to 55. *Biomedicine* 1973; 18:499–506.
35. Mackova E: Changes in functional and efficiency indicators between the 18th and 50th year in man. *Physiol Bohemoslov* 1968; 17:279–287.
36. Bismark HD: On age-dependent changes in adipose tissue layers. *Z Alternforsh* 1967; 20:347.
37. Parizkova J, Eiselt E, Sprynarova S, et al: Body composition, aerobic capacity, and density of muscle capillaries in young and old men. *J Appl Physiol* 1971; 31:323–325.
38. Myhre LG, Kessler WV: Body density and potassium 40 measurements of body composition as related to age. *J Appl Physiol* 1966; 21:1251–1255.
39. Chien S, Peng MT, Chen KP, et al: Longitudinal measurements of blood volume and essential body mass in human subjects. *J Appl Physiol* 1975; 39:818–824.
40. Brozek J: Changes of body composition in man during maturity and their nutritional implications. *Fed Proc Fed Am Soc Exp Biol* 1952; 11:784–793.
41. Young CM, Blondin J, Tensuan R, et al: Body composition of "older" women. *J Am Diet Assoc* 1963; 43:344–348.
42. Parizkova J: Body composition and lipid metabolism. *Proc Nutr Soc* 1973; 32:181–186.
43. Das BC: Linear and curvillinear functional relationships between human blood components and age. *Gerontology* 1967; 13:227.
44. Earney WW, Earney AJ: Geriatric hematology. *J Am Geriatr Soc* 1972; 20:174–177.
45. Shapleigh JB, Mayes S, Moore CV: Hematologic values in the aged. *J Gerontol* 1952; 7:207–219.
46. Garcia JF: Changes in blood, plasma and red cell volume in the male rat, as a function of age. *Am J Physiol* 1957; 190:19–24.
47. Garcia JF: Erythropoietic response to hypoxia as a function of age in the normal male rat. *Am J Physiol* 1957; 190:25–30.
48. Grant WC, LeGrande CM: The influence of age on erythropoiesis in the rat. *J Gerontol* 1964; 19:505–509.
49. Hamilton PJ, Dawson AA, Ogston D, et al: The effect of age on the fribrinlytic enzyme system. *J Clin Pathol* 1974; 27:326–329.
50. Hayes GS, Stinson IV: Erythrocyte sedimentation rate and age. *Arch Ophthalmol* 1976; 94:939–940.

51. Helman N, Rubenstein LS: The effects of age, sex and smoking on erythrocytes and leukocytes. *Am J Pathol* 1975; 63:35.
52. Olbrich O: Blood changes in the aged. *Edinburgh Med J* 1947; 54:306–321.
53. Purcell Y, Brozovic B: Red cell 2,3-diphosphoglycerate concentration in man decreases with age. *Nature (London)* 1974; 251:511–512.
54. Chan K, Kendall MJ, Mitchard M, et al: The effect of aging on plasma pethidine concentration, *Br J Clin Pharmacol* 1975; 2:297–302.
55. Brodie BB, Axelrod J: The fate of acetanilide in man. *J Pharmacol Exp Therap* 1948; 94:29–38.
56. Brodie BB, Axelrod J: The fate of acetophenetidin (phenacetin) in man and methods for the estimation of acetophenetidin and its metabolites in biological material. *J Pharmacol Exp Therap* 1966; 151:133.
57. Calloway N, Foley C, Lagerbloom P: Uncertainties in geriatric data. II. Organ size. *J Am Geriatr Soc* 1965; 13:20–28.
58. Schmucker D, Mooney J, Jones A: Age-related changes in the hepatic endoplasmic reticulum: a quantitative analysis. *Science* 1977; 197:1005–1007.
59. Tauchi H, Sato T: Effect of environmental conditions upon age changes in the human liver. *Mech Aging Dev* 1975; 4:71–80.
60. Pieri C, Nagy ZS, Muzzufferi G, et al: The aging of rat liver as revealed by electron microscopic morphometry. I. Basic parameters. *Exp Gerontol* 1975; 10:291–304.
61. Tauchi H, Sato T: Age changes in size and number of mitochondria of human hepatic cells. *J Gerontol* 1968; 23:454–461.
62. Conney AH: Pharmacological implications of microsomal enzyme induction. *Pharmacol Rev* 1967; 19:317–366.
63. Salem SAM, Rijjayabun P, Shepherd AMM, et al: Reduced induction of drug metabolism in the elderly. *Age Ageing* 1978; 7:68–73.
64. Vestal RE, Wood AJJ, Branch RA, et al: Effects of age and cigarette smoking on propranolol disposition. *Clin Pharmacol Therap* 1979; 26:8–15.
65. Wood AJJ, Vestal RE, Wilkinson GR, et al: Effect of ageing and cigarette smoking on antipyrine and indocyanine green elimination. *Clin Pharmacol Therap* 1979; 26:16–20.
66. Roessle R, Roulet F: *Mass und Zahl in der Pathologic.* Berlin, J. Springer, 1932.
67. Dunnill MS, Halley W: Some observations of the quantitative anatomy of the kidney. *J Pathol* 1973; 110:113–121.
68. Tauchi H, Tsuboi K, Sato K: Histology and experimental pathology of senile atrophy of the kidney. *Nagoya Med J* 1958; 4:71.
69. Moore, RA: Total number of glomeruli in the normal human kidney. *Anat Rec* 1931; 48:153.
70. Kaplan C, Pastenack B, Shaw H, et al: Age-related incidence of sclerotic glomeruli in human kidneys. *Am J Pathol* 1975; 80:227–234.
71. Ljungqvist A, Lagergren C: Normal intrarenal arterial pattern in adult and aging human kidney. A microangiographical and histologic study. *J Anat* 1962; 96:285–298.
72. Ljungqvist A: Structure of the arteriole-glomerular units in different zones of the kidney. *Nephron* 1964; 1:329–337.
73. Takazakura E, Wasabu N, Handa A, et al: Intrarenal vascular changes with age and disease. *Kidney Int* 1972; 2:224–230.
74. Davies DF, Shock NW: Age changes in glomerular filtration rate, effective renal plasma flow and the tabular excretory capacity in adult males. *J Clin Invest* 1950; 29:496–507.
75. Miller JW, McDonald RK, Shock NW: Age changes in the maximal rate of renal tubular reabsorption of glucose. *J Gerontol* 1952; 7:196–200.

76. Ewy GA, Kapadia GG, Yao L, et al: Digoxin metabolism in the elderly. *Circulation* 1969; 39:449–453.
77. Molholm Hansen J, Kampmann J, Lauson H: Renal excretion of drugs in the elderly. *Lancet* 1970; 1:1170.
78. VanZonneveld RJ: Some data on the genito-urinary system as found in old age surveys in the Netherlands. *Gerontol Clin* 1959; 1:167–173.
79. Epstein M, Hollenberg NK: Age as a determinant of renal sodium conservation in normal man. *J Lab Clin Med* 1976; 87:411–417.
80. Weidmann P, DeMyttenaeu-Bursztein S, Maxwell MH, et al: Effect of aging on plasma renin and aldosterone in normal man. *Kidney Int* 1975; 8:325–333.
81. Crane MG, Harris JJ: Effect of aging on renin activity and aldosterone excretion. *J Lab Clin Med* 1976; 87:947–959.
82. Hayduk K, Krause DK, Kaufman W, et al: Age-dependent changes of plasma renin concentrations in humans, *Clin Sci Mol Med* 1973; 45:2735.
83. Flood C, Gherondache C, Pincus G, et al: The metabolism and secretion of aldosterone in elderly subjects. *J Clin Invest* 1967; 46:960–965.
84. Brandfonbrener M, Landown M, Shock NW: Changes in cardiac output with age. *Circulation* 1955; 12:557–566.
85. Jose AD, Collison D: The normal range and determinants of the intrinsic heart rate in man. *Cardiovasc Res* 1970; 4:160–167.
86. Nalefski LA, Brown CFG: Action of atropine on the cardiovascular system in normal persons. *Arch Intern Med* 1950; 86:898–907.
87. Simonson E: Effect of age on the cardiovascular system in recent Russian research. *J Gerontol* 1964; 19:121–127.
88. Conway, J: Effect of age on the response to propranolol. *Int J Clin Pharmacol Ther Toxicol* 1970; 4:148.
89. Wolthuis RA, Froelicher VF, Fischer J, et al: The response of healthy men to treadmill exercise. *Circulation* 1977; 55:153–157.
90. Mittman C, Edelman NH, Norris AH, et al: Relationship between chest wall and pulmonary compliance. *J Appl Physiol* 1965; 20:1211–1216.
91. Rizzato G, Marazzini L: Thoracoabdominal mechanics in elderly men. *J Appl Physiol* 1970; 28:457–460.
92. Muiesan G, Sorbini CA, Grassi V: Respiratory function in the aged. *Bull Physio-Pathol Respir* 1971; 7:973.
93. Michie I: Lung function in the elderly. *Gerontol Clin* 1971; 13:125–135.
94. Chebotarev DF, Korkusko OV, Ivanov LA: Mechanisms of hypoxemia in the elderly. *J Gerontol* 1974; 29:393–400.
95. Gelb AF, Zamel N: Effect of aging on lung mechanics of healthy nonsmokers. *Chest* 1975; 68:538–541.
96. Ashley F, Kannel WB, Sorlie PD, et al: Pulmonary function: Relation to aging, cigarette habit and mortality. *Ann Intern Med* 1975; 82:739–745.
97. Pollock ML, Miller HS, Wilmore J: Physiological characteristics of champion American track athletes 40 to 75 years of age. *J Gerontol* 1974; 29:645–649.
98. Gribbin B, Pickering TG, Sleight P, et al: Effect of age and high blood pressure on baroreflex sensitivity in man. *Circ Res* 1971; 29:424–431.
99. Norris AH, Shock NW, Yiengst MJ: Age changes in heart rate and blood pressure responses to tilting and standardized exercise. *Circulation* 1953; 8:521–526.
100. Caird FI, Andrews GR, Kennedy RD: Effect of posture on blood pressure in the elderly. *Br Heart J* 1973; 35:527–530.
101. Shepherd AMM, Hewick DS, Moreland TA, et al: Age as a determinant of sensitivity to warfarin. *Br J Clin Pharm* 1977; 4:315–320.

102. Reidenberg MM, Levy M, Warner H, et al: Relationship between diazepam dose, plasma level, age, and central nervous system depression. *Clin Pharmacol Therap* 1978; 23:371–374.
103. Castleden CM, George CF, Marcer D, et al: Increased sensitivity to nitrazepam in old age. *Pharmacokinetics* 1977; 1:280–296.
104. Gregory GA, Eger EI, Munson ES: The relationship between age and halothane requirement in man. *Anesthesiology* 1969; 30:488–491.

CHAPTER 4

IMMUNITY AND AGING

WILLIAM B. ERSHLER AND MARC E. WEKSLER

1. INTRODUCTION

The gradual decline in immune function observed with aging has been thought to contribute to the diseases of aging and even the aging process itself.[1] It may be reasonable to relate the increasing incidence of infection or cancer to immune senescence. The precise contribution of changing immune competence to other diseases of aging such as Alzheimer's disease, atherosclerosis, and diabetes must be considered speculative. However, extensive basic and clinical studies of immune function in experimental animals and humans of different ages clearly indicate that the immune system, like many other systems, changes with age. In this chapter we will provide an overview of these studies, particularly as they relate to clinical immune senescence. Methodological problems encountered in immunological studies and aging will be considered and summarized at the end of this chapter.

WILLIAM B. ERSHLER • Division of Hematology and Geriatrics, Department of Medicine, University of Wisconsin, Madison, Wisconsin 53706.   MARC E. WEKSLER • Division of Geriatrics, Department of Medicine, Cornell University Medical College, New York, New York 10021.

2. THYMUS GLAND AND AGING

Among the developments in our increased understanding of the
immune system over the past three decades, the role of the thymus gland
in differentiation and regulation of humoral and cell-mediated immunity
has been most important. Thirty years prior to the studies of Good and
of Miller the striking involution of the thymus that begins about the time
of sexual maturity and is virtually complete by the age of 50 was recognized.[2] This loss of thymic mass was shown to be primarily due to loss
of cortical lymphocytes[3] and thymic epithelial components,[4] a part of
the thymic microenvironment that is essential for T-lymphocyte matu-
ration.[5] The evidence that many of the immunosenescent changes are
related to thymic involution is indirect, but suggestive. The observation
was made that long-lived strains of mice attain peak thymic weight later
and maintain higher thymus : total body weight ratios longer than mice
with shorter life expectancy.[6] The acceleration of age-related decline in
immune function after thymectomy is also compatibile with this thesis.[7,8]
Furthermore, dietary caloric restriction, which extends life expectancy
in mice, delays thymic involution and maintains immune competence
longer.[9,10] A measurable marker of thymic tissue involution is a reduction
in the serum concentration of biologically active peptides (hormones)
known to be produced by thymic epithelium.[11,12] This has been more
clearly demonstrated by bioassay[13] than by analytical assays that detect
the immunoreactive peptide.[14,15] This may reflect an age-acquired functional defect in the secreted hormone rather than a quantitative one.
Hirokawa and Makinodan[16] provided direct evidence that the aging of
the thymus correlates with declining immune function. Thymic glands
were obtained from mice of several different ages (1 day–33 months)
and transplanted into irradiated, neonatally thymectomized, syngeneic
mice that had previously received bone marrow reconstitution. Several
immunological parameters were followed, including antibody response
to a T-dependent antigen, *in vitro* mitogen response, and the development of mature T-cell markers. Those animals which had received young
thymus approximated the responses of the control (sham thymectomy)
animals, whereas recipients of old thymic glands were significantly im-
munologically impaired. These observations and others, including the
restoration of some immune functions noted to be deficient in aged
animals by newborn syngeneic thymus implants,[17,18] indicate a central
role for thymic involution in immunosenescence.

While the age-related immune defects can be related to the involution of the thymus, the failure of transplanting young thymus glands

into old animals to prolong life indicate the difficulty of relating thymic involution to life-span.[19–21] The complexity of immune senescence is illustrated by the work of Twomey and colleagues,[22] who found that the number of immature T cells was elevated in congenital athymic nude mice but not increased in old mice or humans. In fact, they found a reduced number of immature T cells in old subjects. It is apparent, therefore, that thymic involution contributes to, but may not be solely responsible for, immune senescence, as precursor cells may themselves be deficient in number. The combined importance of both bone marrow stem cells and thymic involution to the development of immune senescence is supported by the work of Hirokawa and Makinodan, who found that aged recipients of both young bone marrow and young thymus tissue had greater immune competence than did recipients of either bone marrow or thymus tissue.[23,24] In these studies, successful syngeneic transplantation of bone marrow and thymus resulted in improved immune function in old mice, and this beneficial effect was associated with a statistically longer life-span.

3. IMMUNOSENESCENCE AT THE CELLULAR LEVEL

The function of stem cells, T cells, B cells, and macrophages has been extensively examined in young and old subjects. As expected, based on thymic involution, T-cell function is the primary component altered by aging. The absolute and relative number of T cells changes little, if at all, in humans[25,26] and mice.[27,28] Functional studies indicate that T-cell competence is altered with aging (for review see Ref. 29). Antibody production in response to T-dependent antigens decreases with age and appears to result from deficient T-helper function.[30–33] Other measurements of T-lymphocyte function, such as mixed lymphocyte cultures,[34,35] cell-mediated lymphocytotoxicity,[36] and response to T-cell mitogen,[37,38] have consistently been reported as abnormal. Biochemical[39] and physical properties[40] of the T lymphocyte also change significantly with age. Normal lymphocyte receptors and surface antigens become less dense,[41,42] and new surface antigens appear.[43,44] Lymphokine production, such as Interleukin 2 (IL-2), and γ-interferon have been demonstrated to be reduced in old mice and humans.[45–47] In studies utilizing limiting dilution techniques, Miller and colleagues have demonstrated that IL-2 production per cell was not diminished in cells from old people, whereas the frequency of helper-cell precursors was.[48,49] Of interest, the humoral response to T-independent antigens has also been shown to decline somewhat with age,[50] indicating altered B-cell function with age as well. However, B-cell number,[51] immunoglobulin levels,[52,53] and proliferative response to B-cell mitogens[50] do not change dramatically with age.

It is apparent, therefore, that age-related immune dysfunctions reflect a composite of alterations in function and, for the most part, not number of cell populations. These functional alterations primarily involve T cells but are not restricted to any specific T-cell subset. As mentioned previously, T-helper cells have been demonstrated to be dramatically reduced in function in mice of advanced age. Assays in which antibody response to T-dependent antigens (i.e., requiring T help) is measured have revealed reduced responses with age, and restoration of competent antibody production has been observed with *in vitro* or *in vivo* treatment with thymic hormones.[15,54,55] Other assays in which T cells augment effector cell activity have been demonstrably age-reduced but reconstituted with thymic factors or other T-cell stimulants.[56-58] Thus, there is general agreement that T-helper-cell function is deficient in aged animals. There is no consensus regarding T-suppressor cells, however. The increased incidence of autoantibodies[59,60] and the increased response to chemically altered syngeneic erythrocytes[61] have suggested diminished suppressor-cell function with senescence. Nevertheless, increased T-cell-mediated suppression has been repeatedly demonstrated in a variety of assays from several laboratories.[32,34,62,63] For example, the magnitude of an *in vitro* antibody response to T-dependent and T-independent antigens by lymphocytes from young donors was significantly reduced when T cells from aged mice were cocultured.[64] This age-associated enhancement of antigen-specific suppressor-cell activity contrasts with the observation of increased incidence of autoimmune reactivity in senescence. The latter phenomenon may be explained by an increased resistance to T-suppressor mechanisms in aged animals, and not to impaired T-suppressor-cell activity.[65] Specific autoantiidiotypic suppressor activity is increased with aging, and this may contribute to the impaired immune response.[60] The physiological significance of autoantibodies is far from clear but probably reflects immune dysregulation and not autoimmune disease.

In summary, T cells are diminished in function with age. This deficiency includes most T-cell effector and regulator functions, with the possible exception of T suppression. The resulting deficiency, therefore, is one of cellular immune incompetence, which gradually develops throughout the life-span and may be associated with autoantibody production.

4. METHODOLOGICAL CONCERNS FOR BASIC IMMUNOLOGICAL RESEARCH IN AGING

It is evident from the multitude of investigations to date that normal aging is associated with reduced immune function. Although the deficiency is complex, a central role for thymic involution is probable. Age-related diseases, however, are frequently characterized by their propensity for inducing immune suppression. An example of this is neoplasia and associated immune depression. Despite optimal experimental conditions, mice, rats, and other experimental animals often suffer from age-related diseases that are frequently unrecognized. It is, therefore, imperative to design research protocols that consider and control for the abnormalities that are associated with age-related diseases. One useful suggestion is to do complete necropsies on all experimental animals to rule out these confounding variables.

Pursuing a similar argument, if one is to test, in a nonlethal way, a specific parameter of immune function in young, middle-aged, or old mice, it is possible to reduce the bias produced by doing cross-sectional studies[66] by following each animal to its natural death and then comparing the responses only in mice that reach a prescribed definition of old. A final point in this regard is also suggested. Different strains of laboratory animals have different survival curves, and among the same strains, these vary from laboratory to laboratory. It is important for investigators to be aware of the characteristic survival curves for their experimental animals, to choose animals with appropriate (approaching rectangular) survival curves when investigating age-related (and not disease-related) phenomena, and to report these demographic data with their research methods in their reports. As an arbitrary convention, animals should be considered "old" when they surpass the median natural survival for that strain.

5. CLINICAL CONSEQUENCES

Despite the well-established database outlined here, indicating a significant reduction in immune function with advanced age, the absolute importance of age-acquired immune deficiency has yet to be established. There is no question that infection is more prevalent in elderly populations, especially with viral agents and mycobacteria. Similarly, age is the single most important risk factor for the development of cancer.[68] Nevertheless, several factors other than immune deficiency may be responsible.

With respect to infection, one cannot refute the overwhelming evidence that infections occur more frequently in the elderly, and that they are generally more severe. It is possible, however, that concomitant medical problems, such as malignancy, or cardiorespiratory disease, are at least as important in this increased susceptibility to infection and that healthy, ambulatory, noninstitutionalized senior citizens of any age are not particularly susceptible. In fact, this is the very group that may be more resistant and should be explored in that regard. It is also important to note that despite the sometimes striking immune defects demonstrated by functional assays, the infections common in older people are not the ones usually associated with immune deficiency, such as pneumocystitis or infections due to opportunistic agents, but are influenza and pneumococcal pneumonia.

The concept of immune surveillance as it relates to both aging and the pathogenesis of cancer is one with much theoretical appeal but persistent unresolved issues.[69] The theory is based on the assumption that tumor cells have specific antigens, that immune competent hosts recognize these antigens and destroy tumor cells (primarily T-cell effector mechanisms), and that cancers will develop only if the immune response to tumor antigens is somehow shielded or if the host is immune deficient. Proponents of this theory claim that the increase in cancer prevalence associated with age may be explained by the age-associated immune deficiency. There are, however, persistent problems with the concept. Unlike experimental tumors that are induced by chemical or viral agents, most human tumors have weak or no demonstrable tumor antigens. Furthermore, although tumors occur more frequently in older people, they grow more slowly and metastisize less freqently when compared to histologically similar tumors in the young.[70–73] An alternative hypothesis to explain the increased age-associated cancer prevalence is that carcinogenic agents induce tumors in proportion to their dose and time of exposure. Older people will a priori be exposed to these substances for a longer time than life-style-matched younger controls.[74]

With these caveats in mind, it is still probable that immune deficiency contributes to the pathogenesis of common infections in the elderly and in the development of certain malignancies. With influenza, for example, a reduced antibody response has been observed in healthy, ambulatory elderly volunteers in response to recent vaccines.[54] In fact, in that study, as many as one-fourth of the volunteers did not increase antibody level above preimmunization level. These nonresponders, however, were able to enhance antibody production *in vitro* by the addition of a thymic hormone preparation into the culture supernatant. The significance of this research will become evident if clinical trials of similar thymic peptides administered as adjuvants to historically nonresponding high-risk recipients of influenza or other vaccines are demonstrated to increase their antibody response to the vaccine.

6. METHODOLOGICAL CONCERNS FOR CLINICAL RESEARCH IN IMMUNOLOGY AND AGING

Current population trends indicate that the number of people considered elderly is increasing dramatically. Even with the development of efficient mechanisms for delivering health care, medical expenditure may soon exceed resources. One important area for intense research, therefore, from a socioeconomic and medical perspective, is disease prevention in the elderly. There is now clear evidence that progressive immune deficiency occurs with "normal" aging. The extent that this diminished activity results in disease has not been well defined. It is not difficult to discover abnormalities in a wide variety of responses, but to determine the importance of such findings would be the major focus. If the described defects relate to clinical disease, then research in immune enhancement, by any of a number of mechanisms, may result in disease prevention or treatment.

An additional confounding variable is the heterogeneous population from which "normal" elderly subjects are selected. There are reasons to believe that institutionalized elderly are different in many regards from free-living counterparts. To the extent that these differences will be recongized by immunological testing it is imperative that study populations be thoroughly described and, when comparing old with young, that life-style controls be considered. For example, institutionalized young might serve as adequate controls for studies on nursing-home patients whereas studies of free-living elderly should be compared to studies of similarly free-living young people.

Appropriate studies would be longitudinal in design with cohorts large in number. With reference to our incomplete understanding of immunoregulation with age, any immunological perturbation must be evaluated fully in preclinical studies and very thoroughly in early limited clinical trials designed to determine toxicity prior to larger clinical trials designed to determine efficacy. The challenge in clinical aging research is to separate those effects produced by age-related diseases from those effects produced by normal, physiological involution. Reports of investigations on elderly humans should include careful and well-documented descriptions of the study population. Only by careful subject selection will we be able to realize our goals of separating normal aging from disease.

REFERENCES

1. Walford RL, Gottesman SRS, Weindriech RH, Tam CF: Immunopathology of aging. Biological sciences, in: *The Annual Review of Gerontology and Geriatrics.* 1981, pp 3–48.
2. Boyd E. The weight of the thymus gland in health and disease. *Am J Dis Child* 1932; 43:1162–1214.
3. Yunis EJ, Fernandes G, Teague PO, Stutman O, Good RA: The thymus, autoimmunity and involution of the lymphoid system, in Sigel M, Good RA (eds): *Tolerance Autoimmunity and Aging.* Springfield, IL, Charles C. Thomas Publisher, 1972, pp 62–120.
4. Hirokawa K. The thymus and aging, in Makinodan T, Yunis E (eds): *Immunology and Aging.* New York, Plenum Press, 1977, pp 51–72.
5. Clark SL Jr: Cytological evidence of secretion in the thymus, in: Wolstenholme GE, Porter R (eds): *Thymus-Experimental and Clinical Studies:* CIBA Foundation Symposium. London, Churchill, 1966, pp 3–29.
6. Yunis EJ, Fernandes G, Smith J, Stutman O, Good RA. Involution of the thymus dependent lymphoid system, in Janukovik BD, Isakovic K (eds): *Microenvironmental Aspects of Immunity.* New York, Plenum Press, 1973, pp 301–306.
7. Metcalf D. Delayed effect of thymectomy in adult life on immunological competence. *Nature* 1965; 208:1336.
8. Bach MA, Lymphocyte-mediated cytotoxicity: Effects of aging, adult thymectomy and thymic factor. *J Immunol* 1977; 119:641–647.
9. Weindruch RH, Kristie JA, Walford RL: Dietary restriction imposed in middle age mice: Lifespan, disease and immunologic effects. *Gerontologist* 1980; 20:223.
10. Weindruch RH, Suffin SC: Quantitative histologic effects in mouse of controlled dietary restriction. *J Gerontol* 1980; 35:525–531.
11. Goldstein AL, Low JLK, Thurman GB, et al: Thymosins and other hormone like factors of the thymus gland, in Mihich E (ed): *Immunologic Approaches to Cancer Therapeutics.* New York, John Wiley & Sons, 1980, pp 137–190.
12. Bach J-F, Dardenne M, Pleau JM, Bach M-A: Isolation, biochemical characteristics and biological activity of circulation thymic hormone in the mouse and in the human. *Ann NY Acad Sci* 1975; 249:186–210.
13. Fabris N, Mocchegiani E, Amadio L, Zannotti M, Licastro F, Franceschi C: Thymic hormones deficiency in normal aging and Down's syndrome: Is there a primary failure of the thymus? *Lancet* 1984; 1:983–986.
14. McClure JE, Lameris N, Wara DW, Goldstein AL: Immunochemical studies on thymosin: Radioimmunassay of thymosin α_1. *J Immunol* 1981; 128:368–375.
15. Ershler WB, Moore AL, Hacker MP, Ninomiya J, Naylor PL, Goldstein AL: Specific antibody synthesis *in vitro*. II. Age-associated thymosin enhancement of antitetanus antibody synthesis. *Immunopharmacology* 1984; 8:69–77.
16. Hirokawa K, Makinodan T: Thymic involution: Effect on T-cell differentiation. *J Immunol* 1975; 114:1659–1664.
17. Hirokawa K, Albright JW, Makinodan T: Restoration of impaired immune function in aging animals. I. Effect of syngeneic thymus and bone marrow grafts. *Clin Immunol Immunopathol* 1976; 5:371–376.

18. Fujiwara M, Kishimoto S: IgE antigbody formation and aging. I. Age-related changes in IgE antibody formation and avidity for the NDP-determinant in mice. *J Immunol* 1979; 123:263–268.

19. Metcalf D, Sparrow N, Nakamura K, Ishidate M: The behavior of thymus grafts in high and low leukemia strains of mice. *Aust J Exp Biol* 1961; 39:441–451.

20. Albright JF, Makinodan T: Growth and senescense of antibody-forming cells. *J Cell Physiol* 1966; 67 (suppl. 1):185–206.

21. Kysela S, Steinberg AD: Increased survival of NZB/mice given multiple syngeneic young thymus grafts. *Clin Immunol Immunopathol* 1973; 2:133–136.

22. Twomey JJ, Luchi RJ, Kouttab NM: Null cell senescense and its potential significance to the immunology of aging. *J Clin Invest* 1982; 70:201–204.

23. Kirakawa K. Albright JW, Makinodan T: Restoration of impaired immune function in aging animals. I. Effect of syngeneic thymus and bone marrow grafts. *Clin Immunol Immunopathol* 1976; 5:371–376.

24. Hirokawa K, Sato K, Makinodan T: Restoration of impaired immune function in aging animals. V. Long-term immunopotentiating effects of combined young bone marrow and newborn thymus grafts. *Clin Immunopathol* 1982; 22:297–304.

25. Carosella ED, Mochank K, Braun M: Rosette-forming T cells in human peripheral blood at different ages. *Cellular Immunol* 1974; 12:323–325.

26. Weksler ME, Hütterorth TH: Impaired lymphocyte function in aged humans. *J Clin Invest* 1981; 53:99–109.

27. Stutman O: Cell-mediated immunity and aging. *Fed Proc* 1974; 33:2028–2032.

28. Kay MMB, Mendoza J, Kiven J, Denton T, Lajiness M: Age-related changes in the immune system of mice of eighth medium and long lived strains and hybrids. *Mechanisms Aging Dev* 1979; 11:295–346.

29. Makinodan T, Kay MMB: Age influence on the immune system. *Adv Immunol* 1980; 29:287–330.

30. Callard RE, Basten A: Immune function in aged mice. IV. Loss of T cell and B cell function in thymus-dependent antibody responses. *Eur J Immunol* 1978; 8:552–558.

31. Krogsrud RL, Perkins EH: Age-related changes in T cell function. *J Immunol* 1977; 118:1607–1677.

32. Segre M, Segre D: Humoral immunity in aged mice. I. Age-related decline in the secondary response to DNP of spleen cells propagated in diffusion chambers. *J Immunol* 1976: 116:731–734.

33. Price GB, Makinodan T: Immunologic deficiencies in senescence. I. Characterization of intrinsic deficiencies. *J Immunol* 1972; 108:302–312.

34. Gerbase-Delima M, Meredith P, Walford RL: Age-related changes, including synergy and suppression, in the mixed lymphocyte reaction in long-lived mice. *Fed Proc* 1975; 34:159–161.

35. Hallgren HM, Kersey JH, Dubey DP, Yunis EJ: Lymphocyte subsets and integrated immune function in aging humans. *Clin Immunol Immunopathol* 1978; 10:65–78.

36. Shigemoto S, Kishimoto S, Yamamura Y: Changes of cell-mediated cytotoxicity with aging. *J Immunol* 1975; 215:193–194.

37. Pisciotta AV, Westring DW, DePrey C, and Walsh B: Mitogenic effect of phytohaemagglutinin at different ages. *Nature* 1967; 215:193–194.

38. Callard RE, Basten A: Immune function in aged mice. I. T cell responsiveness using phytohaemagglutinin as a functional probe. *Cell Immunol* 1977; 31:13–25.

39. Tam CF, Walford RL: Cyclic nucleotide levels in resting and mitogen-stimulated cell suspensions from young and old mice. *Mechanisms Aging Dev* 1978; 7:309–320.

40. Ruinay B, Globerson A, Shinitzky M: Visocosity of lymphocyte plasma membrane in

aging mice and its possible relation to serum cholesterol. *Mechanisms Aging Dev* 1979; 10:71–79.

41. Woda BA, Feldman JD: Density of surface immunoglobulin and capping on rat B lymphocytes. I. Changes with aging. *J Exp Med* 1979; 149:416–423.

42. Tada I, Okamura K, Ohmori K, Kaeata M, Goto M: in Orimo H, Shimada K, Iriki M, Shimada K (eds): *Recent Advances in Gerontology*. Amsterdam, Excerpta Medica, 1979, p 442.

43. Gozes Y, Umiel T, Meshorer A, Trainin N: Syngeneic GvH-induced popliteal lymph nodes by spleen cells of old C57B1/6 mice. *J Immunol* 1978; 121:2199–2204.

44. Callard RE, Blanden RV, Basten A: Loss of immune competence with age may be due to a qualitative abnormality in lympocyte membranes. *Nature* 1979; 281: 218–220.

45. Thoman ML, Weigle WO: Lymphokines and aging: Interleukin-2 production and activity in aged mice. *J Immunol* 1981; 127:2102–2106.

46. Gillis S, Kosak R, Durante M, Weksler ME: Immunological studies of aging: Decreased production and response to T cell growth factor by lympocytes from aged humans. *J Clin Invest* 1981; 67:937–942.

47. Miller RA, Stutman O: Enumeration of IL-2 secreting helper T cells by limity dilution analysis and demonstration of unexpectedly high levels of IL-2 production per responding cell. *J Immunol* 1982; 128:2258–2264.

48. Miller RA, Stutman O: Decline, in aging mice, of the anti-2,4,6 trinitrophenyl (TNP) cytotoxic T cell response attributable to loss of Lyt-2-,interleukin 2-producing helper cell function. *Eur J Immunol* 1981; 11:751–756.

49. Miller RA: Age-associated decline in precursor frequency for different T-cell mediated reactions, with preservation of helper or cytotoxic effect per precursor cell. *J Immunol* 1984; 132:63–68.

50. Callard RE, Basten A, Waters LK: Immune function in aged mice. II. B-cell function. *Cell Immunol* 1977: 31:26–36.

51. Diaz-Jouanen E, Strickland RG, Williams RC: Studies of human lymphocytes in the newborn and the aged. *Am J Med* 1975; 58:673–682.

52. Buckely CE, Dorsey FC: Serum immunoglobulin levels throughout the life-span of a healthy man. *Ann Intern Med* 1971; 75:673–682.

53. Radl J, Sepers JM, Skvaril F, Morrell A, Hijmans W: Immunoglobulin patterns in humans over 95 years of age. *Clin Exp Immunol* 1975; 22:84–90.

54. Ershler WB, Moore AL, Socinski MA: Influenza and Aging: Age-related changes and the effects of thymosin on the antibody response to influenza vaccine. *J Clin Immunol* 1984; 4:445–454.

55. Ershler WB, Hebert JF, Blow AJ, Granter S, Lynch J: Effect of thymosin alpha one on specific antibody response and susceptibility to infection in young and aged mice. *Int J Immunopharm* 1985: 7:465–471.

56. D'Agostano G, Frasca D, Garavini M, Doria G: Immunorestoration of old mice by injection of thymus extract: Enhancement of T-cell-T-cell cooperation in the in vitro antibody response. *Cell Immunol* 1980; 53:207–213.

57. Goldstein AL, Low TCK, Thurman GB, et al: in Greep RO (ed): *Recent Progress in Hormone Research*. New York, Academic Press, 1981, vol 37 pp 369–412.

58. Frasca D, Gavavini MI, Doria G: Recovery of T cell functions in aged mice injected with synthetic Thymosin 1. *Cell Immunol* 1982; 72:384–391.

59. Meredith PJ, Walford RL: Autoimmunity, histocompatibility and aging. *Mechanisms Aging Dev* 1979; 9:61–77.

60. Goidl EA, Thorbecke GJ, Weksler ME, Siskind GW: Production of auto-anti-idiotypic antibody during the normal immune response changes in the auto-anti-idiotypic antibody response and the idiotypic associated with aging. *Proc Natl Acad Sci USA* 1980; 77:6788–6798.
61. Naor D, Bonavida B, Walford RL: Autoimmunity and aging: The age-related response of mice of a long-lived strain to trimtrophenylated syngeneic mouse red blood cells. *J Immunol* 1976; 117:2204–2208.
62. Goidl EA, Innes JB, Weksler ME: Immunologic studies of aging. II. Loss of IgG and high avidity in plaque-forming cells and increased suppressor cell activity in aging mice. *J Exp Med* 1976; 144:1037–1048.
63. Doria G, Mancini C, Adonni L: Immunoregulation in senescence: Increased inducibility of antigen specific suppressor T cells and loss of cell sensitivity to immunosuppression in aging mice. *Proc Natl Acad Sci USA* 1982: 79:3803–3807.
64. Walford RL: When is a mouse "old"? *J Immunol* 1976: 117:352–353.
65. Hausman PB, Goidl EA, Siskind GW, Weksler MD: Immunological studies of aging. XI: Age-related changes in idiotypic repertoire of suppressor T cells stimulated during tolerance induction *J Immunol* 1985; 134:3802–3807.
66. Rowe JW, Clinical research on aging. Strategies and directions. *N Eng J Med* 1977; 297:1332–1336.
67. Kohn RR, Causes of death in very old people. *JAMA* 1982; 247:2793–2796.
68. Newall GR, Boutwell WB, Morris DL, Tilley BC, Branyon ES: Epidemiology of cancer, in DeVita VT, Hellman S, Rosenberg SA (eds): *Cancer: Principles and Practice of Oncology.* Philadelphia, J. B. Lipincott Co., 1982.
69. Gatti RA, Good RA: Aging, immunity and cancer. *Geriatrics* 1970; 25:158–168.
70. Ershler WB, Stewart JA, Hacker MP, Moore AL, Tindle BH: B16 murine melanoma and aging: Slower growth and longer survival in old mice. *J Natl Cancer Inst* 1984; 72:161–164.
71. Ershler WB, Gamelli RL, Moore AL, Hacker MP, Blow AJ: Experimental tumors and aging: Local factors that may account for the observed age-advantage in the B16 murine melanoma model. *Exp Gerontol* 1984; 19:367–376.
72. Ershler WB, Moore AL, Shore H, Gamelli RL: Transfer of age-associated restrained tumor growth in mice by old to young bone marrow transplantation. *Cancer Res* 1984; 44:5677–5680.
73. Ershler WB, Socinski MA, Greene GJ: Bronchogenic cancer, metastases and aging. *J Am Geriatr Soc* 1983; 31:673–676.
74. Peto R, Roc FJ, Lee PN, Levy L, Clack J: Cancer and aging in mice and men. *Br J Cancer* 1975; 32:411–416.

CHAPTER 5

CARDIOVASCULAR CHANGES WITH AGING

BRIAN F. JOHNSON AND JANICE C. HITZHUSEN

1. INTRODUCTION

Changes in cardiovascular anatomy and function that are specifically due to aging are often difficult to distinguish because of the high prevalence of heart and blood vessel disease among the elderly in Western cultures.[1-4]

2. PATHOLOGICAL CHANGES

The major pathological changes are illustrated in Fig. 1. Because it is so common in the elderly, atherosclerotic disease tends to obscure a completely separate pathological process that is specific to aging. In that condition, known as arterial ectasia, major arteries suffer diffuse loss of smooth muscle cells and breakdown of the internal elastic membrane, both of which are replaced by collagen fibers.[5,6] The endothelial cells become less regular, and multinucleated giant cells appear in the vessel walls. Connective tissue accumulates in the subendothelial region, lead-

BRIAN F. JOHNSON AND JANICE C. HITZHUSEN • Divisions of Clinical Pharmacology and Geriatrics, University of Massachusetts Medical Center, Worcester, Massachusetts 01605.

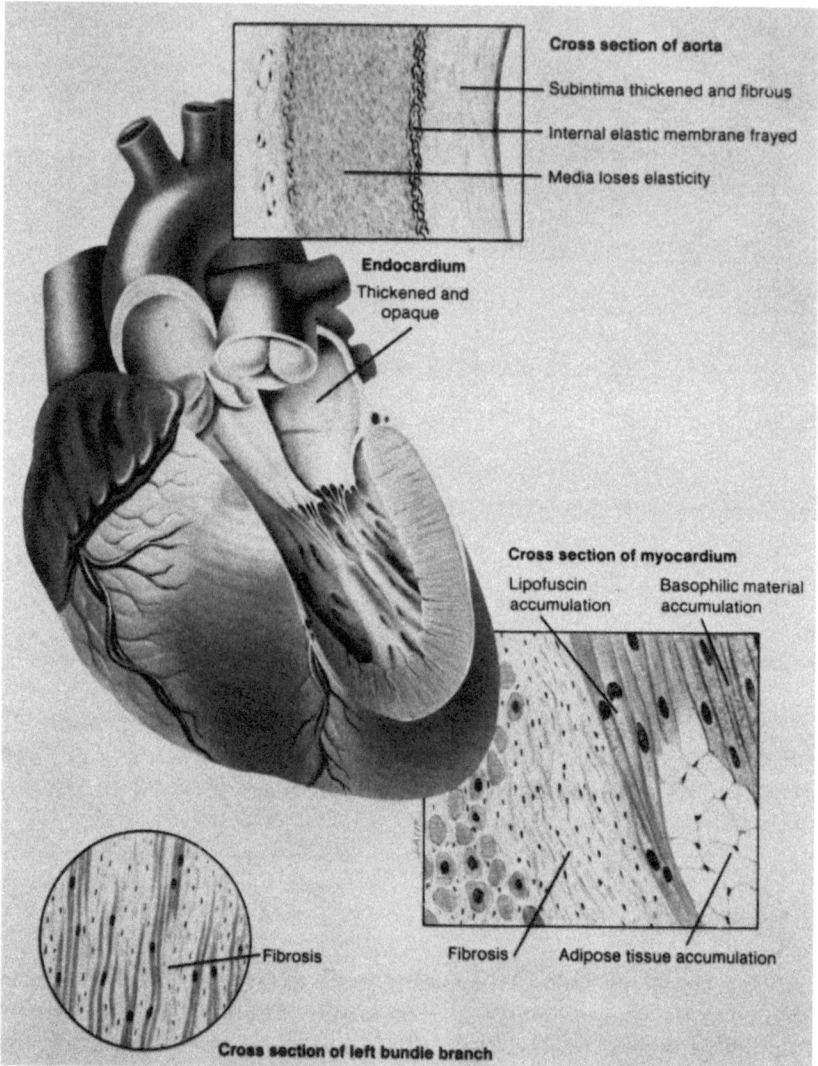

FIGURE 1   Cardiac changes in aging. (From Hitzhusen and Alpert.[17] Reproduced with permission.)

ing to calcium and lipid deposition adjacent to the internal elastic membrane. As a result, the arterial wall becomes slightly thicker, weaker, and substantially stiffer. This process of arterial ectasia becomes more severe in the thoracic aorta, which is about twice as wide and twice as long in a 90-year-old as in a young adult.[7] The distinction between this condition and atherosclerosis was shown by comparisons of aortas from Western and Asian populations.[8,9] Loss of elasticity was at least as common in the Asian aortas, despite a much lower prevalence of atherosclerosis. Similarly, arterial ectasia in the absence of atherosclerosis has been described in aging rhesus monkeys.[10] The reduced aortic distensibility occurs progressively with aging,[11] resulting in loss of the aorta's ability to absorb the suddenly increased pressure during cardiac systole and to recoil with diastole. Reduced distensibility of the major arteries provides a possible explanation for the reduced baroreflex function associated with increased age.[12,13] There is also evidence, however, of loss of autonomic ganglion cells involved with cardiovascular function in the aged.[14]

Heart failure in the elderly is usually associated with multiple abnormalities, including any form of adult heart disease, cardiac amyloidosis, and calcific valvular disease. "Senile heart failure" is not known to occur in structurally normal hearts.[15] Autopsy studies show that heart weight increases with increasing body size but not with advancing age.[16] Histological examination of the myocardium reveals accumulations of lipofuscin, basophilic amorphous material, adipose tissue, and occasionally amyloid material.[17] Empty anatomical spaces left by degenerated myofibers are replaced by connective tissue or interstitial fibrosis, changes that may be distinguishable from the classic changes of myocardial ischemia only by their patchy nature. Microscopic examination of the endocardium also reveals proliferation of elastin fibers, collagen, and mesenchymal cells. On gross inspection these changes cause the endocardium to appear thicker and more opaque in the left atrium and thinner in the ventricles.[18] McMillan and Lev[19] report that the valve leaflets remain thin and delicate with advancing age. However, other investigators have identified sclerosis of valve leaflets, decreased size and number of nuclei in the leaflets, collagen fragmentation, lipid accumulation, and calcification.[20,21] In the conducting system, the sinoatrial node shows an increasing proportion of connective tissue and fewer muscle cells, and the left bundle branch frequently atrophies and becomes fibrosed.[22–24]

3. PHYSIOLOGICAL AND BIOCHEMICAL CHANGES

Total left ventricular mass increases about 20% in the aging rat,[25] which appears to account for the 15% reduction in maximum coronary blood flow per gram dry weight of ventricle in these animals.[26,27]

The mechanisms producing greater contractility with myocardial fiber stretch are not impaired with age.[25] Age apparently has no effect on resting tension or maximal rate of tension development in rat trabecular muscle from the posterior ventricular wall. However, although similar responses of those parameters to increasing concentrations of calcium were seen in muscle from younger and older rats, the latter showed significantly less responsiveness to norepinephrine or isoproterenol.[28] In intact hearts also, the maximal response to catecholamines was reduced with aging,[29] but cyclic AMP and protein kinase responses to catecholamines were unchanged. The deficiency may be at the level of phosphorylation of cell organelles. Cardiac glycosides also increase contractility to a lesser degree in aging myocardium.[30] In rats, the specific activity of the contractile protein (actomyosin) ATPase was shown to fall with increasing age.[31] This may be of importance in view of the reported correlation between the level of activity of this ATPase and myocardial function.[32] Subsequent studies have shown that the age-related loss in enzyme activity in rats can be prevented by continued physical training.[33] It has been suggested that training may minimize the age-related loss of sensitivity of actomyosin ATPase to calcium.[34]

Transmembrane action potential duration is prolonged in loaded right ventricular papillary muscle preparations from aging rats.[35,36] The stimulated senescent heart also requires more time to return to the preexcitation state.[25] However, both contraction duration and myocardial stiffness in chronically exercised rats remain at levels similar to those in younger animals.[37]

During diastole, reuptake of calcium into the sarcoplasmic reticulum has also been found to be slower in older rats than in younger rats.[38] That may explain the slower rate of isometric relaxation of the myocardium in the elderly.

Studies in beagle dogs may explain some of the differing cardiovascular responses to exertion in the elderly. Although no differences between young and old animals were noted in resting left ventricular stroke volume or aortic input impedance, exercise caused a stepwise increase in stroke volume in the young dogs, whereas the old animals showed a major increase in impedance. These differences were abolished by β-adrenergic blockade.[39] It has been postulated that the elderly person has a reduced vasodilator response to catecholamines.

4. ABNORMALITIES OF BLOOD PRESSURE

At least one-third of Americans over the age of 61 years can be considered hypertensive, defined as either a systolic blood pressure level at or above 160 mm Hg or a diastolic pressure at or above 95 mm Hg.[40] As with younger adults, the incidence of cardiovascular disease in these hypertensive patients is more than three times that in elderly normotensives.[41] Hypertension greatly increases the risk of stroke and heart failure in the elderly. Although hypertension increases the risk of coronary disease to a more modest degree, manifestations of ischemic heart disease are the most common complications of hypertension at all ages.

There are two relatively distinct forms of hypertension. Essential hypertension usually has its onset in early adult life and is recognized by increased levels of systolic and diastolic blood pressure. By contrast, predominant systolic hypertension is uncommon before late middle life and becomes much more frequent after the age of 70. It is defined as a systolic blood pressure of 160 mm Hg or greater with an associated diastolic pressure below 95 mm Hg. Isolated systolic hypertension was found in 15% of white men, 31% of white women, 26% of black men, and 39% of black women in the elderly population, according to the United States National Health Examination Survey.[42] Although the pathological bases of the two forms of hypertension are different, both forms often coexist in the same patient. The term disproportionate systolic hypertension may be used for elderly patients with mild elevation of diastolic blood pressure and marked elevation of systolic pressure. Koch-Weser defined disproportionate systolic hypertension as a systolic pressure greater than two times diastolic pressure minus 15.[43] Although little is known about the effects of hypertension on the incidence of cardiovascular disease in the elderly, elevation of systolic pressure appears to be even more important than diastolic pressure as a risk factor for stroke, ischemic heart disease, and heart failure.[40,41,44,45] Isolated elevation of systolic blood pressure is an important risk factor, but whether pharmacological control of isolated systolic hypertension reduces the risk of life-threatening complications is unknown.

In the elderly, as in younger adults, the fundamental hemodynamic abnormality in established essential hypertension is an increase in peripheral vascular resistance. Hyaline thickening of arterioles may contribute to the increased peripheral vascular resistance in the elderly.[46] Both plasma renin activity and aldosterone concentration decrease progressively with age in both normotensive and hypertensive subjects.[47] Increases in plasma renin activity in response to sodium depletion or diuretic treatment may also be reduced in the elderly. By contrast, plasma levels of norepinephrine are often higher in the elderly, although there is no evidence that this is due to an increased level of sympathetic nervous system activity.[48] The elderly also have relatively reduced body fluid volumes, with extracellular fluid volume being most reduced. It would be expected that hemodynamic and fluid volume changes would alter the responsiveness of elderly hypertensives to various antihypertensive agents. However, at present there is little experimental evidence to support these theoretical possibilities.

The pathogenesis of isolated or predominant systolic hypertension in the elderly is only beginning to be understood. As discussed earlier, ectasia of the major arteries is associated with loss of distensibility. Established essential hypertension may also accelerate the aortic stiffening process.[49] As a result, for any level of left ventricular stroke volume the reduced distensibility causes exaggeration of systolic pressure. Reduced baroreceptor sensitivity, resulting from impaired regulation of cardiac output common in the elderly, may contribute to hypertension.

Orthostatic hypotension is also more common in the elderly. Several studies have shown that about 5% of people over the age of 65 years have a drop of at least 40 mm Hg in systolic pressure upon standing.[50] Reduced function of the baroreceptor system is believed to be a major cause of orthostatic hypotension. The elderly have reduced capacity to compensate for either elevated or reduced blood pressure. Orthostatic hypotension is more common in patients with other disorders, such as organic brain disease, heart disease, neurological dysfunction, and varicose veins. It seems likely that those conditions interact with the reduced level of baroreceptor function to produce the clinical disorder.

5. Maximal Work Performance

In their classical studies, Master and Oppenheimer demonstrated that maximal work performance fell at least 20% between the third and eighth decades of life.[51] This age-related reduction in maximal work capacity parallels a similar decline in maximal oxygen consumption, which averages about 1% each year between the third and eighth decades.[52] Efficiency of tissue oxygen extraction is unimpaired in the elderly. A similar decline in maximal oxygen consumption was recorded over a period of 22 years in men aged 18–22 years on initial testing.[53] Some of the drop was found to be prevented by obesity prevention, smoking cessation, and long-term physical conditioning programs, and the decline may be less in women,[54] but some degree of reduction in work performance with age appears inevitable.

6. Invasive Procedures

Cardiac output has been consistently shown to decline with age. In a group of 67 men between 19 and 86 years selected on the basis of absence of symptoms suggestive of cardiovascular disease, a clear inverse correlation was found between resting cardiac output and age. Cardiac output fell about 1% each year. Hemodynamics were based on Evans blue dye dilution techniques. A less striking correlation was that between increasing age and progressively falling stroke volume. Mean resting heart rate also fell with age in these subjects. It is unlikely that the reduction in cardiac output could have been explained by decrease in body size because body-surface-area-adjusted cardiac index remained inversely correlated with age.[55]

In right-heart catheterization studies of 27 healthy men aged 61–83, the Fick method was used to determine cardiac output. The patients were ambulatory and subjected to two consecutive levels of exertion. Compared with data obtained in 23 young men studied earlier by the same investigators under similar circumstances, the older patients had lower mean levels of stroke volume and cardiac output while resting recumbent, but the two groups had similar values while sitting. Increases in levels of exertion were similar in both groups, but there was at least a 25% reduction in the older group in maximal oxygen uptake and maximal cardiac output. Both maximal heart rate and stroke volume were reduced by about 15%. During maximal levels of exertion the older group also showed a 39 mm Hg greater mean rise in systolic blood pressure and about a 40% higher pulmonary wedge pressure and right ventricular end-diastolic pressure.[56]

In another study, significantly lower levels of cardiac output were determined at rest in both recumbent and sitting positions in subjects aged 50 or over without evidence of cardiac disease.[57] In the group as a whole, consisting of 35 men and 19 women aged 18–68, without evidence of cardiac disease, similar increments in cardiac output were seen with exertion, but mean levels of cardiac output remained lower for each level of oxygen consumption in the older subjects. Calculated total peripheral resistance was higher in the older subjects at rest and during exertion. Cardiac output was determined by the indicator-dilution method using indocyanine green. Oxygen consumption was also determined at rest and at progressive levels of exercise on a bicycle ergometer up to the point of maximal voluntary effort.

Although these studies are reasonably consistent, it is intriguing that their conclusions about cardiac function differ somewhat from those obtained using noninvasive methodologies. It has been suggested that differences in preload or afterload may account for the reported aging effect on resting cardiac output.[58] It is possible that in the more recent studies using noninvasive methods, investigators may have more adequately screened out patients with occult coronary artery disease. It is also apparent that invasive methods are more stressful, so that resting determinations may be associated with higher levels of sympathetic nervous system activity. Conceivably, older subjects may respond somewhat differently to the stress of invasive methodology. It must also be admitted that in most invasive studies, few attempts were made to differentiate the effects of age from those of reduced physical fitness so commonly seen in the older population.

7. ECHOCARDIOGRAPHY

The echocardiogram has proved a useful and accurate tool to measure the dimensions of cardiac chambers, septal and myocardial thickness, valve excursion and closure, and aortic root dimension. It is particularly suited to study the effect of age on the heart by providing anatomical information without risk to the patient. M-mode and two-dimensional techniques are routinely used because of their complementary nature: two-dimensional echocardiography provides better imaging of cardiac valves and chambers, whereas M-mode echocardiography is better for recording time motion characteristics and making precise measurements.

Gerstenblith et al. compared M-mode echocardiographic data of 18 men aged 65–84 years with other groups aged 25–44 years and 45–64 years, respectively.[59] They found that the E–F slope of the anterior mitral valve leaflet decreases, the aortic root diastolic dimension increases slightly, and left ventricular diastolic and systolic wall thickness adjusted for body surface area increases with age. Systolic blood pressure was higher than in the two younger groups. Heart rate, left ventricular systolic and diastolic cavity dimensions, velocity of circumferential shortening, ejection fraction, ejection time, and estimated stroke volume did not vary with increasing age. Gardin et al.[60] compared M-mode echocardiographic results of 10 men and two women aged 71–97 years to those of 68 men and 56 women aged 21–70 years and found similar results. However, they noted a decrease in internal dimensions of the left ventricle and a progressive increase in left atrial dimension. In one study of elderly subjects, diminished posterior wall excursion and reduced velocity of shortening of cardiac fibers was found, and it was suggested that the limitation of cardiac performance in the elderly may depend partly on an age-associated decrease in the work of contractile elements.[61] Using both M-mode and two-dimensional echocardiography, Hitzhusen et al.[62] studied 14 women and 20 men aged 70–83. All subjects underwent treadmill thallium-201 exercise perfusion imaging to exclude the presence of coronary artery disease, which had not been performed by previous investigators. The results are shown in Table I. A significant increase in mean aortic root dimension and left ventricular wall thickness and a significant decrease in mean mitral valve E–F slope were again noted in these older subjects when compared with values for younger subjects. In contrast to the findings of Gardin *et al.*, there was no difference in left ventricular internal dimensions between the two groups.

It has been suggested that the decreased E–F slope, which is proportionate to the rate of closure of the anterior leaflet of the mitral valve, is associated with a decreased rate of early diastolic filling of the left ventricle. These abnormalities could be due to intrinsic changes either in the mitral valve or in the left ventricle. Mitral valve changes that may be causative have been described by Sell and Scully,[20] whereas the increased left ventricular myocardial wall thickness may contribute to stiffness, thereby limiting the rate of left ventricular filling in early diastole.[63,64]

The increasing aortic root dimension reflects the findings of published autopsy studies describing increased aortic diameter and volume with age.[65] The resulting increased aortic wall stiffness and the response of the left ventricle to the stiffness could explain the increase in left ventricular wall thickness.

TABLE I   MEAN VALUES FOR ECHOCARDIOGRAPHIC
PARAMETERS

	Mean	SD	n
Mitral valve E-F slope (mm/sec)	67	30	34
Amplitude of motion (mm)	17	4	27
A wave amplitude (mm)	13	4	23
Aortic root (mm)	34	5	34
Aortic root/M2	20	5	30
Left ventricular dimension			
End systole (mm)	26	6	34
End diastole (mm)	45	7	33
Left ventricular wall thickness			
Septum (mm)	12	3	30
Posterior wall (mm)	11	2	34
Left atrium (mm)	36	5	34
Left atrium/M2	21	3	25
Right ventricle (mm)	22	6	33
Right ventricle/M2	13	4	24

8. RADIONUCLIDE TECHNIQUES—THALLIUM-201

The evaluation of myocardial perfusion and cellular viability by thallium-201 is based on its potassium analog characteristic and its radioactive tracer imaging property. Like potassium, thallium-201 accumulates within normal viable myocardial cells. Its distribution is nearly proportional to regional myocardial blood flow and in normal hearts demonstrates homogeneous distribution of radioactivity at rest and with the stress of exercise. A "defect zone" that is present both at rest and with stress suggests myocardial infarction, whereas filling in of a defect zone after rest suggests ischemia. This technique is especially useful in the elderly, in whom the diagnosis or exclusion of coronary artery disease or myocardial infarction is difficult because classic symptoms and prompt request for medical attention are unusual.

The value of stress thallium-201 imaging in detecting coronary artery disease has been well established through comparison to electrocardiography[66,67] and to the combined sensitivity of electrocardiography and coronary arteriography.[68,69]

Tiefenbrunn et al. have documented the sensitivity of thallium-201 imaging at rest in detecting remote myocardial infarction in patients an average of 11 months after documented infarction.[70] Perfusion defects were detected in 90–95% of these patients. Patients with smaller infarct size as defined by serum enzyme levels of less than 20 creatine kinase-gram equivalents failed to exhibit a defect. The incidence of asymptomatic coronary disease in the elderly was studied by Hitzhusen et al.[71] by using treadmill thallium-201 exercise perfusion imaging. Thirty-nine healthy men and women aged 70–83 years were carefully screened for absence of clinical heart disease. Thallium-201 imaging was abnormal in 5 (13%) of the 39 asymptomatic subjects. Electrocardiography during exercise was consistent with ischemia in four of these five subjects. Of the 34 volunteers with normal thallium imaging, 32 had no electrocardiographic changes.

9. RADIONUCLIDE TECHNIQUES—EQUILIBRIUM ANGIOGRAPHY

Equilibrium angiography uses the electrocardiogram for timing the imaging of the entire blood pool of ^{99m}Tc-labeled red cells during cardiac contraction. It is useful as a noninvasive procedure to evaluate regional wall motion and global left ventricular function. Unlike contrast ventriculography and first-pass radionuclide angiography, equilibrium studies allow serial determination of ejection fractions to be made over an extended period of time and provide a useful method for evaluating the cardiac response to exercise.[72–75] Port et al.[76] reported a marked age-related decrease in left ventricular ejection fraction after exercise, but only a limited number of subjects older than 70 years were studied.

Hitzhusen et al.[71] studied 39 volunteers, 22 men and 17 women aged 70–83 years. Twelve of the subjects had minor regional left ventricular wall motion abnormalities during exercise. The female subjects had difficulty performing bicycle exercise and could bicycle only half as long as they could walk on the treadmill. There was no difference in these two exercise protocols for the men. The results are shown in Tables II and III and Fig. 2. As anticipated, the maximal heart rate during bicycle exercise was somewhat lower and the systolic blood pressure somewhat higher than those described for younger subjects. The rate–pressure product achieved was similar to that determined for younger, healthy persons. The data are at variance with the previous findings[76] of marked decrease in left ventricular ejection fraction at maximal exercise. Although there was no significant mean change in exercise ejection fraction in the group, 74% maintained or increased their ejection fraction with exercise. However, this represents a modest decrease in exercise cardiovascular performance given that mean left ventricular ejection fraction failed to increase during maximal exercise, as is routinely seen in younger persons.

TABLE II   EXERCISE TESTING DATA

	Treadmill		Bicycle		
	Rest	Maximal exercise	Rest	Submaximal exercise	Maximal exercise
Men					
Heart rate (beats/min)	66 ± 2	138 ± 4	67 ± 2	108 ± 4	118 ± 3
Blood pressure (mm Hg)	136 ± 4	180 ± 5	136 ± 4	173 ± 5	190 ± 5
	79 ± 2	85 ± 2	80 ± 2	94 ± 2	97 ± 2
Rate–pressure product (mm Hg × beats/min × 10^3)		25			21
Time (min)		7			9
Women					
Heart rate (beats/min)	73 ± 2	134 ± 8	80 ± 4	108 ± 7	117 ± 4
Blood pressure (mm Hg)	140 ± 4	184 ± 5	128 ± 3	163 ± 10	159 ± 4
	78 ± 2	86 ± 3	72 ± 2	91 ± 5	85 ± 2
Rate–pressure product (mm Hg × beats/min × 10^3)		27			22
Time (min)		7			4

TABLE III   CHANGE IN RADIONUCLIDE VENTRICULOGRAPHIC DATA
(% ± STANDARD ERROR OF THE MEAN)[a]

Sex	Rest	Submaximal exercise	Maximal exercise
Men			
Ejection fraction	64 ± 3	71 ± 3	67 ± 3
Relative end-systolic volume		↓ 66 ± 3	↓ 66 ± 3
Relative end-diastolic volume		↓ 1 ± 4	↑ 4 ± 4
Relative cardiac output		↑ 56 ± 12	↑ 85 ± 11
Relative stroke volume		↑ 11 ± 6	↑ 10 ± 4
Women			
Ejection fraction	64 ± 3	76 ± 7	66 ± 3
Relative end-systolic volume		↓ 81 ± 5	↓ 50 ± 19
Relative end-diastolic volume		↓ 1 ± 1	↓ 5 ± 9
Relative cardiac output		↑ 38 ± 12	↑ 65 ± 27
Relative stroke volume		↓ 2 ± 12	↓ 7 ± 7

[a] ↓ = decreases; ↑ = increases.

The influence of age on cardiovascular performance during exercise was discussed earlier in this chapter. The data describing that relationship are similar to the findings from studies of physically active elderly subjects[52,54] and from studies of elderly men and women who have undergone extensive training regimens.[77,78] Young subjects, both untrained and trained, respond to exercise with increased heart rate and myocardial contractility that are presumably mediated by augmented sympathetic nervous system activity.[79–81] Those subjects show decreases in left ventricular end-systolic volume and increases in fractional shortening without change in end-diastolic volume.[80,82] However, untrained elderly subjects usually have increases in heart rate and in end-diastolic pressure and make use of the Frank-Starling mechanism to augment exercise cardiac output.[79–81,83] The vigorous elderly subjects studied by Hitzhusen et al.,[71] all of whom exercised regularly, responded to exertion by increasing heart rate, by decreasing end-systolic volume, and by maintaining an unchanged end-diastolic volume. Their response was similar to that of normal younger persons. Increases in cardiac output and stroke volume with peak exercise, however, were diminished compared with levels in trained and untrained younger subjects. There are many differences between healthy elderly subjects and trained younger subjects, but it is interesting to note that both the athlete and the healthy elderly

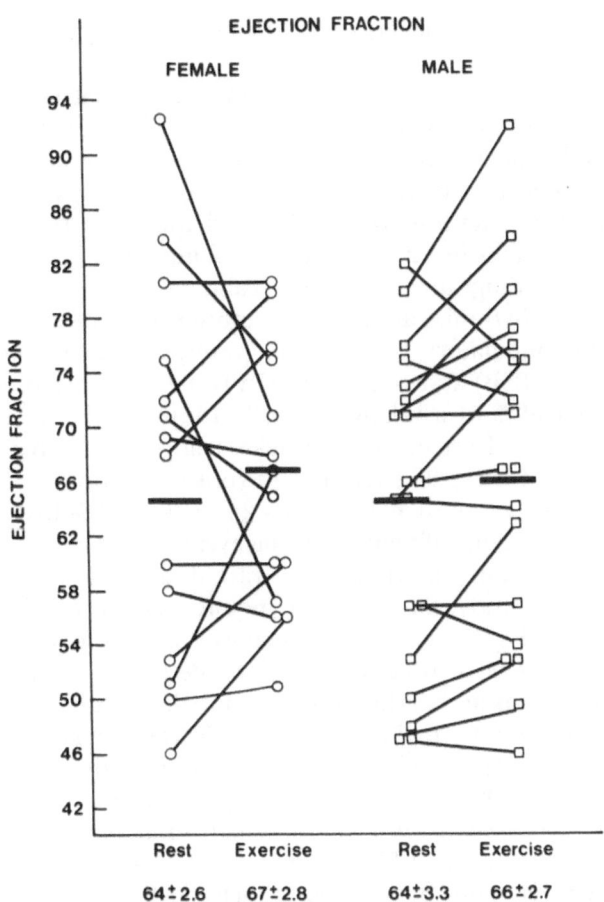

FIGURE 2 Changes in ejection fraction from rest to exercise in the elderly subjects. ○, Female; □, male. (From Hitzhusen *et al.*[71] Reproduced with permission.)

person have increased left ventricular wall thickness.[59,84] The hypertrophied heart may have decreased diastolic compliance and less capacity to increase end-diastolic volume, especially when diastolic filling time is shortened during exercise-induced tachycardia. At peak exercise, lower cardiac output in active elderly subjects may be the result of low maximal heart rate, inability to increase end-diastolic volume, decreased responsiveness to β-adrenergic receptor stimulation, or changes in the myocardial cell that may limit contractility.[33,59,84,85]

10. ELECTROCARDIOGRAPHIC CHANGES

With age, fat may accumulate around the sinoatrial node, which loses many of its pacemaker cells. Valvular calcification may spread, involving and damaging the AV node, bundle of His, or bundle branches. However, a specific degenerative process in the elderly called Lev's disease may be of greater importance. These changes may relate to the age-dependent increases in PR, QRS, and Q–T intervals. Frequent changes in the electrocardiogram are listed in Table IV.

Cardiac arrhythmias are generally more prevalent in the elderly, if recorded by ambulatory electrocardiography. Isolated premature supraventricular (PSB) and ventricular (PVB) beats occurred to some degree in about 80% of a relatively small group of subjects aged 68–85 without evidence of organic heart disease using noninvasive methods.[86] About a third of these 98 subjects had runs of supraventricular tachycardia, 17% had more than 100 PVBs/24 hr, and 15% had ventricular couplets or short runs of ventricular tachycardia. By contrast, evidence of severe conduction abnormality was rare in this study population.

In a recent study, 24-hr Holter monitoring was obtained in 1238 patients over 70 years of age, the majority of whom were inpatients without evidence of cardiac disease.[87] In general, a poor correlation between symptoms and arrhythmia was noted, though 29% of patients with syncopal episodes had associated bradyarrhythmias, and 21% had concurrent tachyarrhythmias. Sinus node dysfunction was again noted to be rare. About 80% had PSB, and brief paroxysms of supraventricular tachycardia occurred in 55%. Paroxysmal atrial fibrillation was rare. Almost three-quarters had PVB, and in the majority they were complex. Eight percent had runs of ventricular tachycardia.

Allowing for the very different composition of those two study samples, the conclusions appear consistent. Unfortunately, because little is known about the prognostic importance of these common rhythm dis-

TABLE IV COMMON ELECTROCARDIOGRAPHIC CHANGES
OF HEALTHY ELDERLY PEOPLE

Leftward shift of the frontal QRS axis	Decrease in QRS duration
Low voltage R, S, and T waves	Right bundle branch block
Isolated left-axis deviation	Premature ventricular beats
Increase in the P-R interval	Premature supraventricular beats
Increase in Q-T interval	

turbances in the elderly, the potential value of therapy remains an important area for research.

11. CONCLUSIONS

In the elderly, the marked stiffness of aortic ectasia has profound effects on systolic blood pressure and on impedance to left ventricular outflow. Isolated systolic hypertension is common and is associated with a high risk of vascular complications.

Accumulations of a variety of substances occur in the aging cardiovascular system, but probably cause no hemodynamic dysfunction unless excessive. At rest, contractility and cardiac output are little affected by aging, though responsiveness to digitalis and catecholamines is reduced. Reduced work performance and oxygen consumption at maximal workloads is associated with a lesser capacity to increase heart rate, cardiac output, and ejection fraction than in young adults. There is no specific clinical cardiomyopathy of aging, and cardiovascular performance can increase manyfold during exertion in the absence of disease, and particularly in physically active individuals. However, the many pathophysiological and biochemical changes that occur may alter responsiveness to both disease and drugs and may include myocardial hypertrophy, reduced myocardial compliance, prolonged contraction, and delayed ventricular filling rate.

Supraventricular and ventricular ectopic activity are frequent, whereas clinical conduction disturbances are rare.

REFERENCES

1. *National Program to Conquer Heart Disease, Cancer, and Stroke*. The President's Commission, Washington, DC, 1964
2. Caird FI, Kennedy RD: Epidemiology of heart disease in old age, in Caird FI, Dall JLC, Kennedy RD (eds): *Cardiology in Old Age*. New York, Plenum Press, 1976, p 1.
3. White NR, Edwards JE, Dry TJ: The relationship of the degree of coronary atherosclerosis with age. *Circulation* 1950; 1:645–654.
4. Ackerman RF, Dry TJ, Edwards JE: Relationship of various factors to the degree of coronary atherosclerosis in women. *Circulation* 1950; 1:1345–1354.
5. Kohn RR: Heart and cardiovascular system, in Finch CE, Hayflick L (eds): *Handbook of the Biology of Aging*. New York, van Nostrand Reinhold Co, 1977, p 281.
6. Wilens SL: The postmortem elasticity of the adult human aorta: Its relation to age and to the distribution of atheroma. *Am J Pathol* 1937; 13:811–815.
7. Wellman WE, Edwards JE: Thickness of the media of the thoracic aorta in relation to age. *Arch Pathol* 1950; 50:183–188.

8. Nakashina T, Tanikawa J: A study of human aortic distensibility with relation to atherosclerosis and aging. *Angiology* 1971; 22:477–490.

9. Avolio AP, Chen S, Wang R, et al: Effect of aging on changing arterial compliance and left ventricular load in a northern Chinese urban community. *Circulation* 1983; 68:50–58.

10. Uno H, Poff B: Coronary arterial ectasia, a predominant type of coronary sclerosis in aged captive rhesus monkeys (Macaca mulatta). *Am J Pathol* 1983; 111:315–322.

11. Hallock P, Benson IC: Studies on the elastic properties of human isolated aorta. *J Clin Invest* 1937; 16:595–602.

12. Gribbin B, Pickering TG, Sleight P, et al: Effect of age and high blood pressure on baroreflex sensitivity in man. *Circ Res* 1971; 29:424–431.

13. McGarry K, Laher MS, Fitzgerald DJ, et al: Baroreflex function in elderly hypertensives. *Br Heart J* 1981; 45:620.

14. Sulkin NM, Kunz A: Histochemical alterations in autonomic ganglion cells associated with aging. *J Gerontol* 1952; 7:533.

15. Pomerance A: Pathology of the heart with and without cardiac failure in the aged. *Br Heart J* 1965; 27:697–710.

16. Smith HL: The relation of the weight of the heart to age and to the weight of the body. *Am Heart J* 1928; 4:79.

17. Hitzhusen JC, Alpert JS: The elderly heart: Special signs and symptoms to watch for. *Geriatrics* 1984; 39:38–51.

18. McMillan JB, Leo M: The aging heart: I. Endocardium. *J Gerontol* 1959; 14:268–283.

19. McMillan JB, Lev M: The aging heart. II. The valves. *J Gerontol* 1964; 19:1–19.

20. Sell S, Scully RE: Aging changes in the aortic and mitral valves: Histologic and histochemical studies with observations on the pathogenesis of calcific aortic stenosis and calcification of the mitral annulus. *Am J Pathol* 1965; 46:345–348.

21. Angrist A: Aging heart valves and a unitary pathology hypothesis for sclerosis. *J Gerontol* 1954; 19:135–143.

22. Davies MJ, Pomerance A: Quantitative study of aging changes in the human sinoatrial node and internodal tracts. *Br Heart J* 1972; 34:150–153.

23. Das DN, Fleg JL, Lakatta EG: Effect of age on the components of atrioventricular conduction in normal humans. *Am J Cardiol* 1982; 49:1031–1032.

24. Lev M, Bharati S: Age-related changes in the cardiac conduction system. *Intern Med* 1981; 6:19–37.

25. Yin FCP, Spurgeon HA, Weisfeldt ML, et al: Mechanical properties of myocardium from hypertrophied rat hearts: A comparison between hypertrophy induced by senescence and by aortic banding. *Circ Res* 1980; 46:292–300.

26. Weisfeldt ML, Loeven WA, Shock NW: Resting and active mechanical properties of trabeculae carneae from aged male rats. *Am J Physiol* 1971; 220:1921–1927.

27. Abu-Erreish GM, Neely JR, Whitmer JT, et al: Fatty acid oxidation by isolated perfused working hearts of aged rats. *Am J Physiol* 1977; 232:E258–E262.

28. Lakatta EG, Gerstenblith G, Angell CS, et al: Diminished inotropic response of aged myocardium to catecholamines. *Circ Res* 1975; 36:262–269.

29. Guarnieri T, Filburn CR, Zitnik G, et al: Contractile biochemical correlates of beta-adrenergic stimulation of the aged heart. *Am J Physiol* 1980; 239:H501–H508.

30. Guarnieri T, Spurgeon H, Froehlich JP, et al: Diminished inotropic response but unaltered toxicity to acetyl-strophanthidine in the senescent beagle. *Circulation* 1979; 60:1548–1554.

31. Chesky J, Rockstein M: Reduced myocardial actomyosin adenosinetriphosphatase activity in the ageing male Rischer rat. *Cardiovasc Res* 1977; 11:242–246.

32. Carey R, Natarajan G, Bove A, et al: Myosin adenosine triphosphatase activity in the volume-overloaded hypertrophied feline right ventricle. *Circ Res* 1979; 45:81–87.

33. Rockstein M, Chesky JA, Lopez T: Effects of exercise on the biochemical aging of mammalian myocardium. I. Actomyosin ATPase. *J Gerontol* 1981; 36:294–297.

34. Rockstein M, Chesky J, Lopez T: Calcium sensitivity of myocardial actomyosin ATPase in young and mature male Fischer rats. A brief note. *Mechanisms Aging Dev* 1978; 8:413–416.

35. Wei JY, Spurgeon HA, Lakata EG: Transmembrane action potential duration and contractile activation are lengthened in cardiac muscle of senescent rats. *Clin Res* 1980; 28:619A.

36. Wei JY, Spurgeon HA, Lakatta EG: Excitation-contraction in rat myocardium: alterations with adult aging. *Am J Physiol* 1984; 246:H784–H791.

37. Spurgeon HA, Steinbach MF, Lakatta EG: Chronic exercise prevents characteristic age-related changes in rat cardiac contraction. *Am J Physiol* 1983; 244:H513–H518.

38. Froehlich JP, Lakatta EG, Beard E, et al: Studies of sarcoplasmic reticulum function and contraction duration in young and aged rat myocardium. *J Mol Cell Cardiol* 1978; 10:472–488.

39. Yin FCP, Weisfeldt ML, Milnor WR: Role of aortic input impedance in the decreased cardiovascular response to exercise with aging in dogs. *J Clin Invest* 1981; 68:28–38.

40. Kannel WB, Gordon T, Schwartz MJ: Systolic versus diastolic blood pressure and risk of coronary heart disease. The Framingham Study. *Am J Cardiol* 1971; 27:335–346.

41. Kannel WB, Gordon T: Evaluation of the cardiovascular risk in the elderly: The Framingham study. *Bull NY Acad Med* 1978; 54:573–591.

42. National Center for Health Statistics: Blood pressure of adults by age and sex: United States, 1960–1962. Public Health Service Publication No 1000, Series 11, No. 4. Washington, DC, Government Printing Office, 1964.

43. Koch-Weser J: Correlation of pathophysiology and pharmacotherapy in primary hypertension. *Am J Cardiol* 1973; 32:499–510.

44. Kannel WB, Wolf PA, Verter J, et al: Epidemiologic assessment of the role of blood pressure in stroke: The Framingham Study. *JAMA* 1970; 214:301–310.

45. Kannel WB, Castelli WP, McNamara PM, et al: Role of blood pressure in the development of congestive heart failure: The Framingham Study. *N Engl J Med* 1972; 287:781–787.

46. Swales JD: Pathophysiology of blood pressure in the elderly. *Age Aging* 1981; 8:104–112.

47. Weidmann P, DeMyttenaere-Burszteins M, DeLima J: Effect of ageing on plasma renin and aldosterone in normal man. *Kidney Int* 1975; 8:325–333.

48. Esler M: Age-dependence of noradrenaline kinetics in normal subjects. *Clin Sci* 1981; 60:217.

49. Hickler RB: Aging and hypertension: Hemodynamic implications of systolic pressure trends. *J Am Geriatr Soc* 1983; 31:421–425.

50. Caird FI, Andrews GR, Kennedy RD: Effect of posture on blood pressure in the elderly. *Br Heart J* 1973; 35:527–530.

51. Master AM, Oppenheimer ET: A simple exercise tolerance test for circulatory efficiency with standard tables for normal individuals. *Am J Med Sci* 1929; 177:223–243.

52. Dehn MM, Bruce A: Longitudinal variations in maximal oxygen uptake with age and activity. *J Appl Physiol* 1972; 33:805–807.

53. Robinson S, Dil DB, Ross JC, et al: Training and physiological aging in man. *Fed Proc* 1973; 32:1628–1634.

54. Hodgson JL, Buskirk ER: Physical fitness and age, with emphasis on cardiovascular function in the elderly. *J Am Geriatr Soc* 1977; 25:385–392.

55. Brandfonbrener M, Landowne M, Shock NW: Changes in cardiac output with age. *Circulation* 1955; 12:557–566.

56. Granath A, Jonsson B, Strandell T: Circulation in healthy old men, studied by right heart catheterization at rest and during exercise in supine and sitting position. *Acta Med Scand* 1964; 176:425–446.

57. Julius S, Amery A, Whitlock LS, et al: Influence of age on the hemodynamic response to exercise. *Circulation* 1967; 36:222–230.

58. Gerstenblith G, Lakatta EG, Weisfeldt ML: Age changes in myocardial function and exercise response. *Prog Cardiovasc Dis* 1976; 19:1–21.

59. Gerstenblith G, Frederiksen J, Yin FCP, et al: Echocardiographic assessment of a normal adult aging population. *Circulation* 1977; 56:273–278.

60. Gardin JM, Henry WL, Savage DD, et al: Echocardiographic measurements in normal subjects: evaluation of an adult population without clinically apparent heart disease. *J Clin Ultrasound* 1979; 7:439–447.

61. Wroblewski T, Szlachcic-Brynczak J: Left ventricular wall motion determined by echocardiography in elderly subjects. *Acta Physiol* 1980; 31:47–51.

62. Hitzhusen JC, Hickler RB, Pape LA, et al: Left ventricular dimensions and function in a healthy elderly population: A 2-dimensional echocardiographic study. *Gerontologist* 1982; 22:119.

63. Sainsbury R, White T, Wray R: Echocardiography in elderly patients with systolic murmurs. *Age Aging* 1981; 10:225–230.

64. Miyatake K, Okamoto M, Kinoshita N, et al: Augmentation of atrial contribution to left ventricular inflow with aging as assessed by intracardiac Doppler flowmetry. *Am J Cardio* 1984; 53:586–589.

65. Learoyd BM, Taylor MG: Alterations with age in the viscoelastic properties of human arterial walls. *Circ Res* 1966; 18:278–292.

66. Bailey IK, Griffith LSC, Rouleau J, et al: Thallium 201 myocardial perfusion imaging at rest and during exercise. Comparative sensitivity to electrocardiography in coronary artery disease. *Circulation* 1977; 55:79–87.

67. Botvinick EH, Tardash MR, Shames DM, et al: Thallium 201 myocardial perfusion scintography for the clinical clarification of normal, abnormal, and equivocal electrocardiographic stress tests. *Am J Cardiol* 1978; 41:43–59.

68. Ritchie JL, Trobaugh GB, Hamilton GW: Myocardial imaging with thallium 201 at rest and during exercise. Comparison with coronary arteriography and resting and stress electrocardiography. *Circulation* 1977; 56:66–71.

69. McGowan RL, Martin ND, Zaret BL: Diagnostic accuracy of noninvasive myocardial imaging for coronary artery disease: an electrocardiographic and angiographic correlation. *Am J Cardiol* 1977; 40:6–10.

70. Tiefenbrunn AJ, Biello DR, Geltman EM, et al: Gated cardiac blood pool imaging and thallium 201 myocardial scintography for detection of remote myocardial infarction. *Am J Cardiol* 1981; 47:1–6.

71. Hitzhusen JC, Hickler RB, Alpert JS, et al: Exercise testing and hemodynamic performance in healthy elderly persons. *Am J Cardiol* 1984; 54:1082–1086.

72. Slutsky R, Karliner J, Ricci D, et al: Response of left ventricular volume to exercise in man assessed by radionuclide equilibrium angiography. *Circulation* 1979; 60:565–571.

73. Borer JS, Bachrach SL, Green MV: Real-time radionuclide cineangiography in the noninvasive evaluation of global and regional left ventricular function at rest and during exercise in patients with coronary artery disease. *N Engl J Med* 1979; 296:839–843.

74. Sorenson SG, Ritchie JL, Caldwell JH, et al: Serial exercise radionuclide angiography: validation of count derived changes in cardiac output and quantitation of maximal

exercise ventricular volume change after nitroglycerin and propranolol in normal men. *Circulation* 1980; 61:600–609.

75. Borer JS, Kent KM, Bacharach SL, et al: Sensitivity, specificity and predictive accuracy of radionuclide cineangiography during exercise in patients with coronary artery disease: Comparison with exercise electrocardiography. *Circulation* 1979; 60:572–580.

76. Port S, Frederick RC, Coleman RE, et al: Effect of age on the response of the left ventricular ejection fraction to exercise. *N Engl J Med* 1980; 303:1133–1137.

77. deVries HA: Physiological effects of an exercise training program upon men aged 52–88. *J Gerontol* 1970; 25:325–336.

78. Adams GM, deVries HA: Physiological effects of an exercise training regimen upon women aged 52–79. *J Gerontol* 1973; 28:50–55.

79. Poliner LE, Dehmer GJ, Lewis SE, et al: Left ventricular performance in normal subjects: A comparison of the responses to exercise in the upright and supine positions. *Circulation* 1980; 62:528–534.

80. Bar-Shiomo B, Druck MN, Morch JE, et al: Left ventricular function in trained and untrained healthy subjects. *Circulation* 1982; 65:484–488.

81. Weiss JL, Weisfeldt ML, Mason SJ, et al: Evidence of Frank-Starling effect in man during severe semisupine exercise. *Circulation* 1979; 59:655–661.

82. Stein RA, Michielli D, Diamond J, et al: The cardiac response to exercise training: Echocardiographic analysis at rest and during exercise. *Am J Cardiol* 1980; 46:219–225.

83. Rodeheffer RJ, Gerstenblith G, Becker LC, et al: Exercise cardiac output is maintained with advancing age in healthy human subjects: Cardiac dilatation and increased stroke volume compensate for diminished heart rate. *Circulation* 1984; 69:203–213.

84. Gilbert CA, Nutter CO, Feiner JV, et al: Echocardiographic study of cardiac dimensions and function in the endurance-trained athlete. *Am J Cardiol* 1977; 40:528–533.

85. Bertel O, Buhler FR, Kiowski W, et al: Decreased beta-adrenoreceptor responsiveness as related to age, blood pressure, and plasma catecholamines in patients with essential hypertension. *Hypertension* 1980; 2:130–138.

86. Fleg JL, Kennedy HL: Cardiac arrhythmias in a healthy elderly population. Detection by 24-hour ambulatory electrocardiography. *Chest* 1982; 81:301–307.

87. Nelson RD, Ezri MD, Denes P: Arrhythmias and conduction disturbances in the elderly, in Messerli FH (ed); *Cardiovascular Disease in the Elderly*. Boston, Martinus Nijhoff Publishing, 1984, p 83.

CHAPTER 6

THE EFFECTS OF AGE ON HEPATIC DRUG METABOLISM

CHO-MING LOI AND ROBERT E. VESTAL

1. INTRODUCTION

Aging is a very complex process that is not completely understood. In the human life cycle, the growth-and-development phase ends near the age of 30 years and is followed by a gradual, often linear decline.[1] Many physiological and biochemical changes occur with "normal" aging. In addition, concomitant pathological conditions, lifelong habits, and environmental exposure also contribute to the biological changes in the elderly. As a result, there are wide differences in the rates of deterioration of organs and enzyme systems with age. Therefore, chronological aging may not necessarily be a true reflection of physiological aging.

1.1. HEPATIC ANATOMICAL CHANGES WITH AGE

A study of organ size in 400 autopsy specimens showed that the liver increases in absolute weight up to age 30.[2] After that, it becomes

CHO-MING LOI • Idaho State University College of Pharmacy, Pocatello, Idaho 83209; and Clinical Pharmacology and Gerontology Unit, Veterans Administration Medical Center, Boise, Idaho 83702.   ROBERT E. VESTAL • Clinical Pharmacology and Gerontology Unit, Veterans Administration Medical Center, Boise, Idaho 83702; and Division of Gerontology and Geriatric Medicine, Department of Medicine, University of Washington School of Medicine, Seattle, Washington 98195.

smaller with uniform progression. However, when the weight of the liver is compared to the total body weight, it bears a relatively constant proportion (2.5%) until about age 50. Thereafter, this proportion declines with age to 1.6% by the tenth decade of life.[3] Subsequent studies[4,5] have yielded similar observations. Thus, the size of the liver, measured by ultrasonic scanning techniques, is smaller in elderly subjects. This age-dependent difference is significant even when the liver volume is related to the concomitant decrease in body surface area.

1.2. HEPATIC PHYSIOLOGICAL CHANGES WITH AGE

Hepatic metabolism of some drugs is shown to be altered in the elderly.[4,6] Based on these observations, several age-related physiological changes of the liver have been suggested. These changes include reduction in hepatic blood flow, alteration in the synthesis of drug binding proteins, and impairment of hepatic drug-metabolizing enzyme activities. However, age-related differences in drug disposition are multifactorial and are infuenced by physiological, pathological, environmental, and genetic factors. Therefore, physiological changes alone do not allow for generalizations regarding the type, magnitude, or importance of age differences in pharmacokinetics and pharmacodynamics.

2. PHYSIOLOGICAL MODEL OF HEPATIC DRUG METABOLISM

Proper evaluation of the literature involving the relationship between age-related differences in hepatic drug metabolism and physiological changes in the liver requires familiarity with the physiological model of pharmacokinetic analysis. Many drugs are removed from the body by hepatic metabolism to less active or inactive metabolites, which are usually excreted by the kidney. The three major biological determinants of this elimination process include (1) the rate at which the drug is delivered to the liver as controlled by the hepatic blood flow; (2) intrinsic activity of the overall elimination process; and (3) the fraction of unbound drug in the blood.[7] A physiological approach to hepatic drug metabolism has been developed that takes into account all these factors.[8] This approach is superior to the classical compartmental pharmacokinetic analysis, in that the parameters of this model correspond to actual physiological parameters, such as hepatic blood flow and the metabolic activity of the liver.[7] This allows the prediction and interpretation of the effects of individual alterations in the biological determinants, disease states, pharmacokinetic drug interactions, and environmental factors on hepatic drug elimination.

2.1. Clearance Concept

Clearance is defined as the volume of blood from which the unmetabolized drug is cleared per unit of time by any pathway of drug elimination. It is a measure of the efficiency of any organ to irreversibly remove a drug from the perfusing blood. For the liver, under steady-state conditions, the hepatic clearance (Cl_H) of a drug is dependent on the total liver blood flow (Q) and extraction ratio (E). The latter is an index of the efficiency of the liver to remove the drug under fixed conditions of blood flow. It is a reflection of the overall activity of the drug metabolizing enzymes or other rate-limiting processes involved in elimination. Thus, hepatic clearance can be estimated from the following relationship:

$$Cl_H = Q \cdot E \qquad \text{(Equation 1)}$$

2.2. Intrinsic Clearance

From Equation (1) it appears that hepatic clearance is directly proportional to liver blood flow. However, this is not the case since extraction ratio is also dependent on hepatic blood flow. This has led to the introduction of the concept of total intrinsic clearance (Cl_i). It is a measure of the maximal intrinsic ability of the liver to irreversibly eliminate the drug in the absence of any flow limitations. The total intrinsic clearance, together with hepatic blood flow, determines the extraction ratio, which is defined as

$$E = \frac{Cl_i}{Q + Cl_i} \qquad \text{(Equation 2)}$$

It follows that the hepatic clearance is

$$Cl_H = Q \cdot E = Q \cdot \frac{Cl_i}{Q + Cl_i} \qquad \text{(Equation 3)}$$

Rearrangement of Equation (2) yields the following relationships:

$$Cl_i = \frac{Q \cdot E}{1 - E} \qquad \text{(Equation 4)}$$

The total intrinsic clearance, therefore, can be viewed as the ratio of V_m and K_m as in classical Michaelis–Menton enzyme kinetics.

2.3. DRUG BINDING

The relationships discussed up to this point are applicable only to drugs that are unbound in the vascular space. However, most drugs are bound to blood constituents such as plasma proteins and red blood cells. They circulate in the blood as both free and bound entities. It has generally been assumed that only the free fraction of the drug will be extracted by the liver. Although this is true for many drugs, there are others whose elimination is not restricted to the free fraction that is delivered to the liver. Propranolol is the prototype of the latter group of drugs.[9] Modification of Equation (3) must be made to take drug binding into consideration:

$$E = \frac{f_b \cdot Cl_{i\ (free)}}{Q + f_b \cdot Cl_{i\ (free)}} \qquad \text{(Equation 5)}$$

where f_b is the free fraction of the drug in the blood and Cl_i (free) is the intrinsic clearance of the free drug. By combining Equations (1) and (5), the hepatic clearance can be described by the following expression:

$$Cl_H = Q \cdot \frac{f_b \cdot Cl_{i\ (free)}}{Q + f_b \cdot Cl_{i\ (free)}} \qquad \text{(Equation 6)}$$

2.4. CLASSIFICATION OF DRUGS ACCORDING TO EXTRACTION RATIO

It is apparent from Equation (5) that hepatic extraction ratio is a function of the intrinsic ability of the liver to remove the drug and the liver blood flow, as well as drug binding. This provides a basis for classification of drugs according to these characteristics. For drugs with a high extraction ratio [i.e., $f_b \cdot Cl_{i(free)} \gg Q$], Equation (6) will be reduced to

$$Cl_H = Q \qquad \text{(Equation 7)}$$

This indicates that the rate of hepatic drug metabolism is limited by the amount of drug delivered to the liver. This, in turn, is determined by liver blood flow. Elimination of these drugs following intravenous adminstration is blood-flow-limited and is nonrestrictive. Both bound and free drug can be functionally removed by the liver. Their disposition will be affected by alterations in blood flow but not changes in metabolic activity or drug binding.

On the other hand, for drugs that have a low extraction ratio [i.e., $f_b \cdot Cl_{i(free)} \ll Q$], Equation (6) reduces to

$$Cl_H = f_b \cdot Cl_{i \ (free)} \qquad \text{(Equation 8)}$$

The hepatic clearance of these drugs is capacity-limited and is restrictive. Only the circulating free drug will be extracted by the liver. The rate of metabolism is dependent on the free fraction of the drug in the blood and the inherent metabolic capacity of the liver enzymes. Changes in drug binding or intrinsic metabolic activity of the liver due to disease states, enzyme induction, or enzyme inhibition will have a direct effect on drug elimination. Alteration in liver blood flow, on the other hand, will not affect the rate of metabolism.

Between these two extremes, there are drugs with an intermediate extraction ratio of 0.3–0.7. They exhibit characteristics of both classes of drugs, with their clearances depending partly on live blood flow and partly on the inherent metabolic activity of the liver.

2.5. SYSTEMIC TOTAL DRUG CLEARANCE AND APPARENT ORAL DRUG CLEARANCE

For practical reasons it is not always possible to measure directly hepatic clearance or its major biological determinants *in vivo*. Instead, a mean total drug clearance (Cl_s) is estimated from the systemic venous blood concentration following intravenous administration of the drug:

$$Cl_s = \frac{D_{i.v.}}{AUC_{i.v.}} \qquad \text{(Equation 9)}$$

where $D_{i.v.}$ is the intravenous dose and $AUC_{i.v.}$ is the total area under the concentration–time curve in the systemic blood. For drugs that are completely metabolized by the liver, the hepatic clearance is equivalent to the total systemic clearance. For drugs that have extrahepatic elimination, if the fraction of the intravenous dose (f) that is eliminated unchanged is known, their hepatic clearance can be estimated by

$$Cl_H = (1 - f) \cdot Cl_s \qquad \text{(Equation 10)}$$

A similar relationship as in Equation (9) has been derived following oral administration of a drug that is completely absorbed and has no extrahepatic metabolism or excretion. The apparent oral total drug clearance (Cl_o) is defined as

$$Cl_o = \frac{D_o}{AUC_o} \qquad \text{(Equation 11)}$$

where D_o is the oral dose and AUC_o is the total area under the concen-tration–time curve and indicates the amount of drug that reaches the systemic circulation.

2.6. First-Pass Metabolism

After oral administration a drug is absorbed from the gastrointestinal tract into the portal circulation, which passes through the liver. During a single transit through the liver, a fraction of the administered dose is removed by "presystemic" or "first-pass" hepatic metabolism. By definition, this is equivalent to the extraction ratio. It follows then the fraction (F) of the orally administered dose that escapes first-pass metabolism into the general circulation is

$$F = 1 - E \qquad \text{(Equation 12)}$$

Adjusting for possible differences in dose, the systemic availability of the drug can also be estimated by the ratio of the area under the curve after oral administration to that after intravenous administration:

$$F = \frac{AUC_o \cdot D_{i.v.}}{AUC_{i.v.} \cdot D_o} = 1 - E \qquad \text{(Equation 13)}$$

Rearrangement of terms yields the following:

$$Cl_o = Cl_i = f_b \cdot Cl_{i \text{ (free)}} \qquad \text{(Equation 14)}$$

Thus, the total intrinsic hepatic clearance (Cl_i) of a drug is equivalent to its apparent oral clearance (Cl_o) assuming that the drug is completely absorbed and is eliminated only by hepatic metabolism.

When Equations (8) and (14) are compared, it is evident that the apparent oral clearance of low-extraction-ratio drugs is the same as their hepatic clearance. The first-pass effect for these drugs is small, with most of the orally administered dose reaching the general circulation. Moreover, this relationship does not hold true for high-extraction-ratio drugs. They exhibit significant first-pass metabolism, and the amount of drug reaching systemic circulation may be substantially less than the dose administered. Alterations in the metabolic capacity of liver enzymes may not affect the disposition kinetics of these drugs after intravenous administration. However, they have profound effects on their pharmacokinetics after oral administration. This hypothesis has been substantiated by Alvan et al.,[10] who found that administration of pentobarbital, a potent hepatic microsomal enzyme inducer, significantly reduced the systemic availability (as reflected by AUC_o) of alprenolol by 78% when administered orally. The disposition of intravenously administered alprenolol, on the other hand, did not show a significant change.

2.7. ESTIMATION OF HEPATIC BLOOD FLOW AND INTRINSIC CLEARANCE

The preceding discussion indicates that for drugs that undergo hepatic metabolism, their disposition kinetics is determined by the physiological parameters of binding, liver blood flow, and intrinsic free clearance. In order to examine the effect of aging on hepatic changes and drug metabolism, it is important to develop methods that allow appropriate estimation of these determinants. Wilkinson and Shand[8] have shown that liver blood flow can be estimated based on the ratio of dose over AUC after intravenous and oral administration:

$$Q = \frac{D_{\text{i.v.}} \cdot D_o}{AUC_{\text{i.v.}} \cdot D_o - AUC_o \cdot D_{\text{i.v.}}} \qquad \text{(Equation 15)}$$

This relationship is valid provided that the drug is completely absorbed after oral administration, that all of the administered dose passes through the liver prior to reaching the systemic circulation, and that there is no extrahepatic elimination.

With these same assumptions, previous analysis [Equation (14)] has demonstrated that the total intrinsic hepatic clearance of a drug is the same as its apparent oral clearance. This parameter can be obtained experimentally by measuring the total area under the concentration–time curve after oral administration. Furthermore, the intrinsic clearance of the free drug can be determined from Equation (14) if the unbound fraction of the drug in the blood is known. The latter parameter is related to the blood/plasma concentration ratio and the fraction of free drug in plasma (f_p):

$$f_b = \frac{f_p \cdot C_p}{C_b} \qquad \text{(Equation 16)}$$

where C_p and C_b are total drug concentration in plasma and in blood, respectively.

3. Physiological Changes in the Liver with Age

The physiological model of hepatic drug metabolism provides a theoretical framework from which an appropriate estimation of the individual hepatic physiological parameters can be determined with relatively noninvasive methods. Based on this approach, investigations with model compounds such as indocyanine green (ICG), antipyrine, and propranolol have been performed to determine the biological determinants of hepatic drug metabolism in relation to age and environmental factors. The findings from these studies provide an insight into the changes in hepatic blood flow, inherent metabolic activity of the liver, and drug binding with age.

3.1. Liver Blood Flow

Early investigations[3,11] showed that regional blood flow to the liver declines with advancing age. The rate of reduction is estimated to range from 0.3% to 1.5%/year. Thus, hepatic blood flow is expected to decrease by 40–45% in a person aged 65 compared with a person aged 25. This decline in hepatic blood flow may be explained partially by a decrease in cardiac output that occurs with aging.[12] It is noteworthy, however, that a recent study, in which coronary artery disease was excluded, suggests that in a very healthy population cardiac output at rest does not significantly differ with age.[13]

More recent studies have supported these earlier investigations. Wood and co-workers[6] studied the disposition of ICG in 24 normal male subjects aged 22–72 years, 11 of whom were smokers. The results showed that the systemic clearance of ICG was 46% greater in the younger group than in the older group. This decline was not influenced by smoking habits. Since ICG has a high hepatic-extraction ratio ($E > 0.85$); its elimination is nonrestrictive and is mostly dependent on hepatic blood flow. In subjects with normal hepatic function, the systemic clearance of ICG following intravenous administration is generally considered to be a valid estimate of liver blood flow. From the observations of this study, it can be inferred that hepatic blood flow diminishes with advancing age.

Vestal and associates[14] conducted a study with propranolol which yielded similar results. The apparent hepatic blood flow was measured indirectly using the technique of simultaneous administration of oral and intravenous propranolol, as described previously [see Equation (15)]. There was a significant negative correlation between age and apparent liver blood flow for all subjects, both smokers and nonsmokers (Fig. 1). In the older subjects (aged 46–73), the apparent hepatic blood flow was

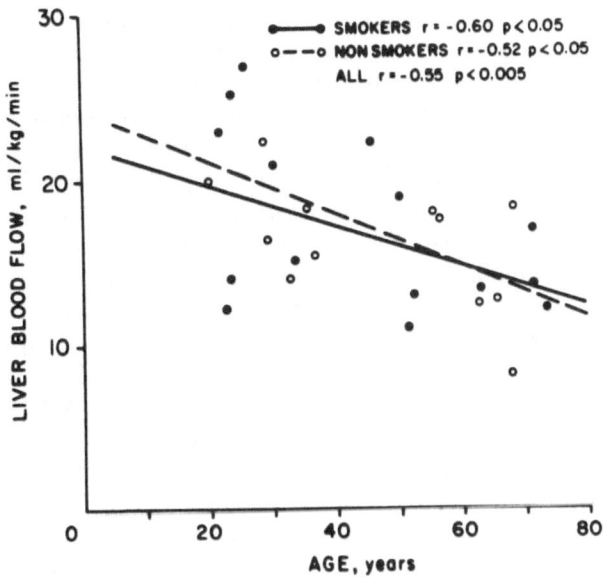

FIGURE 1   Effect of age on hepatic blood flow (Vestal et al.[14]).

24% lower compared with the younger group (aged 21–37). Thus, current evidence supports the concept that liver blood flow diminishes with age. Furthermore, this physiological change does not appear to be influenced by smoking habits.

3.2. DRUG-METABOLIZING ENZYME ACTIVITY

Hepatic drug metabolism generally involves two pathways: (1) microsomal oxidation, mediated by the mixed-function oxidase enzyme systems in the liver, and (2) conjugation, which is independent of microsomal enzyme function. Evidence suggests that the latter pathway is not affected by aging. Farah and co-workers[15] compared the disposition of isoniazid and acetanilid in both young and elderly subjects. Isoniazid is metabolized by acetylation, which is a microsomal enzyme-independent process, whereas acetanilid metabolism is via the microsomal oxidation pathway. There was no observed age difference in metabolism of isoniazid between the elderly and the younger subjects. In contrast, the half-life of acetanilid was significantly increased in the aged group. Further studies on disposition of ethanol,[16] paracetamol,[17] oxazepam,[18] and lorazepam,[19] all of which undergo metabolism via nonmicrosomal enzyme pathways, support this conclusion.

The association between age and alteration in the intrinsic microsomal enzyme activity is less clear. Studies using antipyrine as model compound to assess influence of age on microsomal drug metabolism have yielded conflicting data. Antipyrine is used frequently as a marker in studies involving quantitation of intrinsic microsomal drug-metabolizing enzyme activity. This compound is rapidly absorbed and is distributed in total body water. It is almost completely metabolized in the liver via the microsomal enzyme system. Antipyrine has a low extraction ratio and is minimally bound to plasma protein. Therefore, its hepatic clearance approximates its intrinsic free clearance, which is an index of the microsomal enzyme activity [see Equation (8)]. This property makes antipyrine a useful model drug in studies of the effects of age and environmental factors on microsomal drug-metabolizing enzyme activity.

O'Malley et al.[20] were the first to demonstrate a 45% prolongation of the half-life of antipyrine in elderly subjects (aged 70–100) when compared to young controls (aged 20–50). Further analysis revealed an age-related decrease in total plasma clearance of this drug.[21] In the elderly, there is evidence that liver volume decreases with age.[4,5] The metabolic clearance rate of antipyrine is decreased in the elderly even when the decline in liver volume is taken into consideration. This suggests that the age-related reduction in antipyrine clearance is not only due to a smaller liver, but is also a result of an impairment in microsomal drug-metabolizing enzyme activity.

Observations from the Baltimore Longitudinal Study of Aging,[22] however, suggested an alternate explanation for the age-associated reduction in antipyrine clearance. This study showed that, overall, the total plasma clearance of antipyrine declined with age (Fig. 2). Moreover, there was a wide, sixfold, interindividual variation in the total plasma clearance of this drug. Multivariate analysis revealed that age alone accounted for only 3% of the variance. In comparison, cigarette smoking explained 12% of the variation. The remaining 85% of the variance was unexplained but could be due to other genetic and environmental factors. Thus, age itself appears to have less influence than cigarette smoking on the metabolic clearance rate of antipyrine in adults. Data also indicated that cigarette smoking among young and middle-aged subjects was associated with a faster rate of metabolism of antipyrine (Table I). This same association has also been observed in other studies.[6,23] Tobacco smoke and some of its individual components are inducers of drug-metabolizing enzymes in humans.[24] This could explain the enhanced metabolism of antipyrine demonstrated in these studies. However, in the elderly group, there was no observed difference in antipyrine met-

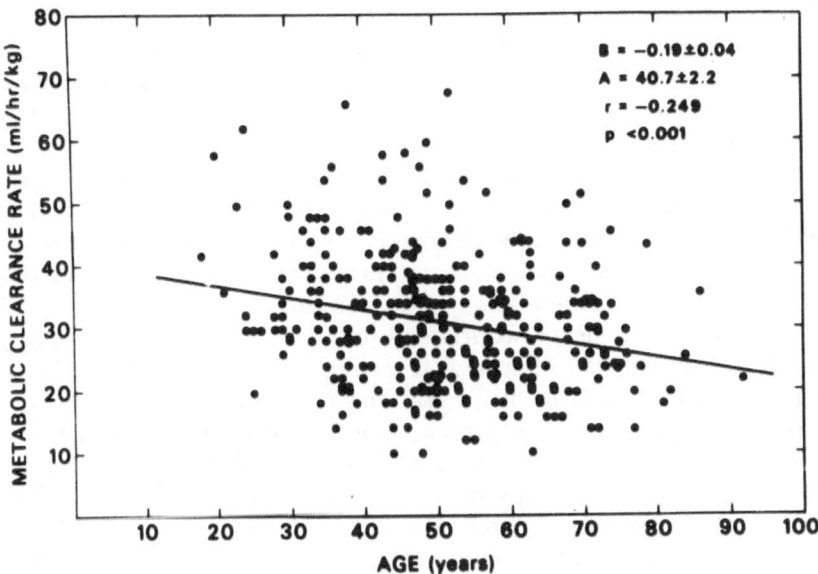

FIGURE 2.   Decline in metabolic clearance rate of antipyrine with age (Vestal et al.[22]).

TABLE I   EFFECT OF AGE AND CIGARETTE SMOKING ON HALF-LIFE($t_{1/2}$) AND
TOTAL SYSTEMIC CLEARANCE (Cl$_s$) OF ANTIPYRINE[a]

Age group	Smoking group[b]	Number of subjects	($t_{1/2}$) (hr)[c]	Cl$_s$ (mg/kg per hour)[c]
Young	Nonsmokers/light	41	14.1 ± 0.61	30.5 ± 1.29
(Age 18–39)	Moderate/heavy	32	10.9 ± 0.73[d]	39.7 ± 2.11[d]
Middle aged	Nonsmokers/light	108	14.6 ± 0.53	28.1 ± 0.86
(Age 40–59)	Moderate/heavy	42	11.5 ± 0.89[d]	37.9 ± 1.89[e]
Old	Nonsmokers/light	72	14.9 ± 0.71	28.0 ± 1.05
(Age 60–92)	Moderate/heavy	12	14.1 ± 1.63	29.6 ± 3.16
All ages	Nonsmokers/light	221	14.7 ± 0.37	28.5 ± 0.60
(Age 18–92)	Moderate/heavy	86	16.6 ± 0.57[e]	37.4 ± 1.32[e]

[a] From data of Vestal et al.[22]
[b] Smoking groups are defined as follows: Nonsmokers did not smoke cigarettes. Light: Smoked once in a while or less than 10 cigarettes/day. Moderate: smoked 10–20 cigarettes/ day. Heavy: Smoked 21 or more cigarettes/day.
[c] Mean ± SEM.
[d] Significantly different from nonsmokers/light group ($p < 0.005$).
[e] Significantly different from nonsmokers/light group ($p < 0.001$).

abolic clearance rate between smokers and nonsmokers. Based on these data, the authors concluded that impairment of antipyrine metabolism in the elderly is not due to age per se, but is related to a diminished response to the drug-metabolizing-enzyme-inducing effect from cigarette smoking.

Other studies have yielded results that are consistent with this postulation. Wood and associates[6] reported a negative correlation between age and antipyrine metabolic clearance rate. A greater antipyrine clearance was observed in smokers than nonsmokers in the younger group. This relationship was not demonstrated in the older subjects.

Another study using propranolol found similar results.[14] Propranolol is a drug with an intermediate extraction ratio. Its elimination is determined by both liver blood flow and enzyme activity. The drug is extensively bound to plasma protein. However, protein binding has little influence on its extraction over the ranges found in normal subjects. Thus, the clearance of propranolol is both flow and capacity limited, but protein binding is nonrestrictive. Following chronic oral dosing (240 mg daily), the mean blood concentration of propranolol was as much as twofold higher in the older subjects. In addition, it was found that the concentration of propranolol in the blood was almost 200% higher in the nonsmokers than in the smokers. The intrinsic clearance of propranolol decreased with age only in subjects who smoked. In nonsmokers, this association was not shown. These data were similar to those reported by Gardner et al.[25] Since intrinisic clearance is a measure of the drug-metabolizing-enzyme activity, it is concluded that age per se has only minor influence on this biological determinant. A diminished response to the enzyme-inducing effects of cigarette smoke in the elderly appears to be a more appropriate explanation for the study results. This same conclusion was also derived from other studies using different enzyme-inducing agents such as rifampin[26] and dichloralphenazone.[27]

However, conflicting results have been presented to challenge these conclusions. Mucklow and Fraser[28] found no correlation between age and antipyrine metabolism. Furthermore, the difference in the rate of antipyrine clearance between smokers and nonsmokers was more significant in the older age group than in the younger group. Cigarette smoking was also associated with an enhanced plasma clearance of theophylline in the elderly as well as in the young.[29] Treatment for 6 days with glutethimide,[30] a potent enzyme inducer, was shown to result in a significant rise in antipyrine plasma clearance in nine elderly nonsmokers. At least with regard to selected substrates, observations from these studies suggest that enzyme induction may occur in the elderly to a similar extent as in the younger subjects.

The reasons for the disparities in findings among these studies are not clear. However, several possible explanations may account for the different results: (1) differences in subject selection, (2) wide interindividual variation related to the inherent differences in genetic drug metabolism phenotype and prior exposure to other environmental influences, (3) differences in enzyme-inducing agents used, and (4) the duration of exposure to the enzyme inducer. Therefore, when conducting studies to determine age-related effects on hepatic microsomal enzyme activity, careful effort must be made to control for these variables.

3.3. DRUG BINDING

Many drugs are reversibly bound to blood constituents such as albumin and α-1-acid glycoprotein. The major factors determining the extent of drug binding to plasma proteins are the following: (1) drug concentration, (2) concentration of the binding proteins, (3) the number of available binding sites, and (4) the binding affinity. Alterations in these factors may lead to a significant change in the plasma free fraction of the drug. For drugs that have a low extraction ratio, the rate of plasma clearance is dependent on the unbound fraction [see Equation (8)]. Therefore, alteration in protein binding will have an impact on the overall elimination of these drugs. This parameter must be taken into account when studying age-dependent changes in hepatic drug metabolism.

Age-related changes in serum albumin concentration have been investigated extensively. Most studies indicate that the serum albumin concentration decreases by 10–20% in old age[31–36] (Fig. 3). This decline is attributed to a reduction in its synthesis rate.[37] Albumin is the major plasma protein to which acidic drugs such as phenytoin,[35,36] salicylic acid,[35] and tolbutamide[34] are bound. *In vivo* and *in vitro* studies[34–36] showed an inverse correlation between the unbound fractions of these drugs and the plasma albumin concentration. Since the latter declines with age, it may explain, in part, the decrease in protein binding of these drugs in elderly subjects.

The association between age and plasma α-1-acid glycoprotein is less well established. α-1-Acid glycoprotein is a plasma glycoprotein with two types of microheterogeneity: the polymorphic forms and the variants.[38] Although some studies[35,39] suggest that this protein increases with age, others fail to confirm this relationship.[33,40] Furthermore, sex difference appears to exert an influence on the age-associated changes in α-1-acid glycoprotein concentration. However, two conflicting observations have been made concerning this association. Verbeeck and

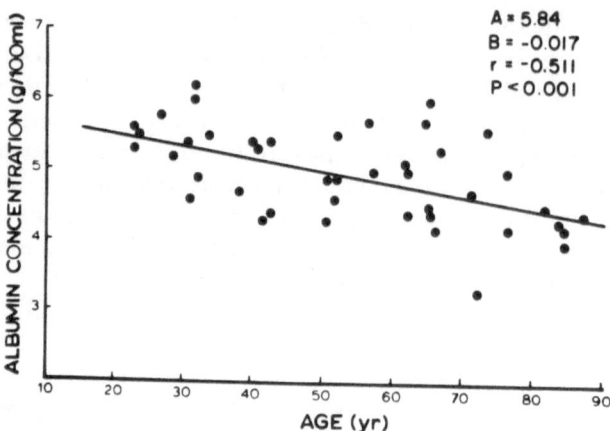

FIGURE 3.   Correlation between plasma albumin concentration and age (Adir et al. [34]).

associates[35] reported an age-dependent elevation of plasma α-1-acid gly-coprotein in males but not in females. Blaine *et al.*,[41] on the other hand, observed an age-related increase in α-1-acid glycoprotein in females but not in males. Besides age and sex, other confounding factors that affect plasma α-1-acid glycoprotein concentration include acute and chronic inflammation,[42] malignancy,[43] myocardial infarction,[44] trauma,[45] and malnutrition.[46]

α-1-Acid glycoprotein is an important binding protein for a number of basic drugs, e.g., propranolol,[33,35] tricyclic antidepressants,[38,39] lido-caine,[47] and verapamil.[48] The extent of propranolol binding varies widely and is directly related to the α-1-acid glycoprotein concentration. Age per se does not affect the binding ratio of propranolol.[35] The same observations have been made with imipramine.[39] Other tricyclic drugs, amitriptyline and nortriptyline, have been shown to bind specifically to the S form of α-1-acid glycoprotein variants.[38] Since the patterns of the variants and the polymorphic forms are genetically determined, this may account for the intersubject variation in protein binding for these drugs. Thus, age per se plays only a minor role, if any, in determining the extent of drug binding to α-1-acid glycoprotein.

4. METHODOLOGICAL CONSIDERATIONS

It is evident, from the preceding discussion, that different studies on age-dependent hepatic physiological changes have not yielded uni-form conclusions. Thus, a clear understanding of the strategies and pitfalls unique to research in clinical gerontology is essential for proper evaluation of the literature and for proper design of future studies. Careful attention must be given to the specific methodologies employed in the studies, such as subject selection, study design, protocol selection, and data analysis.

4.1. SUBJECT SELECTION

The most important factor governing the conduct and evaluation of studies is subject selection.[49] If the study is intended to determine how age may affect pharmacokinetics and pharmacodynamics, every effort must be made to exclude diseased individuals and those who are taking medications. Usually, this can be accomplished through careful history and physical examination along with appropriate laboratory tests to rule out significant occult disease. Furthermore, other factors such as diet, lifelong habits, and unusual occupational exposure should also be noted since they are known to influence drug metabolism in the elderly. On the other hand, studies can also be conducted in the population for which the drugs are intended, namely, elderly patients with one or more diseases of interest. In this instance, the results will reflect the effect of age and disease on drug metabolism. Interpretation of the data in this case will become more difficult.

The importance of how subject selection can influence study outcome is illustrated in a study by Swift *et al.*[4] on antipyrine disposition in relation to liver size. This investigation was carried out in three different groups of subjects: young healthy volunteers, healthy elderly subjects, and elderly hospitalized patients undergoing rehabilitation for locomotor disorders. All subjects were nonsmokers and had normal routine liver function tests and blood count. They were free of medications that affect hepatic enzymes at the time of study or for at least 1 month previously. However, 9 of the 10 hospitalized elderly patients were taking drugs of some type. The study results indicated a significant difference in the plasma clearance, volume of distribution, and plasma half-life of antipyrine between the hospitalized elderly and the normal elderly. These differences were unexplained, but could be attributed to subject selection.

Although careful selection of study subjects may be viewed as "cleaning up" the physiological data, intensive screening of the population may result in a select group of elderly superperformers whose data do not reflect influence of age-related changes.[50] Rowe and Troen[50] illustrate this phenomenon with the following example: In selecting nondiabetic subjects to participate in a study, one might adopt criteria by which individuals with a 2-hr postprandial blood glucose greater than 140 mg/100 ml would be excluded. Since carbohydrate tolerance declines with age, the application of this criterion would result in only a small fraction of individuals in the eighth or ninth decade of life qualifying for the study. Thus, it is important to describe carefully the study population(s) and the selection criteria applied to all age groups.

4.2. STUDY DESIGN

The two general study designs available for use are longitudinal and cross-sectional.[51] In longitudinal studies, one group of subjects is followed over a long period of time. The variables under study are obtained prospectively at specified intervals, and the age-related differences are determined from the slopes for these variables. The advantage of this study design is that the rate of aging is assessed in each individual subject. The major disadvantage is that such studies are expensive and difficult to perform since they require follow-up of a stable population over a long period of time. Another major drawback in longitudinal studies is their particular sensitivity to methodological drift owing to subtle changes in laboratory techniques or equipment over several years. No longitudinal studies of drug metabolism have been performed.

In cross-sectional studies, groups of various ages are observed and age-related differences are measured.[50] These studies are generally easier to perform and are much less expensive and time consuming than longitudinal studies. However, this study design can only determine age differences or effects of age rather than age changes or effects of aging.[53] This distinction is important when interpreting the results of cross-sectional studies, since they may not reflect true age-related changes. For instance, subjects over the age of 75 years represent a sample from a cohort that has experienced at least 75% mortality. If the variable under study is related to survival, a cross-sectional study will seem to show age differences that are due to the progressive loss of individuals with high values rather than aging.[51] This effect of selective mortality can be avoided with the use of longitudinal study design. Thus, even though cross-sectional studies may suggest that advanced age is associated with decreased drug metabolism, it is possible that this apparent difference is due to selective mortality rather than aging per se.

4.3. PROTOCOL SELECTION

Since the effects of age differences in hepatic drug metabolism are difficult to predict, it is possible that differences in the study protocol may influence the study results. Conclusions that appear to be contradictory may, in fact, prove to be quite compatible when differences in the study protocol are taken into consideration.[49] This is exemplified by studies of the effects of age on the pharmacokinetics of propranolol.

After a single 40-mg oral dose of propranolol, it was observed that the elderly subjects had a fourfold increase in plasma propranolol concentrations compared to the young subjects.[53] The higher plasma propranolol levels in the elderly were attributed to a decline in hepatic first-pass extraction and metabolism. A subsequent study by Castleden and George[54] substantiated this hypothesis. All study subjects but one in the elderly group were nonsmokers. The systemic clearance and the first-pass metabolism of propranolol were significantly lower in the elderly than in the young. The mean plasma propranolol concentration after a multiple-dose regimen of 160 mg daily was fourfold higher in the elderly subjects. These data differed from those reported by Vestal et al.,[14] who found no age-related difference in propranolol pharmacokinetics among the nonsmokers. However, the dose used in the latter study was higher than that used in the study of Castleden and George (240 mg daily versus 160 mg daily). It was postulated that the threshold for maximum hepatic extraction might be age-related. A decrease in threshold in the elderly could result in saturation of the extraction process at lower doses than in the young. Such a threshold difference might have been obscured with the use of the higher doses of propranolol.[52] This may account for the observed differences between these two studies.

Evidence to support this postulation was provided by the report of Schneck et al.[55] The influence of dose on the intrinsic clearance of propranolol was investigated in four healthy elderly (mean age 71 years) and six healthy young subjects (mean age 27 years) using single 20-, 40-, and 160-mg doses. All subjects were nonsmokers. The mean intrinsic clearance was lower in the elderly group when compared to the younger population at all doses. In the young group, the intrinsic clearance decreased significantly with increasing dose. In contrast, there were no significant changes in intrinsic clearance among doses in the elderly group. Thus, the effect of age on intrinsic clearance was demonstrated at the lowest dose (20 mg), whereas at the highest dose (160 mg) there was no significant difference between these two groups. These data are consistent with the conclusion that in nonsmokers the effect of age on the hepatic metabolism of propranolol is a dose-related phenomenon.

4.4. DATA ANALYSIS

Valid statistical analysis of the data is critical to appropriate interpretation of the study results. In studies where the age distribution of subjects is bimodal, that is, the subjects are mainly either young or old with few or no middle-aged individuals, the use of linear regression analysis and correlation coefficients to describe the data is questionable.[52] The application of these methods requires a bivariate normal distribution of the dependent and independent variable. If this characteristic is absent from the study sample, other statistical methods such as group comparisons should be used to analyze the data. Furthermore, it should not be assumed that data from the middle-aged group must necessarily be intermediate between the young and the old. They may well be similar to either the young or the old group.

A caution in regard to the interpretation of a significant correlation is warranted. The mere fact of a statistically significant correlation does not establish a cause-and-effect relationship. It only suggests that there is an association between the variables tested in a prospective manner. It is possible, however, that age-related variables, rather than age per se, may account for an apparent effect of age.

5. CONCLUSIONS

Comprehensive review of investigations of the effect of age on drug disposition reveals that age does seem to influence the metabolism and elimination of many drugs.[56] Available studies have provided evidence to suggest several age-associated anatomical and physiological differences in the liver. However, it remains uncertain as to whether these observed differences are due to the phenomenon of aging or whether they result from selective mortality. Differences in the criteria for subject selection, study protocol, and prior environmental exposure may also influence the study outcome. These issues can only be resolved with certainty by careful longitudinal studies with model compounds.

REFERENCES

1. Andres R: Physiological factors of aging significant to the clinician (summary of remarks). *J Am Geriatr Soc* 1969; 17:274–277.
2. Calloway NO, Foley CF, Lagerbloom P: Uncertainties in geriatric data. II. Organ size. *J Am Geriatr Soc* 1965; 13:20–28.
3. Geokas MC, Haverback BJ: The aging gastrointestinal tract. *Am J Surg* 1969; 117:881–892.
4. Swift CG, Homeida M, Halliwell M, et al: Antipyrine disposition and liver size in the elderly. *Eur J Clin Pharmacol* 1978; 14:149–152.
5. Bach B, Hansen JM, Kampmann JP, et al: Disposition of antipyrine and phenytoin correlated with age and liver volume in man. *Clin Pharmacokinet* 1981; 6:389–396.
6. Wood AJJ, Vestal RE, Wilkinson GR, et al: Effect of aging and cigarette smoking on antipyrine and indocyanine green elimination. *Clin Pharmacol Ther* 1979; 26:16–20.

7. Gibaldi M, Perrier D: Pharmacokinetics, ed 2. New York and Basel, Marcel Dekker, Inc., 1982.
8. Wilkinson GR, Shand DG: A physiological approach to hepatic drug clearance. *Clin Pharmacol Ther* 1975; 18:377–390.
9. Kornhauser DM, Wood AJJ, Vestal RE, et al: Biological determinants of propranolol disposition in man. *Clin Pharmacol Ther* 1978; 23:165–174.
10. Alvan G, Piafsky K, Lind M, et al: Effect of pentobarbital on the disposition of alprenolol. *Clin Pharmacol Ther* 1977; 22:316–321.
11. Sherlock S, Bearn AG, Billing BH, et al: Splanchnic blood flow in man by the bromsulpthalein method: the relation of peripheral plasma bromsulpthalein level to the calculated flow. *J Lab Clin Med* 1950; 35:923–937.
12. Bender AD: The effect of increasing age on the distribution of peripheral blood flow in man. *J Am Geriatr Soc* 1965; 13:192–198.
13. Rodeheffer RJ, Gerstenblith G, Becker LC, et al: Exercise cardiac output is maintained with advancing age in healthy human subjects: cardiac dilatation and increased stroke volume compensate for a diminished heart rate. *Circulation* 1984; 69:203–213.
14. Vestal RE, Wood AJJ, Branch RA, et al: Effect of age and cigarette smoking on propranolol disposition. *Clin Pharmacol Ther* 1979; 26:8–15.
15. Farah F, Taylor W, Rawlins MD, et al: Hepatic drug acetylation and oxidation: Effects of aging in man. *Br Med J* 1977; 2:155–156.
16. Vestal RE, McGuire EA, Tobin JD, et al: Aging and ethanol metabolism. *Clin Pharmacol Ther* 1977; 21:343–354.
17. Triggs, EJ, Nation RL, Long A, et al: Pharmacokinetics in the elderly. *Eur J Clin Pharmacol* 1975; 8:55–62.
18. Shull HJ, Wilkinson GR, Johnson R, et al: Normal disposition of oxazepam in acute viral hepatitis and cirrhosis. *Ann Intern Med* 1976; 84:420–425.
19. Kraus JW, Desmond PV, Marshall JP, et al: Effects of aging and liver disease on disposition of lorazepam. *Clin Pharmacol Ther* 1978; 24:411–419.
20. O'Malley K, Crooks J, Duke E, et al: Effect of age and sex on human drug metabolism. *Br Med J* 1971; 3:607–609.
21. Crooks J, O'Malley, K, Stevenson IH: Pharmacokinetics in the elderly. *Clin Pharmacokinet* 1976; 1:280–296.
22. Vestal RE, Norris AH, Tobin JD, et al: Antipyrine metabolism in man: Influence of age, alcohol, caffeine, and smoking. *Clin Pharmacol Ther* 1975; 18:425–432.
23. Hart P, Farrell GC, Cooksley WGE, et al: Enhanced drug metabolism in cigarette smokers. *Br Med J* 1976; 2:147–149.
24. Jusko WJ: Role of tobacco smoking in pharmacokinetics. *J Pharmacokinet Biopharm* 1978; 6:7–39.
25. Gardner SK, Cady WJ, Ong YS: Effect of smoking on the elimination of propranolol hydrochloride. *Int J Clin Pharmacol Ther Toxicol* 1980; 18:421–424.
26. Twum-Barima Y, Finnigan T, Habash AI, et al: Impaired enzyme induction by rifampicin in the elderly. *Br J Clin Pharmacol* 1984; 17:595–596.
27. Salem SAM, Rajjayabun P, Shepherd AMM, et al: Reduced induction of drug metabolism in the elderly. *Age Aging* 1978; 7:68–73.
28. Mucklow JC, Fraser HS: The effects of age and smoking upon antipyrine metabolism. *Br J Clin Pharmacol* 1980; 9:613–614.
29. Cusack B, Kelly JG, Lavan J, et al: Theophylline kinetics in relation to age: The importance of smoking *Br J Clin Pharmacol* 1980; 10:109–114.
30. Pearson MW, Roberts CJC: Drug induction of hepatic enzymes in the elderly. *Age Aging* 1984; 13:313–316.
31. Cammarata RJ, Rodnan GP, Fennell RH: Serum anti-gamma globulin and antinuclear factors in the aged. *JAMA* 1967; 199:115–118.

32. Greenblatt DJ: Reduced serum albumin concentration in the elderly: A report from the Boston Collaborative Drug Serveillance Program. *J Am Geriatr Soc* 1979; 27:20–22.

33. Bendayan R, Pieper JA, Stewart RB, et al: Influence of age on serum protein binding of propranolol. *Eur J Clin Pharmacol* 1984; 26:251–254.

34. Adir J, Miller AK, Vestal RE: Effects of total plasma concentration and age on tolbutamide plasma protein binding. *Clin Pharmacol Ther* 1982; 31:488–493.

35. Verbeeck RK, Cardinal JA, Wallace SM: Effect of age and sex on the plasma binding of acidic and basic drugs. *Eur J Clin Pharmacol* 1984; 27:91–97.

36. Patterson M, Heazelwood R, Smithurst B, et al: Plasma protein binding of phenytoin in the aged: *In vivo* studies. *Br J Clin Pharmacol* 1982; 13:423–425.

37. Gersovitz M, Munro HN, Udall J, et al: Albumin synthesis in young and elderly subjects using a new stable isotope methodology: response to level of protein intake. *Metabolism* 1980; 29:1075–1086.

38. Tinguely D, Baumann P, Conti M, et al: Interindividual differences in the binding of antidepressives to plasma proteins: The role of the variants of alpha-1-acid glycoprotein. *Eur J Clin Pharmacol* 1985; 27:611–666.

39. Abernethy DR, Kerzner L: Age effects on alpha-1-acid glycoprotein concentration and imipramine plasma protein binding. *J Am Geriatr Soc* 1984; 32:705–708.

40. Sager G, Hansteen V, Aakesson I, et al: Effect of heparin on serum binding of propranolol in the acute phase of myocardial infraction. *Br J Clin Pharmacol* 1981; 12:613–620.

41. Blaine PG, Mucklow JC, Rawlins MD, et al: Determinants of plasma alpha-1-acid glycoprotein concentrations in health. *Br J Clin Pharmacol* 1981; 12:275p.

42. Piafsky KM, Borga O, Odar-Cederlof I, et al: Increased plasma protein binding of propranolol and chlorpromazine mediated by disease-induced elevations of alpha-1-acid glycoprotein. *N Engl J Med* 1978; 299:1435–1439.

43. Abramson FP, Jenkins J, Ostchega Y: Effects of cancer and its treatments on plasma concentration of alpha-1-acid glycoprotein and propranolol binding. *Clin Pharmacol Ther* 1982; 32:659–663.

44. Routledge PA, Stargel WW, Wagner GS, et al: Increased alpha-1-acid glycoprotein and lidocaine disposition in myocardial infarction. *Ann Intern Med* 1980; 93:701–704.

45. Edwards DJ, Lalka D, Cerra F, et al: Alpha-1-acid glycoprotein concentration and protein binding in trauma. *Clin Pharmacol Ther* 1982; 31:62–67.

46. Jagadessan V, Krishnaswamy K: Drug binding in the undernourished: a study of the binding of propranolol to alpha-1-acid glycoprotein. *Eur J Clin Pharmacol* 1985; 27:657–659.

47. Routledge PA, Shand DG, Barchowsky A, et al: Relationship between alpha-1-acid glycoprotein and lidocaine disposition in myocardial infarction. *Clin Pharmacol Ther* 1981; 30:154–157.

48. McGowan FX, Reiter MJ, Pritchett ELC, et al: Verapamil plasma binding: Relationship to alpha-1-acid glycoprotein and drug efficacy. *Clin Pharmacol Ther* 1983; 33:485–490.

49. Vestal RE: Pharmacology and aging. *J Am Geriatr Soc* 1982; 30:191–200.

50. Rowe JW, Troen BR: Sympathetic nervous system and aging in man. *Endocr Rev* 1980; 1:167–179.

51. Rowe JW: Clinical research on aging: Strategies and directions. *N Engl J Med* 1977; 297:1332–1336.

52. Vestal RE: Methodological problems associated with studies of drug metabolism in the elderly, in Turner P (ed): *Clinial Pharmacology and Therapeutics: Proceedings of the First World Conference.* London, MacMillan Publisher Ltd, 1980, pp 108–116.

53. Castleden CM, Kayne CM, Parsons RL: The effect of age on plasma levels of propranolol and practolol in man. *Br J Clin Pharmacol* 1975; 2:303–306.

54. Castleden CM, George CF: The effect of aging on the hepatic clearance of propranolol. *Br J Clin Pharmacol* 1979; 7:49–54.
55. Schneck DW, Luderer JR, Pritchard JF, et al: A comparison of the intrinsic clearance of propranolol in young and elderly subjects. *Clin Pharmacol Ther* 1980; 27:284–285.
56. Vestal RE, Dawson G: Pharmacology and aging, in Finch CE (ed): *Handbook of the Biology of Aging*, ed. 2. New York, Van Nostrand, 1985, pp 744–819.

CHAPTER 7

Hematological Effects of Aging
Considerations for Clinical Trials

Elizabeth J. Read and Harvey G. Klein

1. Introduction

In 1951, a standard textbook of hematology reported the virtual absence of satisfactory information about the blood in old age.[1] Although many studies of the effects of aging on peripheral blood cells and bone marrow have been carried out in the intervening years, a great deal of uncertainty and controversy about the nature of these effects remains. Well-designed gerontological studies are difficult to perform. Conclusions have been frequently been drawn on the basis of data generated by studies in which methodology has been insufficient to distinguish between specific hematological effects of aging and effects of disease states that are prevalent in old age. This chapter will critically review the reported effects of aging on the structure and function of the peripheral blood and bone marrow, discuss the effects of drugs on the blood and bone marrow in old age, and present considerations for investigators who plan and carry out clinical trials in the elderly.

ELIZABETH J. READ AND HARVEY G. KLEIN • Department of Transfusion Medicine, Clinical Center, National Institutes of Health, Bethesda, Maryland 20892.

2. Effects of Aging on Blood Cells and Bone Marrow

2.1. Red Blood Cells

The major functions of the red blood cell are to synthesize, transport, and protect hemoglobin, the molecule that plays a critical role in oxygen delivery to the tissues. Normal erythropoiesis is sensitive to the changing physiological demands of the internal environment, such as those which occur in the neonatal period, during growth, and during pregnancy, as well as to less frequently occurring external demands, such as sustained changes in availability of atmospheric oxygen, for example, with changes in altitude. In young subjects, red cell mass appears to be a function of lean body mass, the major determinant of the oxygen requirement.[2,3] Sex differences in red cell mass are reflected in the higher hemoglobin levels in normal adult males compared to females (16.0 ± 2.0 g/dl versus 14.0 ± 2.0 g/dl).[4] These differences are generally attributed to the effect of testosterone, which stimulates red cell production.[5] Racial differences in hemoglobin levels also occur: one large study that controlled for socioeconomic factors and iron status demonstrated that blacks in all age groups in the United States have hemoglobin levels approximately 0.7 g/dl lower than whites.[6]

Most pathological processes have some effects on red cells. Cigarette smoking, for example, has consistently been demonstrated to increase the size of the erythrocyte, as well as the hemoglobin and hematocrit levels.[7,8] Chronic diseases, most notably chronic renal disease, rheumatoid arthritis, chronic infections, and cancer, are frequently accompanied by anemia. Iron deficiency from chronic blood loss, poor dietary intake, or malabsorption is a common cause of anemia. Deficiencies of other nutritional factors, such as vitamin B_{12} and folic acid and, less frequently, copper and pyridoxine, result in anemia. Many drugs may produce anemia by suppressing erythropoiesis, accelerating erythrocyte destruction, or increasing the propensity for blood loss. It is therefore clear that selection of subjects for studies of erythrocyte function in the elderly must take into consideration a variety of factors, some of which may be quite subtle.

Age-related changes in red blood cells are summarized in Table I and are discussed in the next section.

2.1.1. Hemoglobin Levels and Anemia.   A basic but still controversial question concerns the existence of an "anemia of aging." A large number of studies have attempted to define normal values for hemoglobin concentration and other red cell parameters in the elderly.

TABLE I   EFFECTS OF AGING ON RED BLOOD CELLS[a]

Measurement	Effect	Selected ref(s)
Hemoglobin concentration (g/dl)	Unchanged or minimally changed	19, 20, 22
	? Decreased in a subset	30
Mean corpuscular volume	Increased	22, 25, 26
Iron absorption	Unchanged	24
Iron uptake by marrow red cells	Decreased	24
Serum ferritin	Increased	20, 22, 23
Red cell 2, 3-DPG	Slightly decreased	35
Osmotic fragility	Increased	38–40

[a] Effects listed in Tables I–IV are generally accepted as age-related effects.

Evaluation of their results depends on careful scrutiny of selection factors for subjects and controls.

Studies of natural or unselected elderly populations, where subjects with diseases known to cause anemia or other red cell abnormalities have not been excluded, have consistently demonstrated that mean hemoglobin and hematocrit values in the aged are significantly lower than population reference values.[9–14] The prevalence of anemia, defined as a hemoglobin level of <12 g/dl in females and <14 g/dl in males, is variable among these studies. For example, in a Utah population survey with mean subject age of 69, 3% of females and 14% of males were anemic.[11] These figures are at variance with those of a Scottish survey which found 20% of females over age 65 to be anemic[9] and a New York study where 42.5% of the male patients at a geriatric outpatient facility were anemic.[13] One likely explanation for these discrepancies is that differences in mean hemoglobin levels and prevalence of anemia reflect differences in prevalence of disease states associated with anemia. In support of this interpretation is the observation that in the Scottish survey, 45% of the cases of anemia were due to iron deficiency, which was frequently associated with regular salicylate ingestion and gastrointestinal lesions.

Studies of elderly populations selected for absence of diseases known to be associated with anemia have frequently yielded findings similar to those of unselected populations. However, these studies have frequently examined elderly patients in chronic-care facilities,[15–17] which are likely to select populations with less than optimal health status, in spite of attempts to prospectively exclude subjects for diseases known to cause anemia. One such study, which determined hemoglobin values of 92 elderly male patients on chronic-care wards of a Veterans Hospital in Canada, found mean hemoglobin levels of 13.4, 12.9, and 12.2 g/dl in the 65–74, 75–84, and 85–94 age groups, respectively.[16] Although patients were carefully reviewed to exclude conditions known to be associated with anemia, the majority of patients had diagnoses of chronic brain syndrome, cerebral arteriosclerosis, or senility, disorders that are frequently associated with inability to maintain optimal nutritional status. In addition, one-quarter were taking aspirin on a regular or intermittent basis, which raises the possibility of occult gastrointestinal blood loss. In contrast, the progressive decline in hemoglobin level noted with increasing age from 65 to 94 in that population was not observed in a population of 100 selected inpatients in St. Louis.[15] The reason for this discrepancy is not clear.

Decreased mean hemoglobin levels and a high prevalence of anemia unassociated with commonly recognized causes have been reported in a group of healthy ambulatory elderly subjects in a community in Arkansas.[18] These findings suggest that an "anemia of senescence" occurs as a result of an age-related decline in hematopoietic reserve. Similar studies done in communities of higher socioeconomic status[19,20] do not confirm the Arkansas finding. In a study carried out in a Florida religious community, which included a prepaid health plan providing unlimited primary ambulatory care, only 4 of 73 healthy volunteer subjects, aged 63–94, were anemic, despite the fact that other laboratory values commonly associated with anemia were more frequently abnormal: 12 subjects had low serum iron, 10 had elevated total iron-binding capacity, six had decreased iron saturation, 28 had elevated blood urea nitrogen, and 17 had elevated creatinine.[19] Mean hemoglobin values were 15.3 g/dl for men and 14.0 g/dl for women, which are well within the published normal ranges previously cited. A study that compared a selected free-living elderly population aged 60–93 residing in Albuquerque with a young natural (unselected) population aged 20–39 from the New Mexico State Fair found that males had a small but statistically significant decrease in hemoglobin concentration with age (16.3 to 16.0 g/dl), whereas females had a small but statistically significant increase in hemoglobin with age (14.4 to 14.7 g/dl).[20] The prevalence of anemia in these subjects was low: 2.3% in elderly males, 0% in young males, 0% in elderly females, and 1.2% in young females.

Comparison of the latter two studies to the other studies, therefore, suggests that the anemia in elderly populations may be related to a variety of factors, such as socioeconomic status, general health, dietary factors, and racial composition of the study group. A study in Great Britain showed that elderly males living alone had significantly lower hemoglobin levels than males living with their spouses, regardless of socioeconomic status.[21] This suggests that even more subtle differences may influence study results.

Longitudinal data showing no significant changes in hemoglobin levels in 221 selected elderly subjects aged 60–92 at the beginning of the 5-year period of assessment[22] are the best evidence against the occurrence of an "anemia of aging" as a predictable effect of senescence. Although a subset of apparently healthy aged subjects may have anemia without an obvious cause for impaired erythropoiesis, no study to date has shown convincingly a direct detrimental effect of aging on erythropoiesis. This will be discussed further in Section 2.1.4.

2.1.2. IRON ABSORPTION AND METABOLISM. Because of the high prevalence of iron deficiency anemia in the elderly,[9] the kinetics of iron absorption and metabolism in the aged have been investigated. Cross-sectional data,[20] as well as longitudinal data showing a slight increase in ferritin in both males and females over 5 years of follow-up,[22] confirm a previous report of a progressive increase in serum ferritin with age in selected nonanemic subjects aged 20–93 years.[23] One explanation is that increasing ferritin levels reflect increasing storage of iron as a consequence of changes in the handling of iron by the aging reticuloendothelial system. A study of iron absorption and uptake of an orally administered radiolabeled inorganic iron preparation was carried out in healthy elderly subjects aged 65–83, healthy young subjects aged 19–49, and patients with iron deficiency aged 16–84.[24] No decrease was found in iron absorption in the elderly, nor was there a difference in iron absorption between old and young subjects in the face of iron deficiency, where iron absorption was found to be significantly increased. However, there was a significant and marked difference between young and old subjects in bone marrow red cell iron uptake (91% versus 66%), raising the possibility of an increase in ineffective erythropoiesis with age. More detailed ferrokinetic studies are needed to substantiate this observation.

2.1.3. MEAN CORPUSCULAR VOLUME. One consistent, but unexplained, finding of many studies is that after middle age, a progressive increase occurs in red cell size, as measured by the mean corpuscular volume (MCV).[22,25,26] This increase in MCV is present in healthy nonanemic subjects, as well as in many anemic subjects, and is more pronounced in subjects receiving an iron-supplemented diet.[22] The increase in MCV is associated with a proportional increase in mean corpuscular hemoglobin.[25] The increased MCV does not appear to be related to vitamin B_{12} or folate deficiency.[22,26,27] In a group of healthy men aged 20–79 years, the MCV rose and the hemoglobin decreased with age, but the deoxyuridine suppression test, a sensitive method of detecting deficiency of vitamin B_{12} or folate, was found to be normal.[26] In the same report, the ratio of [^{3}H]-thymidine uptake by nucleated bone marrow cells cultured from five subjects aged 70–81 years was significantly higher than that of five young men. The authors speculate that the increased MCV was a consequence of altered nucleic acid metabolism, resembling the state that occurs in megaloblastic anemia.

2.1.4. ERYTHROPOIESIS.   Because lean body mass declines with aging,[28] one might expect an age-related decline in total erythropoiesis on the basis of decreased tissue oxygen requirement, a direct function of lean body mass. This decline in erythropoiesis would be reflected in a decreasing red cell mass. This expected age-related effect has not been convincingly demonstrated to be either present or absent. Piomelli *et al.* found that the correlation between total red cell volume and lean body mass observed in young men was much weaker in 91 octogenarian male subjects, and this led to the conclusion that in the elderly, erythropoiesis varies, independent of oxygen demands of the lean tissues.[29] Because the subjects had a variety of medical ailments, the results of this study may simply demonstrate the variability of erythropoiesis in relationship to pathological factors.

The notion that anemia associated with a decline in erythropoiesis occurs as a result of aging was explored in a study of 18 healthy elderly subjects with unexplained mild anemia, 15 healthy nonanemic elderly subjects, and 15 healthy nonanemic young subjects.[30] All subjects were white males, and the two sets of elderly subjects were similar except for the presence of unexplained mild anemia. The anemic elderly, when compared to the nonanemic elderly and young control subjects, had a decrease in marrow normoblast number and a decrease in number of the erythropoietin-responsive progenitor cells (CFU-E), but no difference in the more primitive erythroid progenitor cells (BFU-E). Decreased peripheral granulocyte and lymphocyte counts, decreased marrow myeloid precursors, and decreased macrophage/granulocyte precursor cells (CFU-C) were also found. The nonanemic elderly consistently had values for marrow progenitor cells intermediate between the young controls and elderly anemic subjects. One might conclude from this study that some elderly individuals may have anemia associated with a decline in erythropoietin-sensitive erythroid progenitor cells, which may be due to aging itself or to another age-related phenomenon, not as yet defined.

The demonstration of normal testosterone levels in elderly anemic subjects,[30] and the finding of unchanged testosterone, testosterone-binding protein, and cell responsiveness to testosterone with aging,[31] suggest that the differential effects of testosterone on hematopoiesis are not responsible for the unexplained hypoproliferative anemia observed in some elderly subjects. Although reduced responsiveness to erythropoietin has been demonstrated *in vitro* and *in vivo* in aged mice[32,33] and *in vitro* in aged rats,[34] there is not yet evidence for this phenomenon in humans.

A study of erythropoietic function and the heme synthetic pathway in rats has shown decreased activity of Δ-aminolevulinic acid synthetase, the rate-limiting enzyme in heme biosynthesis, increased activity of heme oxygenase, the rate-limiting enzyme in heme degradation, decrease in [^{14}C]glycine incorporation into heme, and reduction of [^{14}C]leucine and [^{3}H]uridine incorporation into protein and nucleic acid synthesis, in senescent rat marrow.[34] It is possible that the decrease in CFU-E growth seen in senescent rat bone marrow cells is due to these age-associated changes in the balance between heme biosynthesis and degradation.

2.1.5. ERYTHROCYTE METABOLISM AND FUNCTION. Several studies have examined the effect of age on various metabolic and functional properties of erythrocytes. Erythrocyte 2,3-diphosphoglycerate (DPG) decreases slightly but significantly with age.[35] However, the observed difference (14.9 μmol/g hemoglobin at age 18–24 versus 13.9 μmol/g hemoglobin at age 75–84) is of doubtful physiological significance because it would result in very little change in hemoglobin affinity for oxygen. In elderly anemic subjects, DPG levels are higher than in nonanemic elderly subjects, but not significantly higher than in young controls and lower than levels usually found in young subjects as a physiological response to the same degree of anemia.[30] Red cell glucose-6-phosphate dehydrogenase (G-6-PD) activity was found to be decreased in selected elderly hospitalized subjects aged 80–90 compared to normal young subjects aged 20–30.[36] This was not related to a shift in the age of the red cell population, because age-unbiased erythrocyte samples were obtained. It is not clear whether this finding represents decreased enzyme synthesis in erythroid precursor cells, some other age-related effect on enzyme level, or an effect of other factors present in the elderly subjects, most of whom were taking medications. It is also not known whether the observed decrease in G-6-PD level is associated with any increase in susceptibility to oxidant hemolysis.

One consistent finding of several studies is the increase in red cell osmotic fragility with human aging.[37–40] Araki and Rifkind, using healthy subjects from the National Institute on Aging Longitudinal Study, demonstrated that aging is associated with an increase in the midpoint, breadth, and asymmetry of the osmotic fragility curve.[38] They attributed these changes to the increase in the MCV. The rate of osmotic hemolysis of erythrocytes was increased in older subjects, and this may be attributed to a change in erythrocyte membrane properties. In a subsequent study, the rate of osmotic hemolysis was related inversely to cholesterol content of the membrane, and cholesterol content of the membrane was related to age, suggesting a causal relationship between increased cholesterol content of the membrane and the decreased rate of osmotic hemolysis in elderly subjects.[39] The increase in membrane cholesterol content may reflect a change in plasma cholesterol distribution in the elderly, rather than a change in total plasma cholesterol.

Other studies of age-related changes in human erythrocyte membranes have demonstrated a variety of effects, many of which have uncertain physiological significance. An age-related decline in sialic acid content of glycophorin,[41–43] a major structural glycoprotein of erythrocyte membranes, has been reported. Because experimental removal of sialic acid from rabbit erythrocytes has been associated with rapid red cell removal from the circulation,[44] decreased sialic acid content in red cell membranes might be expected to result in decreased red cell survival. However, red cell survival in elderly human subjects appears to be no different from that in younger subjects.[45] This is in contrast to data in rats, which show a significant decline in red cell survival with aging.[46] Vomel has speculated that the possibility of early red cell removal related to loss of sialic acid and possibly all glycoproteins from the membrane surface might be counteracted by reduced activity of the humoral immune system in elderly subjects,[43] but specific data demonstrating this interaction are not available. One study of the fluidity of liposomes prepared from erythrocytes of young and old subjects has suggested a decrease in membrane lipid fluidity with aging.[41] Another study, however, showed no age-related change in lipid fluidity in erythrocyte ghost membranes, but found a change in the physical state of membrane proteins, as demonstrated by a spin labeling technique.[47]

Several studies have reported changes in the sodium, potassium-dependent adenosine triphosphatase (ATPase) activity of the red cell membrane with aging. A decline in erythrocyte ATPase activity with increasing age has been reported.[48] However, Simat et al. have recently elucidated the methodological difficulties of assaying the enzyme, and using carefully controlled techniques, they found no variation in erythrocyte ATPase activity with increasing age in males but a modest decrease in females.[49] The values for elderly females, however, were not significantly different from those of elderly males. The finding of a decline in the enzyme in postmenopausal females was also reported by Naylor et al.,[50] who further demonstrated that this phenomenon appears to be due to a reduction in the number of pump sites (ATPase molecules) per cell, rather than decreased activity at each pump site.[51] Although the etiology of this change in postmenopausal females is not known, the fact that the phenomenon is sex-related suggests that erythrocyte ATPase activity is not the basis for the age-related decline in basal oxygen consumption[52] or for the age-related changes in the MCV and osmotic fragility discussed previously.

2.2. LEUKOCYTES

The total and differential white blood counts are commonly used in clinical practice because of widely recognized changes in these counts in a variety of infectious, metabolic, inflammatory, and neoplastic diseases.[53] Examples of conditions associated with abnormalities in the leukocyte count that are less widely appreciated but are potential causes of variability in studies of elderly populations are vitamin B_{12} and folate deficiency, which may cause neutropenia, acute exercise or emotional stress, which may cause neutrophilia, and congestive heart failure, which may be associated with lymphopenia. Abnormalities of neutrophil function are seen in a wide variety of disorders including alcoholism, rheumatoid arthritis, uremia, and diabetes mellitus. Medications frequently alter both leukocyte number and function; these are discussed in Section 3.1.

It is well documented that a leukocytosis, consisting of proportional increases in granulocytes, lymphocytes, and monocytes, occurs in cigarette smokers.[8,54] The magnitude of leukocytosis increases in relationship to the number of cigarettes smoked per day,[8] but the etiology is unknown. Racial variability in the leukocyte count is also well described. Absolute neutrophil counts are significantly lower in black populations than in white populations. Whereas African and West Indian blacks have average neutrophil counts of 2100–2600/μl,[55] American blacks have neutrophil counts of 3000–4000/μl and American whites have neutrophil counts of 4400–4600/μl.[56,57] A subset of American blacks with neutrophil counts of 2000/μl or less can be identified and are said to have "benign" neutropenia because it is not associated with an increase in infectious complications.[58] The observed decrease in peripheral blood neutrophil count appears to result from a difference in marrow release of granulocytes rather than a decrease in granulocyte production.[58]

Age-related changes in leukocytes are summarized in Table II and are discussed in the next section.

2.2.1. LEUKOCYTE COUNT. A large number of reports concerning the effect of aging on the peripheral total and differential leukocyte count have been published. The wide variability in results suggests that many of these studies have been confounded by methodological prob-

TABLE II EFFECTS OF AGING ON WHITE BLOOD CELLS

Measurement	Effect	Selected ref(s)
Total leukocyte count	Unchanged	59–61
Lymphocyte count	Unchanged	61, 70–72
	? Decreased	67
Granulocyte function	Variable defects measured *in vitro*	70, 75–77, 79–81
Monocyte function	Variable defects measured *in vitro*	77, 82
Lymphocyte function	See Chapter 4	

lems similar to those of studies of red blood cell parameters in the elderly. Several cross-sectional studies of aged subjects in nursing homes or hospitals,[15,17] as well as aged subjects living in the community, selected from relatively healthy employee or clinic populations,[59–61] have shown no significant change in total or differential leukocyte count with aging. Other studies have concluded that a decrease in the total leukocyte count related to decreases in both the lymphocyte and neutrophil counts occurs in females, but not males, over age 50.[62,63] Caird reported that leukopenia occurred owing to a decrease in absolute lymphocyte count, in a stratified random sample of the elderly population of a town in Scotland.[64]

Many other studies have looked specifically at the effect of aging on the absolute lymphocyte count, with some showing a decline[66–69] and others showing no change.[61,69–72] In a well-designed cross-sectional and longitudinal study that used a large selected male population from the Normative Aging Study in Boston, there were no significant differences in lymphocyte count among the three age groups (23–44, 45–54, and 55 +), nor did significant changes occur in lymphocyte counts at three points measured over a 10-year period.[72] This study, however, could have been inadequate in detecting an aging effect on lymphocyte count because of the small number of subjects above age 69 at the start of the 10-year study period. A Mayo Clinic Study of 507 healthy male and female subjects aged 15 to greater than 70 years of age found no age or sex differences in absolute lymphocyte count, but it too is flawed by the relatively small number of subjects in the over-70 age group.[61] A study of 1684 inpatients and 884 outpatients at the University of Wisconsin Hospitals showed a downward trend in absolute lymphocyte count from age 30 to age 90, with statistically significant differences appearing in the 70–79 and 80–89 age groups when compared to the 20–29 age group.[67] However, because exclusions were made only for patients on the hematology or oncology services and patients with abnormal differential counts, rather than specific disease states, it is possible that the difference in the white counts between old and young reflected differences in general health status. To date, there has been no clear demonstration of changes in the total or differential leukocyte count that are directly related to aging.

2.2.2. LEUKOCYTE FUNCTION. The well-documented increase in incidence and morbidity of infections in the elderly[73,74] has prompted numerous investigations of the effects of aging on host defense mechanisms. The majority of these studies have concerned changes in lymphocyte-mediated immune functions; these are reviewed in detail in Chapter 4. Neutrophilic granulocytes, or polymorphonuclear leukocytes (PMNs), are critical to host defense against bacterial infections, and the effect of aging on their *in vitro* function has been the subject of several studies. Granulocyte adherence was found to be increased in aged subjects in two studies,[75,76] but was not different from that of younger controls in another study.[77] Elderly subjects in two studies had decreased chemotaxis;[76,77] however, only 5 of 70 elderly subjects in another study demonstrated this abnormal function, and it was not associated with a decrease in bacterial killing.[70] Age-related defects in bacterial phagocytosis[77,79] and *Candida* killing[76] have been reported; however, one study found that impaired bacterial killing in 8 of 70 elderly subjects was a transient phenomenon.[70] Quantitative reduction of nitroblue tetrazolium (NBT) dye, which measures oxidative metabolism of PMNs, decreased progressively up to age 79, but then increased after age 80 in one study,[76] but was decreased in all elderly subjects in other studies.[77,79] Chemiluminescence response of PMNs, another measure of PMN oxidative metabolism, was decreased in subjects over age 70.[80] It is likely that differences in subject selection factors, which in some studies are not well described, are responsible for these discrepant results. In a study of *in vitro* PMN function in well-characterized subjects in the Baltimore Longitudinal Study, NBT reduction and bacterial killing did not change with age, but a subtle defect in superoxide production was present in aged subjects.[81]

The relationship between *in vitro* PMN functional abnormalities and infection in the elderly host has not been explored in detail, although a longitudinal study by Phair *et al.* examined measures of several components of the immune system and clinical status in 70 healthy aged subjects.[70] Granulocyte function, *in vivo* and *in vitro* measures of cell-mediated immunity, and immunoglobulin levels were minimally different in the old subjects compared to young controls. Over a 2-year period,

12 subjects had infections, most of which were relatively minor. Four of nine subjects who had urinary tract or respiratory infections had decreased *in vitro* bacterial killing measured on one occasion, but repeat studies 6 months later were normal in all.

Several studies of monocyte function in aging have shown decreased phagocytosis,[77,82] decreased chemotaxis to leukocyte-derived chemotactic factor[77] but normal chemotaxis to zymosan-activated serum,[77,82] and no change in intracellular killing of *Candida albicans*.[82] Monocyte leukocytic pyrogen (LP) production after *in vitro* stimulation by *Staphylococcus epidermidis* was not significantly decreased in healthy elderly subjects compared to young controls, suggesting that the decreased febrile response to infection observed in elderly patients is not due to an intrinsic defect in monocyte LP production.[83]

These studies of *in vitro* leukocyte function in the elderly raise several issues. First, assays of granulocyte and monocyte function may have a wide distribution of normal results, and it may not be possible to determine what minimum response would be adequate to provide adequate *in vivo* function. Furthermore, these assays do not test functional reserve, which would be required for optimal defense against infection. The issue of whether selective mortality, i.e., death of individuals with inadequate leukocyte function, may have yielded populations of healthier survivors, who do not have significant abnormalities of leukocyte function but do not represent a "normal" elderly population, has been raised by some authors.[70,81] Despite these considerations, current data do not suggest that aging is associated with any severe or widespread impairment of granulocyte or monocyte function which is easily measured or which is clearly associated with susceptibility to specific infections.

2.3. PLATELETS

Age-related changes in platelets are summarized in Table III and are discussed in this section.

2.3.1. PLATELET COUNT.   No age-related changes in the platelet count have been reported in studies of selected ambulatory or hospitalized elderly subjects.[15,19] However, a group of healthy elderly subjects with unexplained anemia and leukopenia had significantly lower platelet counts than did normal young controls.[30] These subjects had evidence of decreased stem cell proliferation, but it is uncertain whether this represents an age-related phenomenon.

2.3.2. PLATELET FUNCTION.   There have been relatively few studies of platelet function in aged subjects. Certainly, there has been no clinical or laboratory data suggesting an age-associated bleeding diathesis

TABLE III   EFFECTS OF AGING ON PLATELETS

Measurement	Effect	Selected ref(s)
Platelet count	Unchanged	15, 19
Platelet function	Increase in epinephrine-induced aggregation	98–100
	Increase in collagen-induced aggregation	103
Platelet survival	Slightly decreased	92, 94, 95

on the basis of abnormal platelet function. The majority of studies have explored the possibility of an age-related increase in platelet aggregation and consumption, perhaps because of increasing recognition of the role of platelets in the pathogenesis of atherosclerosis and thrombotic disease,[84–86] which are common in aged populations. Standard *in vitro* tests of platelet function, such as platelet adhesion, aggregation, release and procoagulant activity, do not correlate well with *in vivo* states of abnormal platelet deposition.[87,88] Newer methods, such as determination of plasma concentrations of platelet-specific proteins released during aggregation, e.g., platelet factor 4 (PF4) and β-thromboglobulin (BTG), as well as quantitation of circulating platelet aggregates and measurement of platelet life-span, have been used increasingly, but these too are subject to a number of limitations in their predictive value for *in vivo* abnormalities.[89]

Elevated BTG levels[90–92] and PF4 levels[90] have been observed in selected elderly subjects when compared to young normal controls. However, although the difference between mean values for young and old subjects was statistically significant, there was considerable overlap of the range for older subjects with the range for younger subjects in each of these studies. Furthermore, the finding in one study[90] that both these platelet-specific proteins are elevated proportionally in elderly subjects raises the possibility of an *in vitro* (i.e., postsampling) platelet release reaction, because PF4 released *in vivo* quickly adheres to vascular endothelium, and an *in vivo* platelet release reaction should be characterized by an increased ratio of BTG to PF4.[93] Other findings suggesting an increased *in vivo* platelet release reaction in elderly subjects are an increase in the percentage of low-density platelets and decreased platelet granule content.[92]

Platelet survival, assessed by [51]Cr- or ^{111}In-labeled autologous platelets[94,95] or by measurement of the platelet production time,[92] appears to decrease slightly with increasing age. In one study, this finding did not appear to be related to the degree of vascular disease in the elderly subjects,[92] confirming previous observations that even severe widespread atherosclerosis decreases platelet survival to only a minimal degree.[96,97]

An age-associated increase in epinephrine-induced platelet aggregation *in vitro* has been reported by several investigators.[98–100] This sensitivity of platelets to aggregation by epinephrine was related to low-density lipoprotein (LDL) and total cholesterol in the subjects' plasma, but not to high-density lipoproteins (HDL) or very-low-density lipoproteins (VLDL).[101] Although LDL and cholesterol tended to increase with age, the correlation between age and epinephrine concentrations required to produce secondary platelet aggregation was weak. Another study explored the effect of purified lipoproteins on aggregation of and serotonin release from gel-filtered platelet preparations. VLDL increased thrombin-induced platelet aggregation as well as epinephrine-, ADP-, and thrombin-induced serotonin release from platelets, whereas HDL inhibited these platelet functions.[102] It appears, therefore, that these lipoproteins have a profound effect on *in vitro* platelet function and could be responsible for the observed changes in epinephrine-induced platelet aggregation with aging. The effects of these lipoproteins *in vivo* on platelet aggregation are not known. Another possible explanation for the reported increase in epinephrine-induced platelet aggregation in aged subjects is the recent finding of an age-associated increase in human platelet α-adrenergic receptor capacity.[100] Despite these provocative findings, the questions of whether *in vitro* platelet sensitivity to epinephrine reflects *in vivo* platelet sensitivity to endogenous catecholamines and whether this observation has clinical significance have yet to be answered.

A study of the effects of aging and smoking on plasma proteins and platelet functions, using well-characterized subjects aged 40–69 years from the Normative Aging Study in Boston, revealed an age-associated increase in collagen-induced aggregation of platelets *in vitro*.[103] This study did not examine epinephrine-induced platelet aggregation. The physiological significance of this finding is not known.

2.4. BONE MARROW: CELLULARITY, CYTOGENETICS, AND FUNCTION

Age-related changes in the bone marrow are presented in Table IV and are discussed in this section.

2.4.1. BONE MARROW CELLULARITY. In the clinical setting, hematopoietic function is often assessed by examination of the bone marrow, in conjunction with data from other clinical evaluations. Bone marrow aspirates and core biopsies are usually obtained from the posterior

TABLE IV   EFFECTS OF AGING ON THE BONE MARROW

Measurement	Effect	Selected ref(s)
Bone marrow cellularity (measured conventionally)	Decreased	104, 105
Bone marrow cytogenetics	Loss of Y chromosome	107–110
Bone marrow granulocyte reserve	Slightly decreased	145

iliac crest in adults and are examined microscopically for quantitative and qualitative abnormalities of hematopoietic cells and their stroma. Quantitative assessment of bone marrow cellularity is usually carried out by visual estimation of the relative areas covered by bone marrow cells and adipose tissue in the marrow space, but can be measured by more elaborate methods such as the point-counting method of Hartsock *et al.*[104] Bone marrow cellularity of a given specimen is expressed as the percentage of (bone marrow cells)/(bone marrow cells + adipose tissue) in a given area and does not take into account the quantitative relationship of bone marrow to bone mass or the possibility of variability in sampling. Therefore, although estimation of bone marrow cellularity is useful in many clinical situations, it cannot be regarded as an accurate measure of total-body bone marrow.

One study of 100 unselected autopsy cases found that marrow cellularity of the rib, sternum, and vertebra declined gradually from age 30 to age 70.[105] Because it is likely that the underlying diseases and causes of death were not similar in all age groups, and relatively small numbers of cases were examined in each age group, it cannot be concluded that these changes are simply an age-related effect. Furthermore, the sternal marrow cellularity of approximately 40% at age 70 falls within the range of 35–47.5% for sternal marrow cellularity in four normal adult subjects (ages unspecified) reported by Beutler *et al.*[106] In a study of 177 cases of sudden death, the majority of which were caused by trauma or arteriosclerotic heart disease, Hartsock *et al.* observed a progressive decline in anterior iliac crest marrow cellularity from 80% in the first decade of life to 45% by age 30, but very little change from age 30 to age 70.[104] However, subjects aged 70–79 had a mean cellularity of 30%. The significance of this data is difficult to assess because the over-70 age group may not be comparable to the under-70 age group with regard to circumstances of death. A more important consideration in evaluating any of these studies of bone marrow cellularity is that a decrease in marrow cellularity, if assessed by conventional techniques that do not take bone mass into account, may simply reflect an increase in adipose tissue resulting from decreased bone mass in elderly subjects.

2.4.2. BONE MARROW CYTOGENETICS. Hypodiploidy due to the loss of the Y chromosome, yielding a 45,X karyotype in bone marrow cells of aging males, is an intriguing but unexplained phenomenon.[107–110] In one study, 41 of 165 males aged 3 months–94 years, including 17 of the 21 males over age 80, had 45,X cell lines on direct cytogenetic preparations of their bone marrow.[110] The incidence of Y chromosome loss increases with increasing age, but the degree of hypodiploidy, i.e., the percentage of cells in an individual that manifest this abnormality, does not correlate with age.[110] The clinical significance of this phenomenon is uncertain, because the Y chromosome loss in the marrow cells is not associated with a significant degree of Y-chromosome loss in peripheral blood lymphocytes[107,108,110] and is not associated with impairment of hematopoiesis.[110]

2.4.3. BONE MARROW FUNCTION: HEMATOPOIESIS. Adult mammalian hematopoietic cells are considered to be "continuous replicators"—that is, they normally divide throughout the life-span of the host.[111] In the current model of hematopoiesis, a complex system of pluripotential, multipotential, and unipotential progenitor cells, or stem cells, is able to maintain a continuous supply of mature functional blood cells.[112] These stem cells are capable of self-renewal, proliferation, and differentiation, and these functions are subject to regulation by complex humoral and cellular factors. A great deal of controversy has centered on whether bone marrow stem cells have an intrinsic limitation on their self-renewal proliferative capacities, and whether normal aging is associated with a measurable and physiologically significant decline in these functions.

Hayflick and Moorehead observed that human lung fibroblasts have a limited number of population doublings *in vitro* before failure of the culture, and that fewer subcultivations were obtained with increasing *in vivo* age of the donor fibroblasts.[113,114] Hayflick hypothesized that this finite *in vitro* life-span of diploid cells might be an expression of aging at the cellular level.[114] This hypothesis, which implies that aging may be associated with a decline in hematopoiesis due to the exhaustion of bone marrow stem cells, has been tested in numerous studies. Serial transplantation of syngeneic bone marrow into irradiated mice cannot be performed indefinitely, suggesting that cells responsible for repopulation of the bone marrow may not possess unlimited self-renewal capacity.[115,116] Because serial transplantation studies require stem cells to proliferate at a rate more rapid than in the natural setting, it can be argued that studies of the functional capacity of transplanted bone marrow would be a more reasonable approach to investigating the effects of senescence.[117–119] Experiments of serial transplantation of syngeneic bone marrow into irradiated congenitally anemic W/W^v mice, an animal model for an abnormality of the pluripotent stem cell, have demonstrated that marrow cells from old donors and young donors are equally successful in supporting normal hematopoiesis.[119–121] Failure of the transplanted marrow to function after five transplants occurred with both old and young donor cells, with some old cell lines functioning for up to 100 months, which is more than 2.5 times the mean life-span of the donor animals.[118–120] A corollary study of transplantation of normal young marrow into aged mice demonstrated defective red cell production following bleeding, suggesting that the defective erythropoiesis of aged mice was not intrinsic to their erythropoietic stem cells.[122] Although these studies do not disprove the notion that bone marrow cells have an intrinsic proliferative limit, it appears that any such proliferative limit would not be realized in the normal life-span of the organism.

Although the *in vivo* hematopoietic capacities of stem cells obtained from young or old mice do not appear to be different, there have been conflicting results on the reconstitution of immunological functions of stem cells transplanted from young or old mice.[123–133] These will be discussed in Chapter 4.

The distinction between an intrinsic defect of stem cells and a defect in stromal cells was explored in a study of proliferative capacity of long-term bone marrow cultures from old and young mice, using a technique that allows separation of stroma from hematopoietic stem cells.[134] Young hematopoietic stem cells had greater proliferation than did old, and young stromal cells conferred a growth advantage on old hematopoietic cells. Although *in vitro* culture data must be interpreted with caution because they may not reflect *in vivo* phenomena, these findings deserve further investigation.

It is unclear how these complex data on hematopoiesis in aged animals may be applied to aged humans. There are no data to suggest any donor age-related differences in the success of human marrow transplants, but no truly aged donors have been used. Invasive transplant studies on normal humans similar to those on animals are obviously not possible. However, clinical data from the human studies reviewed in the foregoing sections, examined in the light of these animal data, suggest that the bone marrow is able to support normal hematopoiesis throughout the human life-span. Reports that some subsets of elderly subjects may have unexplained hypoproliferative anemia should prompt further basic and clinical studies on the function of aging hematopoietic tissue and stroma.

3. EFFECTS OF DRUGS ON THE BLOOD AND BONE MARROW

3.1. GENERAL CONSIDERATIONS

Drugs exert a wide range of effects on the blood and bone marrow, which may be manifest clinically as quantitative or functional changes in one or more of the peripheral blood cell lines. Examples of these effects, and drugs frequently implicated with each effect, are presented in Table V. More extensive catalogs of the hematological effects of drugs are available in reference texts.[4,135,136]

True adverse effects of drugs should be distinguished from predictable drug effects on the blood and bone marrow. Predictable effects occur in the majority of subjects exposed to a drug given at a specified dose and schedule, often in the therapeutic range. One example is the predictable myelosuppression leading to pancytopenia following administration of certain cytotoxic cancer chemotherapeutic agents. Another example is the granulocytosis produced by various agents, including epinephrine, corticosteroids, and lithium. Abnormalities of platelet ag-

TABLE V   DRUG EFFECTS ON THE BLOOD AND BONE MARROW

Clinical effect	Frequently associated drugs
Pancytopenia	
Aplastic anemia: idiosyncratic, irreversible	Chloramphenicol, sulfonamides, phenybutazone, gold salts, indomethacin
Dose-related in susceptible hosts, usually reversible	Chloramphenicol, phenytoin
Predictable, dose-related, reversible	Alkylating agents
Pure red cell aplasia	Gold salts
Agranulocytosis	Chlorpromazine, propylthiouracil
Granulocytosis	Epinephrine, corticosteroids, lithium
Thrombocytopenia	Quinine, quinidine, sulfonamides
Hemolytic anemia	
Immune	Penicillin, cephalothin, methyldopa
Associated with G-6-P-D deficiency	Antimalarials, sulfonamides, nitrofurantoin
Megaloblastic anemia	Trimethoprim, triamterene, phenytoin, phenobarbitol, ethanol
Methemoglobinemia	Acetanilid, naphthalene, amyl nitrate, nitroglycerin, sulfonamides
Granulocyte dysfunction	Corticosteroids, ethanol
Platelet dysfunction	Aspirin, other nonsteroidal antiinflammatory agents

gregation are routinely observed following the administration of aspirin and other nonsteroidal antiinflammatory agents.

In contrast to these predictable effects, true adverse drug effects occur in only a fraction of the population at doses that are well tolerated by the majority. This phenomenon implies that certain individuals have enhanced susceptibility to certain drugs. Although the etiology of this susceptibility is well defined in some instances, such as increased sensitivity of G-6-PD-deficient individuals to oxidant drug hemolysis, the vast majority of adverse drug effects on the blood are not associated with a well-defined genetic or acquired cause for increased susceptibility.

There are two general categories of mechanisms for adverse hematological effects of drugs: immunological and toxic (nonimmunological). A wide variety of mechanisms within these two general categories have been implicated in adverse hematological reactions. For many drugs, however, the mechanisms are poorly understood.

3.2. ARE THE ELDERLY MORE SUSCEPTIBLE TO DRUG EFFECTS ON THE BLOOD AND BONE MARROW?

Very few data support, or even address, the notion that aging might be associated with an inherent increase in susceptibility of the blood and bone marrow to the effects of drugs. Several studies, however, suggest an increase in hematological toxicity of some drugs in elderly subjects. As with the majority of studies on the effects of aging on the blood and bone marrow discussed previously, it is not possible to distinguish direct aging effects from effects of diseases associated with aging.

Aplastic anemia is a heterogeneous clinical entity that is drug-associated in one-third to one-half of cases. Mortality rates for aplastic anemia compiled from U.S. vital statistics are higher in the over-60 population than in the younger population.[137] However, these data may not reflect the incidence rates of aplastic anemia because of inaccuracies in diagnosis and coding of causes of death. Furthermore, aplastic anemia may be reversible or of limited severity in some patients, and factors associated with age may influence morbidity or mortality.

In a study that analyzed mortality from phenylbutazone-induced aplastic anemia and agranulocytosis in relationship to the number of prescriptions written in England and Wales, it appeared that the elderly, particularly females over the age of 65, were more susceptible to phenylbutazone toxicity.[138] Because of reliance on mortality figures, one cannot conclude from these data that increasing age alone is associated with increased susceptibility of the bone marrow to this drug effect. Delayed clearance of phenylbutazone, which has been observed in patients who have recovered from phenylbutazone-induced bone marrow aplasia,[139] might be a possible explanation for these data if elderly patients were more likely to have delayed clearance of this drug.

Chlorpromazine is associated with two effects on white blood cells, as reviewed by Pisciotta.[140] In about 10% of patients, continuous chlorpromazine therapy causes transient leukopenia, which reverses spontaneously even when the drug is continued. A much smaller number of patients develop severe, life-threatening agranulocytes after receiving cumulative doses of 10–20 g over periods of greater than 20 days. About 70% of cases are found in patients over age 40, and females are more commonly affected than males. In view of this age association and *in vitro* data suggesting that recovered patients have limited proliferative potential of granulocyte precursors, Pisciotta has postulated that chlorpromazine-induced agranulocytosis might result from an age-related decrease in granulocyte proliferation. Because decreased granulocyte proliferation in retrospectively identified chlorpromazine-sensitive individuals may be an effect rather than a cause, further investigation is needed to explain this apparent age-related effect.

Data on the outcome of elderly patients with cancer or leukemia have frequently been cited as clinical evidence that bone marrow reserve declines with aging. Hellman *et al.* have suggested that the elderly are more sensitive to cytotoxic agents because their bone marrow stroma may be less able to support hematopoiesis.[141] The concept that a series of toxic exposures occurs throughout the normal human life-span and may render the bone marrow more susceptible to drug effects deserves exploration. However, there are no specific data demonstrating an inherently increased sensitivity of the bone marrow in the elderly. Several host factors have been defined as prognostic factors for development of increased bone marrow suppression with cytotoxic drugs: prior cytotoxic therapy, performance status (functional status, which reflects overall health), bone or bone marrow disease, age (greater than 65), liver function, and renal function.[142] Although these factors may appear to be clinically useful in determining cytotoxic drug doses and schedules, no specific studies have isolated the effect of age as an independent prognostic factor. In acute leukemia, increased age is considered to be the most consistently important predictor of a poor response to therapy.[143] However, a study demonstrating similar remission rates and survivals in young and old leukemia patients treated with identical intensive chemotherapy regimens at a large referral center suggests that age does not directly influence outcome.[144] It also provides indirect evidence that the elderly are not intrinsically more susceptible to the effects of cytotoxic drugs on the functional reserve of the bone marrow.

4. Methodological Considerations for Clinical Drug Trials in the Elderly

From the foregoing review of studies on the hematological effects of aging, it may be concluded that there are few well-documented and clinically significant direct effects of aging on the blood and bone marrow. However, because age is associated with an increased prevalence of chronic disease states known to affect the hematopoietic system, it is critical that clinical drug trials in the elderly take into consideration the complex interactions between chronic diseases, drugs, and the hematopoietic system. Study populations for geriatric trials must be well defined in terms of sex, race, underlying chronic diseases, nutritional factors, socioeconomic status, chronic drug therapy, and previous drug exposures, because all these factors may be associated with quantitative and functional changes in the blood and bone marrow. It is important that each of these factors be considered in both the patient selection process and the interpretation of hematological data during clinical drug trials, so that a specific effect is not simply attributed to aging alone.

Available data suggest that aging itself is not associated with an increased susceptibility of the bone marrow to the effects of drugs. Any observed increase in the susceptibility of the elderly to adverse hematological effects of drugs should therefore prompt investigation of other possible factors present in elderly subjects. Such factors may include alterations in absorption, distribution, metabolism, and excretion of drugs, which may lead to increased risk of exposing the bone marrow to potentially toxic levels of the drug or its metabolites. Other factors include possible drug interactions in elderly subjects, who are frequently taking several drugs simultaneously, and the probability in elderly subjects of prolonged or repetitive drug exposure leading to cumulative drug toxicity.

Studies of drugs with known myelotoxicity in the elderly pose a problem, because the risk for increased susceptibility to predictable myelosuppression may be difficult to assess in elderly patients with one or more chronic diseases. Estimates of bone marrow granulocyte reserves can be obtained in a relatively simple fashion by measuring the increment in peripheral blood granulocyte count after oral or parenteral corticosteroid administration.[145] The granulocyte increment correlates well with postmitotic marrow granulocytes (the "reserve" granulocyte pool) in subjects with normal-sized or absent spleens.[78] In one study, prednisolone-induced granulocyte increments in 11 healthy elderly subjects, aged 56–90, were somewhat lower than those in 35 younger subjects, aged 18–55, but the values for all but one of the elderly subjects fell in the normal range.[145] However, 27 of 31 subjects with disease states expected to produce impairment of marrow function had abnormally low granulocyte increments. Measurement of marrow granulocyte reserve by this simple method might, therefore, be of use in the clinical trial setting to define myelosuppression prior to or after drug therapy in elderly subjects with multiple underlying disease states. Additional methods, either to evaluate specific cell lines or to measure overall marrow function, are clearly needed.

REFERENCES

1. Wintrobe MM: Clinical Hematology, ed 3. Philadelphia, Lea & Febiger, 1951.
2. Muldowney FP: The relationship of total red cell mass to lean body mass in man. *Clin Sci* 1957; 16:163–169.
3. Muldowney FP, Crooks J, Bluhm MM: The relationship of total exchangeable potassium and chloride to lean body mass, red cell mass and creatinine excretion in man. *J Clin Invest* 1957; 36:1375–1381.
4. Wintrobe MM: Clinical Hematology, ed 8. Philadelphia, Lea & Febiger, 1981.
5. Shahidi NT: Androgens and erythropoiesis. *N Engl J Med* 1973; 289:72–80.
6. Garn SM, Ryan AS, Owen GM, Abraham S: Income matched black-white hemoglobin differences after correction for low transferrin saturations. *Am J Clin Nutr* 1981; 34:1645–1647.
7. Okuno T: Smoking and blood changes. *JAMA* 1973; 225:1387–1388.
8. Helman N, Rubenstein LS: The effects of age, sex and smoking on erythrocytes and leukocytes. *Am J Clin Pathol* 1975; 63:35–44.
9. McLennan WJ, Andrews GR, MacLeod C, Caird FI: Anaemia in the elderly. *Q J Med* 1973; 165:1–13.
10. Kelly A, Munan L: Haematologic profile of natural populations: Red cell parameters. *Br J Haematol* 1977; 35:153–160.
11. Fisher S, Hendricks DG, Manoney AW: Nutritional assessment of senior rural Utahns by biochemical and physical measurements. *Am J Clin Nutr* 1978; 31:667–672.
12. O'Neal RM, Abrahams OG, Kohrs MB, Eklund DL: The incidence of anemia in residents of Missouri. *Am J Clin Nutr* 1976; 29:1158–1166.
13. Htoo MS, Kofkoff RL, Freedman ML: Erythrocyte parameters in the elderly: An argument against new geriatric normal values. *J Am Geriatr Soc* 1979; 27:547–551.
14. *Hemoglobin and Selected Iron-Related Findings of Persons 1–74 Years of Age: United States 1971–1974.* Hyattsville, MD, DHEW, PHS Publication No 46, 1979.
15. Shapleigh JB, Mayes S, Moore CV: Hematologic values in the aged. *J Gerontol* 1952; 6:207–219.
16. Smith JS, Whitelaw DM: Hemoglobin values in aged men. *Can Med Assoc J* 1971; 105:816–818.
17. Zaino EC: Blood counts in the nonagenarian. *NY State J Med* 1981; 81:1199–1200.
18. Lipschitz DA, Mitchell CO, Thompson C: The anemia of senescence. *Am J Hematol* 1981; 11:47–54.
19. Jernigan JA, Gudat JC, Blake JL, Bowen L, Lezotte DC: Reference values for blood findings in relatively fit elderly persons. *J Am Geriatr Soc* 1980; 28:308–314.

20. Garry PJ, Goodwin JS, Hunt WC: Iron status and anemia in the elderly: New findings and a review of previous studies. *J Am Geriatr Soc* 1983; 31:389–399.

21. Hobson W, Blackburn EK: Haemoglobin levels in a group of elderly persons living at home alone or with spouse. *Br Med J* 1953; 1:647–649.

22. Garry PJ, Goodwin JS, Hunt WC: Longitudinal assessment of iron status in a group of elderly, in Chandra RK (ed): *Nutrition, Immunity, and Illness in the Elderly*. Elmsford, NY, Pergamon Press, 1985, pp 77–83.

23. Casale G, Bonora C, Migliavacca A, Zurita IE, De Nicola P: Serum ferritin and ageing. *Age Aging* 1981; 10:119–122.

24. Marx JJM: Normal iron absorption and decreased red cell iron uptake in the aged. *Blood* 1979; 53:204–211.

25. Okuno T: Red cell size and age. *Br Med J* 1972; 1:569–570.

26. Okabe T, Ishizawa S, Ishii T, Kataoka K, Matsuki S: Erythrocyte aging changes evaluated by the deoxyuridine suppression test. *J Am Geriatr Soc* 1982; 30:626–631.

27. Garry PJ, Goodwin JS, Hunt WC: Folate and vitamin B_{12} status in a healthy elderly population. *J Am Geriatr Soc* 1984; 32:719–726.

28. Forbes GB: The adult decline in lean body mass. *Hum Biol* 1976; 48:161–173.

29. Piomelli S, Nathan DG, Cummins JF, Gardner FH: The relationship of total red cell volume to total body water in octogenarian males. *Blood* 1962; 19:89–98.

30. Lipschitz DA, Udupa KB, Milton KY, Thompson CO: Effect of age on hematopoiesis in man. *Blood* 1984; 63:502–509.

31. Harman SM, Tsitouras PD: Reproductive hormones in aging men. I. Measurement of sex steroids, basal luteinizing hormone, and Leydig cell response to human chorionic gonadotropin. *J Clin Endocrinol Metab* 1980; 51:35–40.

32. Udupa KB, Lipschitz DA: Erythropoiesis in the aged mouse: I. Response to stimulation *in vivo*. *J Lab Clin Med* 1984; 103:574–580.

33. Udupa KB, Lipschitz DA: Erythropoiesis in the aged mouse: II. Response to stimulation *in vitro*. *J Lab Clin Med* 1984; 103:581–588.

34. Ibraham NG, Lutton JD, Leuere RD: Erythroid colony development as a function of age: The role of marrow cellular heme. *J Gerontol* 1983; 38:13–18.

35. Purcell Y, Brozovic B: Red cell 2,3-diphosphoglycerate concentration in man decreases with age. *Nature* 1974; 251:511–512.

36. Rodgers GP, Lichtman HC, Sheff MF: Red blood cell glucose-6-phosphate dehydrogenase activity in aged humans. *J Am Geriatr Soc* 1983; 31:8–11.

37. Detraglia M, Cook FB, Stasiw DM, Cerny LC: Erythrocyte fragility in aging. *Biochim Biophys Acta* 1974; 345:213–219.

38. Araki K, Rifkind JM: Age dependent changes in osmotic hemolysis of human erythrocytes. *J Gerontol* 1980; 35:499–505.

39. Araki K, Rifkind JM: Erythrocyte membrane cholesterol: An explanation of the aging effect on the rate of hemolysis. *Life Sci* 1980; 26:2223–2230.

40. Bowdler AJ, Dougherty RM, Bowdler NC: Age as a factor affecting erythrocyte osmotic fragility in males. *Gerontology* 1981; 27:224–231.

41. Hegner D, Platt D, Heckerse H, Schloeder U, Breuninger V: Age-dependent physiochemical and biochemical studies of human red cell membranes. *Mech Aging Dev* 1979; 10:117–130.

42. Platt D, Norwig P: Biochemical studies of membrane glycoproteins during red cell aging. *Mech Aging Dev* 1980; 14:119–126.

43. Vomel T: Properties of ATPases and energy-rich phosphates in erythrocytes of young and old individuals. *Gerontology* 1984; 30:22–25.

44. Jancik J, Schauer R: Sialic acid—A determinant of the life-time of rabbit erythrocytes. *Hoppe-Seyler's Z Physiol Chem* 1974; 355:395–400.
45. Hurdle ADF, Rosin AJ: Red cell volume and red cell survival in normal aged people. *J Clin Pathol* 1962; 15:343–345.
46. Glass GA, Gershon H, Gershon D: The effect of donor and cell age on several characteristics of rat erythrocytes. *Exp Hematol* 1983; 11:987–995.
47. Butterfield DA, Ordaz FE, Markesbery WR: Spin label studies of human erythrocyte membranes in aging. *J Gerontol* 1982; 37:535–539.
48. Gambert SR, Duthie GH: Effect of age on red cell membrane sodium–potassium dependent adenosine triphosphatase ($Na^+ - K^+$ ATPase) activity in healthy men. *J Gerontol* 1983; 38:23–25.
49. Simat BM, Morley JE, From AHL, et al: Variables affecting measurement of human red cell Na^+, K^+ ATPase activity: Technical factors, feeding, aging. *Am J Clin Nutr* 1984; 40:339–345.
50. Naylor GJ, Dick DAT, Worrall EP, Dick P, Boardman L: Changes in the erythrocyte sodium pump with age. *Gerontology* 1977; 23:256–261.
51. Naylor GJ, Dick EG, Smith AHW, Dick DAT, McHarg AM, Chambers CA: Changes in erythrocyte membrane cation carrier with age in women. *Gerontology* 1980; 26:327–329.
52. Keys A, Taylor HL, Grande F: Basal metabolism and age of adult man. *Metabolism* 1973; 22:579–587.
53. Dale DC: Abnormalities of leukocytes, in Petersdorf RG, Adams RD, Braunwald E, Isselbacher KJ, Martin JB, Wilson JD (eds): *Principles of Internal Medicine*, ed 10. New York, McGraw-Hill, 1983, pp 304–310.
54. Corre F, Lellouch J, Schwartz D: Smoking and leukocyte counts: Results of an epidemiologic survey. *Lancet* 1971; 2:632–634.
55. Ezeilo GC: Non-genetic neutropenia in Africans. *Lancet* 1972; 2:1003–1004.
56. Karayalcin G, Rosner F, Sawitoky A: Pseudoneutropenia in Negroes. A normal phenomenon. *NY State J Med* 1972; 72:1815–1817.
57. Caramihai E, Karayalcin G, Aballi AJ, Lanzkowsky P: Leukocyte count differences in healthy white and black children 1 to 5 years of age. *J Pediatr* 1975; 86:252–254.
58. Mason BA, Lessin L, Schechter GP: Marrow granulocyte reserve in black Americans. Hydrocortisone-induced granulocytosis in the "benign" neutropenia of the black. *Am J Med* 1979; 67:201–205.
59. Sanders C, Orr RG, Ecans RJ: *Blood Counts on Radiation, Non-radiation, and New-Entry Employees.* U.K.A.E.A. Research Group Report, AERE-R, 1968. London, H.M.S.O.
60. Helman N, Rubenstein LS: The effects of age, sex, and smoking on erythrocytes and leukocytes. *Am J Clin Pathol* 1975; 63:35–44.
61. Zacharski LR, Elveback LR, Linman JW: Leukocyte counts in healthy adults. *Am J Clin Pathol* 1971; 56:148–150.
62. Allan RN, Alexander MK: A sex difference in the leukocyte count. *J Clin Pathol* 1968; 21:691–694.
63. Cruickshank JM, Alexander MK: The effect of age, sex, parity, haemoglobin level, and oral contraceptive preparations on the normal leukocyte count. *Br J Haematol* 1970; 18:541–550.
64. Caird FI, Andrews GR, Gallie TB: The leucocyte count in old age. *Age Aging* 1972; 1:239–244.
65. Diaz-Jouanen E, Strickland RG, Williams RC: Studies of human lymphocytes in the newborn and the aged. *Am J Med* 1975; 58:620–628.

66. Ferguson T, Chrichton DN, Price WH: Lymphocyte counts in relation to age. *Lancet* 1977; 2:35.
67. MacKinney AA: Effect of aging on the peripheral blood lymphocyte count. *J Gerontol* 1978; 33:213–216.
68. Polednak AP: Age changes in differential leukocyte count among female adults. *Hum Biol* 1978; 50:301–311.
69. Davey FR, Huntington S: Age-related variation in lymphocyte subpopulations. *Gerontology* 1977; 23:381–389.
70. Phair JP, Kauffman CA, Bjornson A, Gallagher J, Adams L, Hess EV: Host defenses in the aged: Evaluation of components of the inflammatory and immune responses. *J Infect Dis* 1978; 138:67–73.
71. Reddy MM, Goh K: B and T lymphocytes in man IV. Circulating B, T, and "null" lymphocytes in aging population. *J Gerontol* 1979; 34:5–8.
72. Sparrow D, Silbert JE, Rowe JW: The influence of age on peripheral lymphocyte count in men: A cross-sectional and longitudinal study. *J Gerontol* 1980; 35:163–166.
73. Gladstone JL, Recco R: Host factors and infectious diseases in the elderly. *Med Clin North Am* 1976; 60:1225–1240.
74. Gardner ID: The effect of aging on susceptibility to infection. *Rev Infect Dis* 1980; 2:801–810.
75. Silverman EM, Silverman AG: Granulocyte adherence in the elderly. *Am J Clin Pathol* 1977; 67:49–52.
76. Corberand J, Ngyen F, Laharrague P, et al: Polymorphonuclear functions and aging in humans. *J Am Geriatr Soc* 1981; 29:391–397.
77. Antonaci S, Jirillo E, Ventura MT, Garofalo AR, Bonomo L: Non-specific immunity in aging: Deficiency of monocyte and polymorphonuclear cell-mediated functions. *Mech Aging Dev* 1984; 24:367–375.
78. Deubelbeiss KA, Roth P: Postmitotic marrow neutrophils and neutrophil mobilization in man: Role of the spleen. *Blood* 1978; 52:1021–1032.
79. Charpentier B, Fournier C, Fries D, Mathieu D, Noury J, Bach JF: Immunological studies in human ageing I. *In vitro* functions of T cells polymorphs. *J Clin Lab Immunol* 1981; 5:87–93.
80. VanEpps DE, Goodwin JS, Murphy S: Age-dependent variations in polymorphonuclear leukocyte chemiluminescence. *Infect Immunity* 1978; 22:57–61.
81. Nagel JE, Pyle RS, Chrest FJ, Adler WH: Oxidative metabolism and bactericidal capacity of polymorphonuclear leukocytes from normal young and aged adults. *J Gerontol* 1982; 37:529–534.
82. Nielson H, Blom J, Larsen SO: Human blood monocyte function in relation to age. *Acta Path Microbiol Immunol Scand Sec C* 1984; 92:5–10.
83. Jones PG, Kauffman CA, Bergman AG, Hayes CM, Kluger MJ, Cannon JG: Fever in the elderly. Production of leukocytic pyrogen by monocytes from elderly persons. *Gerontology* 1984; 30:182–187.
84. Ross R, Glomset JA: The pathogenesis of atherosclerosis. *N Engl J Med* 1976; 295:369–377, 420–425.
85. Schafer AI, Handin RI: The role of platelets in thrombotic and vascular disease. *Prog Cardiovasc Dis* 1979; 22:31–52.
86. Mustard JF, Packham MA, Kinlough-Rathbone RL: Platelets and thrombosis in the development of atherosclerosis and its complications. *Adv Exp Med Biol* 1978; 102:7–30.
87. Lowe GDO: Laboratory evaluation of hypercoagulability. *Clin Haematol* 1981; 10:407–442.

88. Packham MA: Methods for detection of hypersensitive platelets. *Thromb Haemostas* 1978; 40:175–195.
89. Turpie AGG, DeBoer AC, Genton E: Platelet consumption in cardiovascular disease. *Sem Thromb Hemostas* 1982; 8:161–185.
90. Zahavi J, Jones NAG, Leyton J, Dubiel M, Kakkar VV: Enhanced *in vivo* platelet "release reaction" in old healthy individuals. *Thromb Res* 1980; 17:329–336.
91. Ludlam CA: Evidence for the platelet specificity of β-thromboglobulin and studies on its plasma concentration in healthy individuals. *Br J Haematol* 1979; 41:271–278.
92. Sie P, Montagut J, Blanc M, et al: Evaluation of some platelet parameters in a group of elderly people. *Thromb Haemostas* 1981; 45:197–199.
93. Kaplan KL, Owen J: Plasma levels of beta thromboglobulin and platelet factor 4 as indices of platelet activation *in vivo*. *Blood* 1981; 57:199–202.
94. Abrahamsen AF: Platelet survival studies in man with special reference to thrombosis and atherosclerosis. *Scand J. Haematol* 1968; suppl 3.
95. Sinzinger H, Jager E, Leithner CL, Hofer R: Altersabhangigkeit der thrombozyten-halbwertszeit. *Akt Gerontol* 1982; 12:213–216.
96. Harker LA, Slichter SJ: Arterial and venous thromboembolism: Kinetic characterization and evaluation of therapy. *Thrombos Diathes Haemorrh* 1974; 31:188–203.
97. Najean Y, Dassin E, Renner C, Wacquet M: Cinetique plaquettaire au cours des maladies arterielles. *Nouv Presse Med* 1979; 46:3813–3816.
98. Johnson M, Ramey E, Ramwell PW: Sex and age differences in human platelet aggregation. *Nature* 1975; 253:355–357.
99. Yokoyama M, Kawashima S, Sakamoto S, et al: Platelet reactivity and its dependence on alpha-adrenergic receptor function in patients with ischaemic heart disease. *Br Heart J* 1983; 49:20–25.
100. Yokoyama M, Kusui A, Sakamoto S, Fukuzaki H: Age-associated increments in human platelet α-adrenoceptor capacity. Possible mechanism for platelet hyperactivity to epinephrine in aging man. *Thromb Res* 1984; 34:287–295.
101. Hassall DG, Forrest LA, Bruckdorfer R, et al: Influence of plasma lipoproteins on platelet aggregation in a normal male population. *Arteriosclerosis* 1983; 3:333–338.
102. Aviram M, Brook JH: The effect of blood constituents on platelet function: Role of blood cells and plasma lipoproteins. *Artery* 1983; 11:297–305.
103. Chao FC, Tullis JL, Alper CA, Glynn RJ, Silbert JE: Alternation in plasma proteins and platelet functions with aging and cigarette smoking in healthy men. *Thromb Haemostas* 1982; 47:259–264.
104. Hartsock RJ, Smith EB, Petty CS: Normal variations with aging of the amount of hematopoietic tissue in bone marrow from the anterior iliac crest. *Am J Clin Pathol* 1965; 43:326–331.
105. Custer RP, Ahfeldt FE: Studies on the structure and function of bone marrow. II. Variations in cellularity in various bones with advanced years of life and their relative response to stimuli. *J Lab Clin Med* 1932; 17:960–962.
106. Beutler E, Drennan W, Block M: The bone marrow and liver in iron-deficiency anemia. *J Lab Clin Med* 1954; 43:427–439.
107. O'Riordan ML, Berry EW, Tough IM: Chromosome studies on bone marrow from a male control population. *Br J Haematol* 1970; 19:83–90.
108. Pierre RV, Hoagland HC: 45, X cell lines in adult men: Loss of Y chromosome, a normal aging phenomenon? *Mayo Clinic Proc* 1971; 46:52–55.
109. Walker LMS: The chromosomes of bone-marrow cells of haematologically normal men and women. *Br J Haematol* 1971; 21:455–461.

110. Pierre RV, Hoagland HC: Age-associated aneuploidy: Loss of Y chromosome from human bone marrow cells with aging. *Cancer* 1972; 30:889–894.

111. Post J, Hoffman J: Cell renewal patterns. *N Engl J Med* 1968; 279:248–258.

112. Quesenbery PJ: The concept of the hemopoietic stem cell, in Williams WJ, Beutler E, Erslev AJ, Lichtman MA (eds): *Hematology.* New York, McGraw-Hill, 1983, pp 129–143.

113. Hayflick L, Moorehead PS: The serial cultivation of human diploid cell strains. *Exp Cell Res* 1961; 25:585–621.

114. Hayflick L: The limited *in vitro* lifetime of human diploid cell strains. *Exp Cell Res* 1965; 37:614–636.

115. Cudkowitz G, Upton AC, Shearer GM: Lymphocyte content and proliferative capacity of serially transplanted mouse bone marrow. *Nature* 1964; 201:165–167.

116. Simonovitch L, Till JE, McCulloch EA: Decline in colony-forming ability of marrow cells subjected to serial transplantation into irradiated mice. *J Cell Comp Physiol* 1964; 64:23–32.

117. Harrison DE: Normal function of transplanted mouse erythrocyte precursors for 21 months beyond donor lifespans. *Nature (New Biol)* 1972; 237:220–222.

118. Harrison DE: Normal production of erythrocytes by mouse marrow continuous for 73 months. *Proc Natl Acad Sci USA* 1973; 70:3184–3188.

119. Harrison DE: Normal function of transplanted marrow cell lines from aged mice. *J Gerontol* 1975; 30:279–285.

120. Harrison DE: Mouse erythropoietic stem cell lines function normally 100 months. Loss related to number of transplantation. *Mech Aging Dev* 1979; 9:427–433.

121. Boggs DR, Saxe DF, Boggs SS: Aging and hematopoiesis. II. The ability of bone marrow cells from young and aged mice to cure and maintain cure in W/W^v. *Transplantation* 1984; 37:300–306.

122. Harrison DE: Defective erythropoietic responses of aged mice not improved by young marrow. *J Gerontol* 1975; 30:286–288.

123. Farraer JJ, Longhman BE, Nordin A: Lymphopoietic potential of bone marrow cells from aged mice. *J Immunol* 1974; 112:1244–1249.

124. Albright JF, Makinodan T: Decline in growth potential of spleen-colonizing bone marrow cells of long-lived mice. *J Exp Med* 1976; 144:1204–1213.

125. Kishimoto S, Shigemoto S, Yamamura Y: Immune responses in aged mice. Change of cell-mediated immunity with aging. *Transplantation* 1973; 15:455–459.

126. Tyan ML: Impaired thymic regeneration of lethally irradiated mice given bone marrow from aged donors. *Proc Soc Exp Biol Med* 1976; 152:33–35.

127. Tyan ML: Age-related decrease in mouse T cell progenitors. *J Immunol* 1977; 11:846–851.

128. Tyan ML: Effect of age on the intrinsic regulation of murine hematopoiesis. *Mech Aging Dev* 1982; 19:15–20.

129. Gozes Y, Umiel T, Trainin N: Selective decline in differentiating capacity of immunohematopoietic stem cells with aging. *Mech Aging Dev* 1982; 18:251–259.

130. Makinodan T, Perkin EH, Chen MG: Immunologic activity of the aged. *Adv Gerontol Res* 1971; 3:171–198.

131. Harrison DE, Doubleday JW: Normal function of immunologic stem cells from aged mice. *J Immunol* 1975; 114:1314–1317.

132. Harrison DE, Astle CM, Doubleday JW: Stem cell lines from old immunodeficient donors give normal responses in young recipients. *J Immunol* 1977; 118:1223–1227.

133. Harrison DE, Astle CM, Delaittre JA: Loss of proliferative capacity in immuno-

hemopoietic stem cells caused by serial transplantation rather than aging. *J Exp Med* 1978; 147:1526–1531.

134. Mauch P, Botnick LE, Hannon EC, Obbagy J, Hellman S: Decline in bone marrow proliferative capacity as a function of age. *Blood* 1982; 60:245–252.

135. Horler AR: Blood disorders, in Davies DM (ed): *Textbook of Adverse Drug Reactions,* ed 2. New York, Oxford University Press, 1981, pp 503–517.

136. Swanson M, Cook R: *Drugs Chemicals and Blood Dyscrasias.* Hamilton, IL, Hamilton Press, 1977.

137. Alter BP, Potter NU, Li FP: Classification and aetiology of the aplastic anemias. *Clin Haematol* 1978; 7:431–465.

138. Inman WHW: Study of fatal bone marrow depression with special reference to phenylbutazone and oxyphenbutazone. *Br Med J* 1977; 1:1500–1505.

139. Cunningham JL, Leyland MJ, Delamore IW, Price Evans DA: Acetanilide oxidation in phenylbutazone-associated hypoplastic anemia. *Br Med J* 1974; 3:313–317.

140. Pisciotta AV: Immune and toxic mechanisms in drug-induced agranulocytosis. *Sem Hematol* 1973; 10:279–310.

141. Hellman S, Reincke U, Botnick L, Mauch P: Functional organization of the hematopoietic stem cell compartment: Implications for cancer and its therapy. *J Clin Oncol* 1983; 1:277–284.

142. Lokich JJ: *Primer of Cancer Management.* Boston, G. K. Hall & Co., 1978, pp 110–114.

143. Wiernik PH: Acute leukemias of adults, in DeVita VJ, Hellman S, Rosenberg SA (eds): *Cancer: Principles and Practice of Oncology.* Philadelphia, Lippincott, 1982, pp 1402–1426.

144. Foon KA, Zighelboim J, Yale C, Gale RP: Intensive chemotherapy is the treatment of choice for elderly patients with acute myelogenous leukemia. *Blood* 1981; 58:467–470.

145. Cream JJ: Prednisolone-induced granulocytosis. *Br J Haematol* 1968; 15:259–267.

CHAPTER 8

The Aging Brain

Neal R. Cutler

1. Introduction

This chapter discusses the relationship between functional alterations and brain metabolism in aging animals and humans and in the pathological conditions of Alzheimer's disease (AD).

The normal aging process in animals and humans involves many alterations in both morphology and neurochemistry. In pathological conditions such as AD, those alterations have been shown to be greatly accelerated; however, the functional significance of these changes remains to be determined.

Neuronal loss has been reported in specific brain regions, such as superior frontal and temporal regions and striatum. Hippocampal pyramidal and granular cells and cerebellar Purkinje cells decline whereas a majority of brain stem nuclei are age invariant.[1]

Age-related morphological changes, such as a 10% decrease in brain weight, occur between the ages of 20 and 80, and there may be neuronal dropout of up to 25% in some areas of the brain.[2] There is a loss of both gray and white matter with age. The most affected areas of brain are the frontal and parietal regions.[3] Gray and white matter differ proportionally throughout life. It has been suggested that gray matter is

NEAL R. CUTLER • Department of Geriatrics, Cedars–Sinai Medical Center, University of California–Los Angeles School of Medicine, Los Angeles, California 90048.

lost in early life (20–50 years) and white matter later (70–90 years). Neurochemical changes are also reported to occur, such as a decrease of 62–90% in neocortical choline acetyltransferase enzyme activity (found at autopsy) in the frontal, temporal cortex, and hippocampal regions[4] and decrease in tyrosive hydroxylase in basal ganglia and related structures,[4] but no age-related changes have been found in neuropeptides (somatostatin, cholecystokinin, substance P).[5]

The question remains, however, whether these senescent changes are accompanied by age-related declines in brain oxidative metabolism, an indicator measure of brain function.[6,7] Results to date have been mixed; some indicate no change in cerebral blood flow and cerebral metabolic rates for oxygen during the normal aging process in the brain,[6] and others report decrements in those parameters as well as in cerebral metabolic rate for glucose (CMR_{glc}).[7–9]

Using the Sokoloff deoxyglucose technique,[10] regional cerebral metabolic rates for glucose ($rCMR_{glc}$) in both the normal aging rat and humans were studied, as a prelude to studies on disease states associated with aging.

2. BRAIN METABOLISM AND AGING IN THE RAT

The $rCMR_{glc}$ can be measured for specific brain regions of awake rats with the use of ^{14}C-2-deoxy-D-glucose (^{14}C-DG).[10] The ^{14}C-DG is injected intravenously as a bolus, and plasma radioactivity and glucose concentration are sampled at fixed times thereafter, until the animal is killed and brain radioactivity, due to trapped ^{14}C-DG-6-phosphate, is determined. A three time-constant equation is used to estimate $rCMR_{glc}$.

Initial studies suggested that $rCMR_{glc}$ declined in many regions of the brains of Sprague-Dawley rats between 4–6 months and 14–16 months of age but not thereafter.[11] Declines were evident in the parietal cortex and in parts of the visual, auditory, and extrapyramidal motor systems, as well as in the inferior olive and the gracile and cuneate nuclei. Values for whole-brain CMR_{glc}, which are given in Table I, summarize the time courses of the regional measures. Because glucose utilization fell in the first 16 months of life, the decline was considered to reflect senescence.[11]

This conclusion was reexamined using the ^{14}C-DG technique in awake Fischer-344 rats.[12] As illustrated in Table II, $rCMR_{glc}$ rose in many brain regions between 1 and 3 months of age (when the rat brain continues to grow), fell between 3 and 12 months, but did not change thereafter. The decline after 3 months was not considered to reflect

TABLE I   AVERAGE CEREBRAL GLUCOSE UTILIZATION IN
AWAKE SPRAGUE-DAWLEY RATS AT THREE AGES[a]

Age of rats (months)	Number of animals	Weighted average cerebral glucose utilization (μmol/100 g per min)
4–6	4	77 ± 2[b]
14–16	7	61 ± 2[c]
26–36	6	64 ± 4[c]

[a] Data obtained from Smith et al. [11]
[b] Mean ± S.E.M.
[c] Significantly different from mean at 4–6 months ($p < 0.05$).

TABLE II   REGIONAL CEREBRAL GLUCOSE UTILIZATION IN
AWAKE FISCHER-344 RATS AT DIFFERENT AGES[a]

Brain region	Age (months)				
	1	3	12	24	34
	rCMRglc, μmol/100 g/min				
Olfactory bulb	49 ± 3[b,c]	66 ± 4[c]	56 ± 2	56 ± 4	58 ± 5
Frontal pole	69 ± 5	77 ± 4	59 ± 2[c]	61 ± 4	63 ± 5
Sensorimotor cortex	61 ± 4	74 ± 4[c]	60 ± 3[c]	59 ± 3	61 ± 5
Temporalparietal–occipital cortex	60 ± 5	70 ± 4	55 ± 2[c]	54 ± 4	56 ± 5
Hypothalamus + thalamus	42 ± 4	53 ± 4	41 ± 3	42 ± 4	56 ± 5
Putamen + head of caudate nucleus	68 ± 5	73 ± 4	55 ± 3[c]	52 ± 4	58 ± 4
Nucleus accumbens	—	54 ± 6	45 ± 3	50 ± 6	49 ± 5
Septum	38 ± 4	48 ± 5	46 ± 3	47 ± 4	40 ± 4
Hippocampus	44 ± 4	56 ± 3[c]	46 ± 3[c]	43 ± 3	46 ± 4
Inferior colliculus	71 ± 5	94 ± 5[c]	68 ± 3[c]	57 ± 4	56 ± 4
Superior colliculus	46 ± 3	64 ± 3[c]	50 ± 3[c]	47 ± 4	40 ± 4
Midbrain basis + tegmentum	44 ± 4	56 ± 3[c]	43 ± 2[c]	39 ± 3	40 ± 4
Medulla	40 ± 3	46 ± 2[c]	36 ± 3[c]	34 ± 3	32 ± 3
Corpus callosum	—	32 ± 3	28 ± 4	26 ± 4	28 ± 3

[a] Data obtained from London et al.[12]
[b] Mean ± S.E.M. ($n = 10$, except at 34 months, when $n = 7$).
[c] Differs significantly from mean at preceding month ($p < 0.05$).

senescence, however, because it occurred within the first one-third to one-half of the rat's life-span. The maximum survival time of Fischer-344 rats is approximately 36 months, and mean survival is 29 months.

3. THE STRESSED NERVOUS SYSTEM

Despite the finding that $rCMR_{glc}$ remains constant in the unstimulated rat, the senescent brain may respond differently from the young brain to pharmacological or physiological stimulation. Table III shows $rCMR_{glc}$ in two representative brain regions (the visual cortex and striatum) in response to oxotremorine, a cholinergic muscarinic agonist.[13] Without the drug, mean $rCMR_{glc}$ at 12 months was similar to the value at 24 months in both regions. However, the 12- and 24-month means were less than the means at 3 months. In response to oxotremorine, the 3- and 12-month means for $rCMR_{glc}$ approached each other and were higher than the means at 24 months.

Glucose utilization by other parts of the nervous system may actually increase during senescence. In awake Fischer-344 rats, glucose utilization by the superior cervical ganglion rose from a mean of 16 μmol 100 g^{-1} min^{-1} at 3 months of age to 23 μmol 100 g^{-1} min^{-1} at 30 months. The increase was accompanied by an elevated plasma norepinephrine concentration, growth of paraaortic paraganglia replete with catechol-

TABLE III   EFFECT OF OXOTREMORINE ON REGIONAL CEREBRAL GLUCOSE UTILIZATION IN AWAKE FISCHER-344 RATS OF THREE AGES[a]

| Brain region | Age (months) | $rCMR_{glc}$, μmol.100 g^{-1}.min^{-1} | |
		Control	Oxotremorine (0.1 mg.kg^{-1}, i.p.)
Visual cortex	3	92 ± 3[b]	101 ± 6
	12	77 ± 7	110 ± 9[d]
	24	74 ± 3[c]	84 ± 7
Striatum	3	73 ± 5	77 ± 5
	12	53 ± 5[c]	78 ± 7[d]
	24	47 ± 4[c]	63 ± 6

[a] Data from Dam et al.[14]
[b] Mean ± S.E.M. for six to eight rats.
[c] Significantly different from $rCMR_{glc}$ in 3-month rats.
[d] Significantly different from $rCMR_{glc}$ in control rats ($p \leq 0.05$).

amines, and a decrease in heart rate.[15,16] Functional activity of the aged sympathetic nervous system may be chronically activated as a partial response to decreased sensitivity of myocardial and vascular tissues to adrenergic agonist.[17]

In summary, even though multiple morphological and neurochemical sequelae, such as a decrease in the density of dendritic spines in cortical pyramidal neurons, decreases in cerebellar Purkinje cells, and reductions in regional neurotransmitter synthetic enzymes and receptor densities,[18] occur in the rat after 12 months of age, no change was found in brain metabolism between 12 and 34 months in the resting Fischer-344 rat. However, under pharmacological stimulation with oxotremorine, a muscarinic agonist, a differential responsivity in various brain regions was demonstrated.

4. CEREBRAL METABOLIC FUNCTION AND AGING IN HUMANS

The well-known morphological and neurochemical changes with age have been addressed earlier. There have been attempts to quantify those changes *in vivo* through computed tomographic analysis in aging individuals, and we have found age-related gray matter atrophy and ventricular dilatation.[19,20] However, measurement of brain metabolism is one way to assess the implications of morphometric changes.

Early studies measuring brain metabolism and cerebral blood flow found reduction with age.[21] However, the subjects were not completely free of all disease, a fact that could account for the age-related changes found.

In lieu of the previous studies that used positron emission tomography (PET) and [18F]fluoro-2-deoxy-D-glucose ([18FDG), we again sought to determine whether brain metabolic function changes with age in healthy humans. All the subjects selected for our study were male and free of medical and neurological disease. All subjects were normotensive and had had complete physical and neurological examinations. To date, 40 subjects in good health between ages 21 and 83 years[22,23] have been studied. PET scanning methodology and analysis have been previously described.[22,23]

No change in brain metabolic function and age has been found in any of the 59 regions examined ($p < 0.05$) (Table IV and Figs. 1 and 2).

The finding of age invariance in brain metabolic function may reflect the use of careful health screening criteria, which distinguishes our study

TABLE IV   MEAN VALES FOR THE RIGHT HEMISPHERE
AND THE FOUR LOBES THAT WERE EXAMINED—NO
STATISTICALLY SIGNIFICANT CORRELATION
WAS FOUND $(p < 0.05)^a$

Region	rCMRglc[b] mg/100g per min	r^c
Hemisphere (right)	4.60 ± 1.08	0.01
Frontal lobe	5.41 ± 1.35	−0.05
Parietal lobe	5.45 ± 1.32	0.01
Temporal lobe	4.48 ± 1.25	0.03
Occipital lobe	5.39 ± 1.23	0.02

[a] Data obtained from Duara et al.[22,23]
[b] Mean ± S.D. ($n = 40$).
[c] r is correlation with age. No correlation was statistically significant
($p < 0.05$).

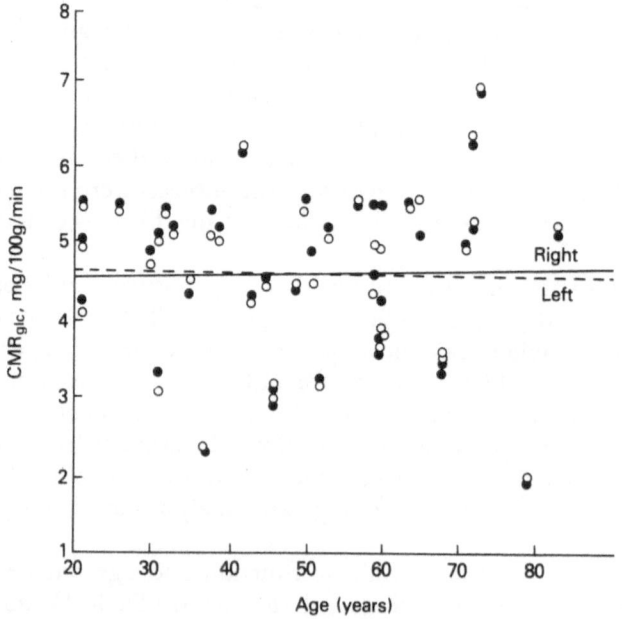

FIGURE 1   Hemisphere cerebral metabolic rates for glucose (CMRglc) for 40 healthy subjects between 21 and 83 years of age. No correlation was statistically significant ($p > 0.05$).
○, Right; ●, left. (Adapted from Duara et al.[23])

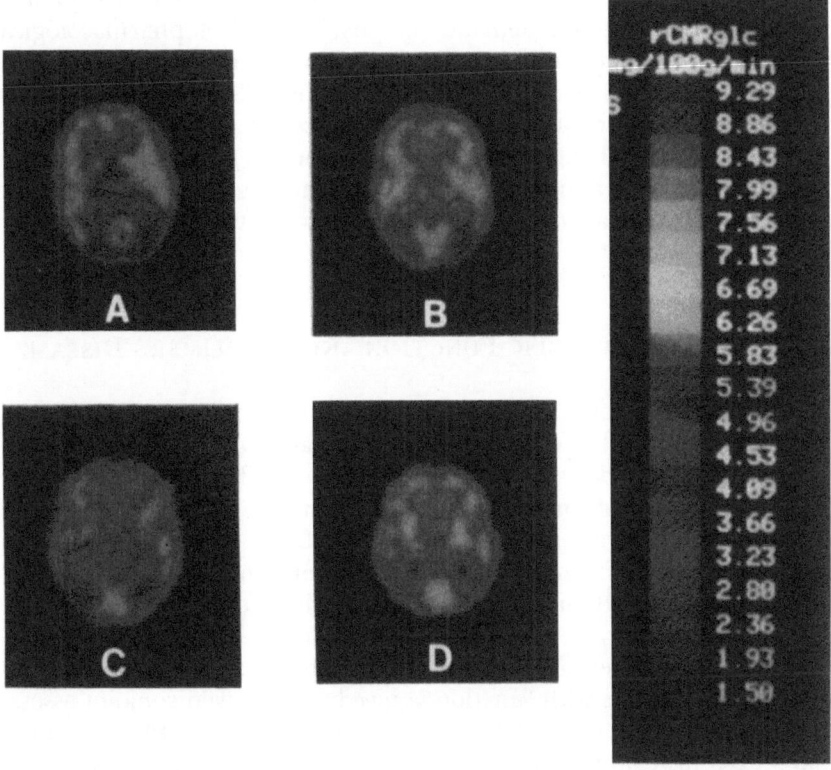

FIGURE 2   Representative PET scans (50 mm above the inferior orbitomeatal line) of individual healthy subjects from four different age groups. (A) 21-year-old normal; (B) 49-year-old normal; (C) 67-year-old normal; (D) 83-year-old normal. Qualitatively there is no overall change in brain metabolic function and age.

from other previous studies.[8,9] Our findings are in agreement with some data from investigators who have carefully screened for exclusion of disease.[6]

A neuronal compensatory mechanism such as neuronal plasticity and increased dendritic growth could explain the finding of age invariance in brain metabolism. Buell and Coleman[24] have shown that in postmortem brain tissue, dendrites grow branches and develop a segmental elongation mechanism, possibly as a compensation for the neuronal loss that occurs in the aging process. In the resting state, that is, with eyes covered and ears plugged, during the PET scanning procedure, the subject is not under a stressed state. However, if the subject were stressed, compensatory mechanisms might not be able to cope with the increased metabolic demands of physiological or pharmacological stimulation, and reductions in metabolism might result.[13] An example of this reaction has been shown in this chapter (Section 3), where the aged rat did not show the same metabolic response to oxotremorine as the young or middle-aged rat. This phenomenon certainly needs to be considered in central nervous system drug studies in the elderly where subjects may show a varying behavioral response to different drugs.

5. Cerebral Metabolic Function and Alzheimer's Disease

The inclusion of AD in this discussion of the aging brain is appropriate owing to the large number of individuals over the age of 65 years affected with this disease process—at least 4 million people. AD is a slowly progressive disorder that includes both intellectual decline and personality deterioration. Autopsies of the AD brain have revealed neurofibrillary tangles in the cerebral cortex, preferential loss of large cortical neurons in midfrontal and temporal regions, marked reductions in the presynaptic enzyme choline acetyltransferase, and reductions in adrenergic, serotoninergic, and muscarinic receptors.[25] Attempts to correlate those changes with function *in vivo* have involved conjoint assessment of clinical severity and brain metabolic function. PET scanning and [18]FDG have accurately assessed metabolic changes in various brain regions that occur in AD. We have examined and previously reported[26-28] on PET scan measures of rCMRglc in patients with mild-to-moderate and severe AD.

The following is a description of those studies. All our patients were given a diagnosis of presumptive AD by DSM-III criteria and agreement of two physicians trained in diagnosis of dementias and were free of all other medical and neurological diseases. They had a mean age of 65 years (range 49–81). Two were classified as severe and 10 as mild-to-moderate as determined by severity-rating scales.

The mild-to-moderate AD patients were assessed with PET scanning and [18]FDG with methods similar to those previously described[22,23] and had no significant reductions in hemisphere or lobar metabolism as compared to controls (Figs. 3 and 4). The severe group had significant reductions throughout both hemispheres and three of the four lobes examined.

Study limitations that need to be considered are the need for con-

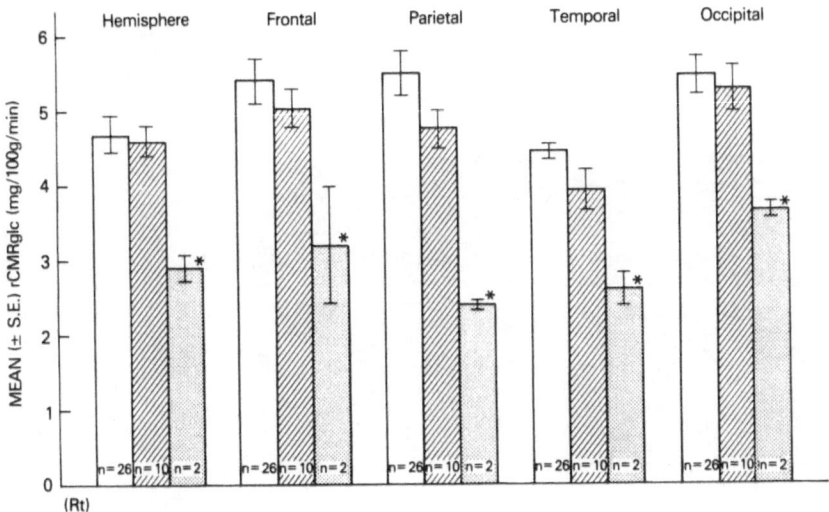

FIGURE 3. Brain metabolic function for the right hemisphere and the four lobes for healthy controls and mild-to-moderate and severe Alzheimer's disease (AD) patients. No differences were found in metabolic function in the hemisphere or the four lobes in the mild-to-moderate AD patients ($p > 0.05$). The severe AD patients revealed marked reductions in brain metabolism throughout the brain ($p < 0.05$).☐ , Controls (mean age = 62 years);▨, AD (mild-mod; mean age = 64 years);▩, AD (severe; mean age = 70 years); *, p < 0.05.

firmation of diagnosis, which can only be made at autopsy or by cerebral biopsy. On clinical grounds our diagnoses are only 70% accurate.[29] The PET scan also has a number of limitations such as spatial resolution and partial volume effects.[22,23]

In an attempt to explain our findings, it has been shown by studies of Buell and Coleman that, at least in AD, the compensatory, dendritic arborization of the normal aging process is lost.[24,30]

Alternatively, the decrements found in late-severe AD may reflect the "threshold principle" that Roth and others have proposed for cognitive functioning in dementia.[31] According to that principle, changes in metabolic function may not become manifest until a threshold level of damage is reached. That level of brain damage has been quantified for clinical findings of dementia to 12 or more senile plaques per visual field and is found in greater than 90% of individuals with AD. In addition, loss of neurotransmitters and associated receptors in AD adds to the "threshold capacity."[32]

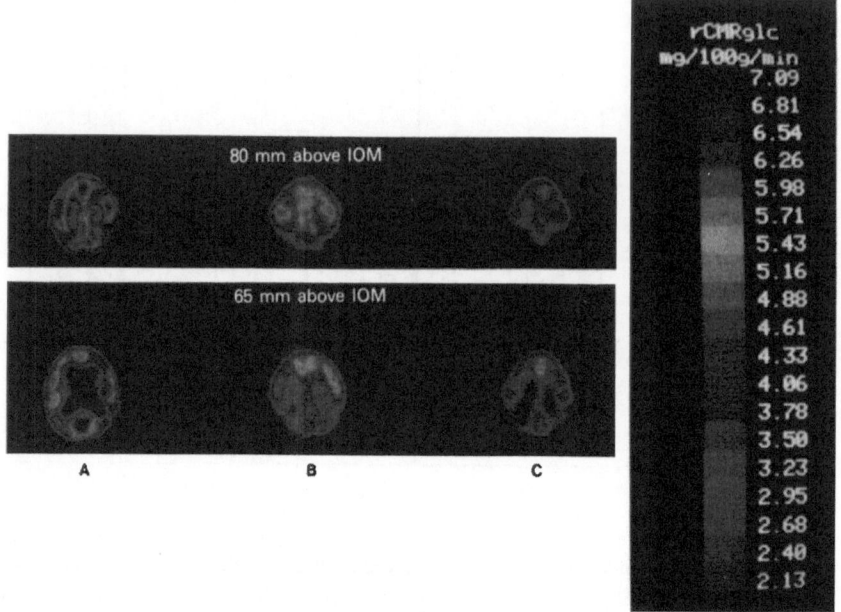

FIGURE 4   Positron emission tomography scans at 65 and 80 mm above the inferior or-bitomeatal line (IOM): (A) Alzheimer's disease patient with mild severity; (B) 49-year-old AD patient of moderate severity; (C) 69-year-old AD patient with the severe form. Qual-itatively there are gradual reductions in brain metabolism (as represented by darker areas on the scans) as the patients' severity increases.

6. RECEPTOR FUNCTION AND THE AGING BRAIN

The development of PET scanning technology has made it possible to delineate various neurotransmitter receptor subtypes. PET scanning with 3-N-methylspiperine has revealed an age-related decline in dopamine receptor numbers[33,34] that agrees with postmortem studies of brain tissue in animals and man[35,36] (Fig. 5). Moreover, another dopaminergic receptor analog such as [76Br]-bromospiroperidol has been examined in animals, and it is now being developed for use in humans.[37] Other receptor brain systems such as the benzodiazepine receptors are also being examined.[38]

Eventually analogs for other receptor systems and appropriate radioactive tracers will be developed to examine *in vivo* changes of the aging brain. Muscarinic antagonists may be used to map the brain cholinergic system in pathological conditions such as the dementias in which a brain cholinergic dysfunction has been found.[39,40] A muscarinic an-

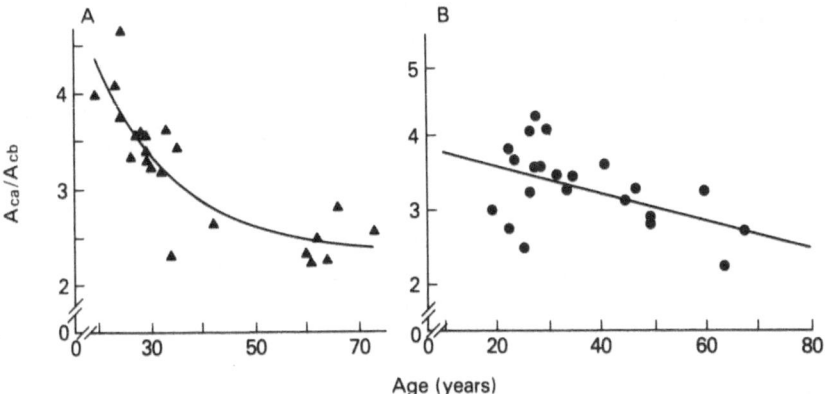

FIGURE 5   Positron emission tomography scanning using 3-*N*-methylspiperine, a neuroleptic liquid, in 44 healthy subjects [(A) 22 male and (B) 22 female] in the age range of
19–73 years. Age-related declines were found in the dopamine receptors in the basal
ganglia of both male and female subjects. The ratio Aca/Acb is an estimate of the dopamine
receptor. The ratios represent an average value for the right and left sides of the brain.
(From Wang et al.[34]; used with permission; copyright 1984 by the AAAS.)

tagonist has been used to assess muscarinic cholinergic receptors in the
brain of an AD patient.[41]

The implications of studies with these various receptor agents will
lead to tests of specific drugs that may have therapeutic value in the
various pathological brain dysfunctions that may develop during the
normal aging process.

7. SUMMARY

The brain in animals and humans undergoes many age-related alterations in both morphometric and neurochemical assessments. In the
AD brain, those morphological and neurochemical changes are greatly
accelerated. The functional significance of those changes has been determined by measuring brain metabolic function with PET scanning and
[18]FDG in healthy aging and AD. Our findings demonstrate that brain
metabolic function did not change with age in 40 healthy men between
21 and 83 years of age.

This finding of age invariance in both animals and humans may be
due to compensatory mechanisms of the aging brain such as dendritic
arborization. However, under various stressed states, there are different
pharmacological responses between young and older subjects.

In the pathological condition of AD, significant metabolic changes were found throughout the brains of the late-severe disease patients but not in the brains of mild-moderate AD patients. The findings may reflect that a threshold amount of brain damage, either morphological or neurochemical, is required for dementia or reduction in metabolism to occur.

As more radioactive tracers are developed, an *in vivo* exploration of the human brain in various normal and pathological conditions will increasingly provide a more accurate understanding of the mechanisms responsible for them and for the normal aging process.

REFERENCES

1. Ball MJ: Neuron loss, neurofibrillary tangles, and granulovacuolar degeneration in the hippocampus with aging and dementia. A quantitative study. *Acta Neuropathol (Berl)* 1977; 37:111–118.
2. Tomlinson BE, Henderson G: Some quantitative cerebral findings in normal and demented old people, in Terry RD, Gershon S (eds): *Neurobiology of Aging (Aging, vol 3)*. New York, Raven Press, 1976, pp 183–204.
3. Anderson JM, Hubbard BM, Coghill GR, et al: The effect of advanced old age on the neurone content of the cerebral cortex. Observations with an automatic image analyser point counting method. *J Neurol Sci* 1983; 58:235–246.
4. McGeer E, McGeer PL: Neurotransmitter metabolism in the aging brain, in Terry RD, Gershon S (eds): *Neurobiology of Aging (Aging, vol 3)*. New York, Raven Press, 1976, p 389.
5. Buck SH, Deskmukh PP, Burks TF, et al: A survey of substance P, somatostatin, and neurotensin levels in aging in rat and human central nervous system. *Neurobiol Aging* 1981; 2:257–264.
6. Dastur DK, Lane MH, Hansen DB, et al: Effects of aging on cerebral circulation and metabolism in man, in: Birren JE, Butler RN, Greenhouse, SW, Sokoloff L, Yarrow MR (eds): *Human Aging, A Biological Behavioral Study*. USPHS Publication 986. Washington, DC, US Government Printing Office, 1963, p 57.
7. Gottstein U, Held K: Effects of aging on cerebral circulation and metabolism in man. *Acta Neurol Scand* 1979; 60(suppl 72):54–55.
8. Frackowiak RSJ, Lenzi GL, Jones T, et al: Quantitative measurement of regional cerebral blood flow and oxygen metabolism in man using ^{15}O and positron emission tomography: Theory, procedure and normal values. *J Comput Assist Tomogr* 1980; 4:727–736.
9. Kuhl DE, Metter EJ, Riege WH, et al: Effects of human aging on patterns of local cerebral glucose utilization determined by the [^{18}-F]fluorodeoxyglucose method. *J Cerebral Blood Flow Metab* 1982; 2:163–171.
10. Sokoloff L, Reivich, M, Kennedy C, et al: The ^{14}C-deoxyglucose methods for the measurement of local cerebral glucose utilization: Theory, procedure, and normal values in the conscious and anesthetized albino rat. *J Neurochem* 1977; 28:897–916.
11. Smith CB, Goochee C, Rapoport SI, et al: Effects of aging on local rates of cerebral glucose utilization in the rat. *Brain* 1980; 103:351–365.

12. London ED, Nespor SM, Ohata M, et al: Local cerebral glucose utilization during development and aging of the Fischer-344 rat. *J Neurochem* 1981; 37:217–221.

13. London ED, Mahone P, Rapoport SI, et al: Effect of age on oxotremorine-induced stimulation of local cerebral glucose utilization. *Fed Proc* 1982; 41:1323.

14. Dam M, Rapoport SI, London ED: Aging and cholinergic system: A 2-deoxyglucose study in the rat brain. *Monogr Neural Sci* 1984; 11:62–67.

15. Partanen M, London ED, Rapoport SI: Glucose utilization in sympathetic ganglia of male Fischer-344 rats at different ages. *J Auton Nerv Syst* 1982; 5:391–398.

16. Partanen M, Chiueh CC, Rapoport SI: Age-related increase in catecholamine-containing paraganglia in male Fischer-344 rats. *Anat Rec* 1981; 201:563–566.

17. Lakatta EG, Gerstenblith G, Agell CS, et al: Diminished inotropic response of aged myocardium to cathecholamines. *Circ Res* 1975; 36:262–269.

18. Weiss B, Greenberg L, Cantor E: Age-related alterations in the development of adrenergic denervation supersensitivity. *Fed Proc* 1979; 38:1915–1921.

19. Cala LA, Thickbroom GW, Black JL et al: Brain density and cerebrospinal fluid space size: CT of normal volunteers. *AJNR* 1981; 2:41–47.

20. Schwartz M, Creasey H, Grady CL, et al: Computed tomography analysis of brain morphometrics in 30 healthy men, aged 21–83 years. *Ann Neurol* 1985; 17:146–157.

21. Kety SS: Human cerebral blood flow and oxygen consumption as related to aging. *Res Publ Assoc Res Nerv Ment Dis* 1956; 35:31–45.

22. Duara R, Margolin RA, Robertson-Tchabo EA, et al: Cerebral glucose utilization, as measured with positron emission tomography, in 21 resting healthy men between the ages of 21 and 83 years. *Brain* 1983; 106:761–775.

23. Duara R, Grady C, Haxby J, et al: Human brain glucose utilization and cognitive function in relation to age, *Ann Neurol* 1984; 16:702–713.

24. Buell SJ, Coleman PD: Quantitative evidence for selective dendritic growth in normal human aging but not in senile dementia. *Brain Res* 1981; 214:23–41.

25. Terry RD, Davies P: Some morphologic and biochemical aspects of Alzheimer's disease, in Samuel D, Algeri S, Gershon S, Grimm VE, Toffano G (eds): *Aging of the Brain (Aging,* vol 22). New York, Raven Press, 1983, p 47.

26. Cutler NR, Schapiro MB: Brain metabolism as measured with positron emission tomography: Alzheimer's disease and Down syndrome. *Clin Neuropharmacol* 1984; 7(suppl 1): S177–S178.

27. Cutler NR: Brain metabolism in Alzheimer's disease, in Cutler NR (moderator): Brain imaging: Aging and dementia. *Ann Intern Med* 1984; 101:360–362.

28. Cutler NR, Haxby JV, Duara R, et al: Clinical history, brain metabolism and neuropsychological function in Alzheimer's disease. *Ann Neurol* 1985; 18:298–309.

29. Jellinger K. Neuropathological aspects of dementia resulting from abnormal blood and cerebrospinal fluid dynamics. *Acta Neurol Belg* 1976; 76:83–102.

30. Schiebel AB, Tomigasu U: Dendritic sprouting in Alzheimer's presenile dementia. *Exp Neurol* 1978; 60:1–8.

31. Roth M, Tomlinson BE, Blessed G: Correlation between scores for dementia and counts of "senile plaques" in cerebral grey matter of elderly subjects. *Nature* 1966; 209:109–110.

32. Rossor MN. Neurotransmitters in CNS disease: Dementia. *Lancet* 1982; 2:200–204.

33. Wagner HN, Burns HD, Dannals RF, et al: Imaging dopamine receptors in the human brain by positron tomography. *Science* 1983; 221:1264–1266.

34. Wang DF, Wagner HN, Jr., Dannals RF, et al: The effects of age on dopamine and serotonin receptors measured by positron emission tomography in the living human brain. *Science* 1984; 226:1393–1396.

35. Severson JA, Marcusson J, Winblad B, et al: Age-correlated loss of dopaminergic binding sites in human based ganglia. *J Neurochem* 1982; 39:1623–1631.
36. Thal LJ, Horowitz SG, Dvorkin B, et al: Evidence of loss of brain [³H]spiroperidol and [³H]ADTN binding sites in rabbit brain with aging. *Brain Res* 1980; 192:185–194.
37. Maziere B, Loch C, Hantroye P, et al: ⁷⁶Br-Bromospiroperidol: A new tool for quantitative *in-vivo* imaging of neuroleptic receptors. *Life Sci* 1984; 35:1349–1356.
38. Hantroye P, Kaijima M, Prenant C, et al: Central type benzodiazepine binding sites: A positron emission tomography study in the baboon's brain. *Neurosci Lett* 1984; 48:115–120.
39. Whitehouse PJ, Price DL, Struble RG, et al: Alzheimer's disease and senile dementia: Loss of neurons in the basal forebrain. *Science* 1982; 215:1237–1239.
40. Whitehouse PJ, Price DL, Clark AW, et al: Alzheimer's disease. Evidence for selective loss of cholinergic neurons in the nucleus basalis. *Ann Neurol* 1983; 10:122–126.
41. Holman BL, Gibson RE, Hill TC, et al: Muscarinic acetylcholine receptors in Alzheimer's disease. *JAMA* 1985; 254:3063–3066.

PART II

Pharmacokinetics and Pharmacodynamics in the Elderly

CHAPTER 9

THE TRICYCLIC ANTIDEPRESSANTS

MATTHEW V. RUDORFER AND WILLIAM Z. POTTER

1. INTRODUCTION

Depression is a common, albeit probably underdiagnosed and under-treated, disorder of late adult life.[1] This illness affects an estimated 7–11% of individuals over 65 years of age.[2] When depression is properly diagnosed and treated, the prognosis for recovery is good, even in older patients. Pharmacotherapy remains the mainstay of treatment of affective illness in all age groups.

The nature of the disorder and of the treatment in depression are comparable for the older and the younger patient. However, there are special concerns in administering antidepressant drugs to the elderly. Psychological and physiological changes associated with aging, as reviewed in Part I, influence every aspect of antidepressant prescribing for the geriatric person: differential diagnosis, drug distribution and metabolism, and end-organ receptor response. Moreover, the frequent presence of concomitant physical illness and multiple medication use in the elderly complicate all these factors.[3]

These issues confront not only the clinician but also the investigator designing and executing drug trials in the elderly. In this chapter we will explore the application of current knowledge of diagnostic, phar-

---

MATTHEW V. RUDORFER AND WILLIAM Z. POTTER • Section on Clinical Pharmacology, Laboratory of Clinical Science, National Institute of Mental Health, Bethesda, Maryland 20892.

macokinetic, and pharmacodynamic factors necessary for optimal research methodology. Although our focus will be on the tricyclic antidepressants, other drugs used for treatment of affective illness in geriatric populations, including monoamine oxidase inhibitors and lithium, will also be addressed.

2. DIAGNOSIS

Homogeneity of study groups is essential in clinical research. Thus, accurate diagnosis of psychiatric illness is a necessary prerequisite for a successful antidepressant drug trial. A detailed discussion of differential diagnosis is beyond the scope of this chapter, but several general principles should be borne in mind.

The previously impressionistic diagnosis of affective illness has been replaced in the last decade and a half by reliable, standardized instruments utilizing operational criteria for diagnosis, including those by Feighner et al.,[4] Research Diagnostic Criteria,[5] and the third edition of the Diagnostic and Statistical Manual.[6] Specific age-related depressive syndromes, such as "involutional melancholia," have been eliminated from the nomenclature. Thus, a diagnosis of depression in an elderly individual would require the same criteria as in a younger person.

However, the borderline between affective illness and the expected decline in functioning, on the one hand, and frank organic disease, on the other, tends to be particularly blurred in the elderly.[7] In clinical practice the diagnosis of depression is often missed as deficits in memory or concentration may be inappropriately attributed to "normal" forgetfulness or, if severe, to "senility." In addition to or even in lieu of apparent dysphoria, the elderly depressive may present with apathy and vague physical complaints, which may be difficult to weave together as a depressive syndrome.

Even more challenging for the clinical investigator is making the distinction between "pure" depressive illness and that secondary to, or superimposed on, gross organic disease or dysfunction. Although initial reports were encouraging, use of the dexamethasone suppression test[8] or other biological variables to distinguish the "pseudodementia" of profound depression in the elderly from true organic brain syndrome has not proven consistently reliable. Careful medical screening is necessary to rule out the many physical disorders that may present as depression,[9] such as thyroid abnormalities, pernicious anemia, and Cushing's or Parkinson's disease, and which are more common in geriatric populations. Furthermore, it is increasingly uncommon to encounter an older patient who is medication-free at presentation. Moreover, a variety of frequently prescribed drugs have been implicated in inducing or aggravating mood changes in vulnerable individuals; these include antihypertensives (reserpine, propranolol, methyldopa, clonidine), antiparkinsonian agents with L-dopa, and corticosteroids.[3] A high index of suspicion combined with appropriately detailed history taking and medical evaluation is required to identify true affective illness.

One further comment is necessary in planning drug trials in depressed elderly patients. Depression often occurs in people suffering from other illnesses. Although it is necessary and preferable to study otherwise healthy psychiatric patients and normal volunteers, in addition it is important to understand how antidepressant drugs act in the physically impaired body and how they interact with other medications that cannot be discontinued. These aspects of antidepressant pharmacotherapy will be dealt with in this chapter.

3. PHARMACOKINETIC FACTORS

Tricyclic antidepressants (TCAs) are virtually completely absorbed on oral administration. Being very lipophilic drugs, they are widely distributed in the body. In the blood they are approximately 90% bound to plasma proteins, primarily α_1 acid glycoprotein. Metabolism is mainly through oxidative demethylation and hydroxylation by the hepatic cytochrome P-450 enzyme system. The hydroxy metabolites are then subject to renal excretion. Figure 1 illustrates the metabolic pathways for the prototypical TCA, imipramine (IMI). Alteration from the norm at any of these steps—whether endogenously or exogenously induced—can affect the amount of drug presenting to various receptors and thus the resulting degree of clinical response or toxicity.[10,11]

For example, let us look at the pharmacokinetic parameter volume of distribution (V). The already high V of TCAs is increased further in the elderly owing to the rising ratio of fat to muscle with advancing age.[12] Moreover, V is also dependent on the degree of plasma protein binding of drug. Insufficient dietary protein intake may result in a drop in plasma protein concentration. Or, concomitant somatic disease such as inflammation or malignancy may increase α_1 acid glycoprotein, causing a decreased free fraction and increased total plasma steady-state concentration (Css) of the tricyclic.[11] It should be noted that in the latter instance clearance and the unbound (pharmacologically active) drug concentration will not change (see Potter et al.[10] for discussion).

Turning from theory to the observed data, it does not appear that

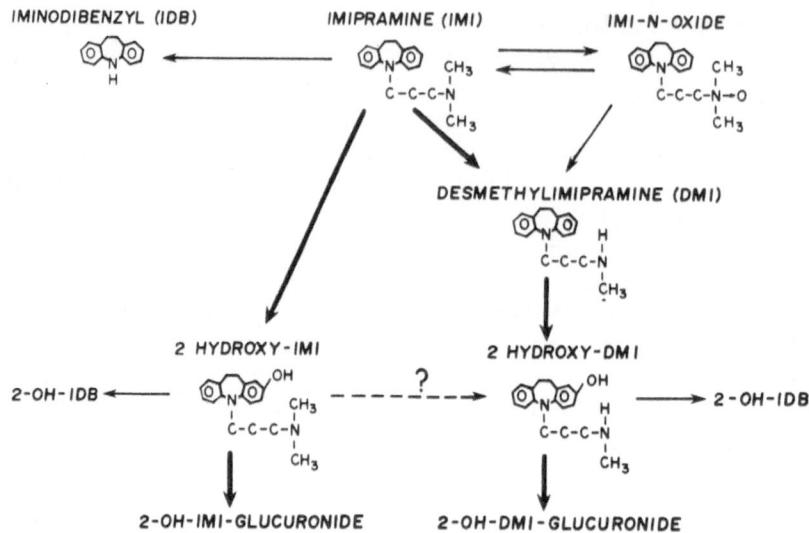

FIGURE 1   Major metabolic pathways for the tricyclic antidepressant imipramine.

TCA plasma protein binding is affected by age per se (Table I). Young and elderly male volunteers showed comparable binding after a single dose of amitriptyline.[13] In 69 patients between the ages of 20 and 97 years treated with imipramine, no significant relationship was found between age and plasma TCA binding.[14] Despite an increase in plasma α_1 acid glycoprotein concentration with advancing age, free imipramine concentrations were similar. On the other hand, *in vitro* protein binding of imipramine did correlate with α_1 acid glycoprotein concentrations when the drug was added to plasma samples from healthy volunteers and cardiac patients with two-to-three times higher protein levels.[15] Cutler *et al.*[16] found normal α_1 acid glycoprotein levels in a group of elderly depressed women undergoing desipramine (DMI) treatment. Interestingly, these patients displayed equivalent DMI plasma concentrations per unit dose as younger subjects. Thus medical illness may be more of a confounding factor than age per se in terms of plasma protein binding.

A much more commonly evaluated pharmacokinetic parameter than volume of distribution is the plasma concentration of a drug after single or chronic dosing. This would reflect the dose and clearance (Cl) of the drug. A related measure is the elimination half-life ($t_{1/2}$), which = 0.693 V/Cl. Thus the $t_{1/2}$ can reflect changes in either V or Cl.

Cl of antidepressant drugs occurs primarily in the liver, with subsequent elimination of metabolites by the kidney. A decline in activity

TABLE I   REPORTED EFFECTS OF ADVANCING AGE ON PHARMACOKINETICS
OF HETEROCYCLIC ANTIDEPRESSANTS

TCA	Kinetic parameter[a]						Ref.
	V	Cl	t1/2	Css	Protein binding	OH-metabolite plasma conc.	
All	↑						12
Amitriptyline		?↓	↑		0		13
				0			21, 22
				↑			24
		?↓		?↑			26
						↑	48
Nortriptyline	↓	↑					17
				0			20, 21, 22
				↑			23
						↑	48–50
Imipramine					0		14
			↑	↑			24
	↓	↑					19
				↑			27
						0	31
Desipramine				0	0		16
		0	0	0			18
						↑	45
						?↑	51
Clomipramine				↑			28, 29
Maprotiline	0	0					19
Mianserin				0			36

[a] 0 = no age effect; ? = nonsignificant trend.

of liver oxidative enzymes and decrease in hepatic blood flow, reducing the "first-pass" metabolism after drug absorption from the stomach and small intestine, are associated with aging.[12] The complexity and interaction of the multiple factors determining TCA kinetics in the elderly are illustrated in two studies comparing single doses of antidepressants in old versus young subjects. Using amitriptyline (AT), Schulz et al.[13] observed a slightly (10%) but not significantly lower Cl in elderly as compared to young male volunteers. Mean $t_{1/2}$, however, was considerably (34%) elevated in the elderly, based not only on the reduction in Cl, but on an increase in V. Cl was slower and $t_{1/2}$ prolonged after single-dose nortriptyline (NT) in a group of elderly depressed inpatients,[17] but the presence of significant medical illness and concomitant polypharmacy compared to the young healthy volunteer group limited interpretation. These confounding variables, of great importance in conducting antidepressant trials in the elderly, will be discussed further below. Probably due in part to the multitude of potential confounding variables, at present there is considerable disagreement in the literature regarding the presence and extent of pharmacokinetic changes referable to age during antidepressant treatment, as outlined in Table I.

Patients in their sixties showed no abnormalities in single-dose DMI kinetics.[18] However, five women in their seventies and eighties demonstrated prolonged $t_{1/2}$, slower Cl, reduced rate of demethylation, and lower first-pass removal following a single dose of IMI compared to published norms in younger subjects.[19] Interestingly, kinetic parameters of a single dose of the tetracyclic compound maprotiline in the same subjects were similar to those of IMI, albeit more in line with reports in younger individuals.

Åsberg and her colleagues[20] administered NT to 49 psychiatrically healthy individuals, most of them relatives of patients known to achieve unusually high plasma NT steady-state concentrations. No effect of age (ranging from 16 to 75 years) on NT kinetics could be demonstrated. Other studies in naturalistic patient settings have yielded mixed results. No significant effect of age on steady-state concentrations has emerged in some studies of AT and NT.[21,22] On the other hand, Kragh-Sørensen and Larsen[23] found higher NT levels over age 70, and Nies et al.[24] reported higher TCA concentrations in old versus young patients when AT was prescribed; for NT there was no difference. In a relatively young group of patients[25] age was unrelated to plasma TCA concentrations after overdose. Using a nomogram constructed from single-dose data, Dawling et al.[26] found a mean AT routine daily dose of only 62 mg sufficient to produce therapeutic Css in 15 depressed elderly inpatients.

In the frequently cited study of Nies et al.,[24] compared to young patients, those over 65 years of age showed an almost threefold increase in Css of IMI and DMI, with longer DMI $t_{1/2}$s. As noted, Potter et al.[18] saw no age effects on DMI kinetics. Although the elderly women depressives in the study of Cutler et al.[16] did not show disproportionately high DMI concentrations per 100 mg dose, it is noteworthy that most showed clinical response at low doses (25 mg b.i.d.) with correspondingly low Css, indicating that simple kinetic factors alone could not account for any age difference in DMI effects. Gram et al.[27] found generally higher IMI levels with advancing age, particularly in men.

Two fixed-dose clomipramine (CMI) studies also yielded higher drug concentrations in older patients. This relationship was nonlinear in the subjects of Träskman and associates,[28] with a steep rise of CMI levels in the upper age group. The desmethyl metabolite did not change in older subjects. In a similar investigation[29] endogenously depressed patients older than 65 years developed higher Css of CMI and desmethyl CMI than those less than 40 years old. There was even an age differential among the elderly: those above 75 years had higher drug levels than patients between 65 and 75 years of age.

The metabolism of antidepressant drugs generally follows first-order (linear) kinetics, with drug elimination proportional to dose.[10] A provocative report from Odense, Denmark[30] described dose-dependent kinetics of therapeutic doses of IMI in elderly depressed patients. In these individuals elevations of the demethylated metabolite, i.e., DMI plasma levels, were disproportionately greater than the linear rises in parent drug concentrations with increases in IMI dose. NT, on the other hand, showed first-order kinetics at all dosages. These authors hypothesized saturability of IMI and DMI hydroxylation but not demethylation pathways to account for their findings.[31] At least for DMI levels up to 150 ng/ml, our group[32] has not found dose dependence of DMI hydroxylation. Anecdotal reports of disproportionately high Css of DMI in older patients have continued to appear in the literature. Recently, for example, Dugas and Bishop[33] documented a case of a 65-year-old man with linear DMI kinetics in the 100 to 250 mg/day range and nonlinear kinetics at higher doses. This was interpreted to represent saturation of metabolic pathways at the higher dose levels, although metabolite concentrations were not reported. In nine young adult depressives, nonlinear kinetics of DMI could not be accounted for by the hydroxy metabolite, which did increase proportionately to dose.[34] Thus, the question of quantitative and qualitative differences of TCA metabolism in the geriatric population remains open and subject to further study.

The newer "second-generation" antidepressants[35] are metabolized much like the standard TCAs. For most, a specific age effect in kinetics has not been investigated. Mianserin has generally produced similar Css across age groups.[36] Prior to its withdrawal from the clinical and research markets because of suspected neurotoxicity,[35] the serotonin reuptake inhibitor zimelidine was associated with prolonged $t_{1/2}$s and increased Css for both the parent drug and the demethylated metabolite in older versus younger patients.[37] Longer $t_{1/2}$s and reduced Cl were seen in elderly men after a single dose of the triazolobenzodiazepine alprazolam.[38]

The "marker drug" debrisoquine has been used in genetic studies to identify a subpopulation of slow hydroxylators[39,40] but not as a probe for altered metabolism in aging. It would be interesting to see whether the distribution of debrisoquine hydroxylation was the same in an aged group as in the younger populations studied thus far. Using such a phenotype indicator, one could attempt to factor out the role of oxidative hepatic processes per se in contributing to slower drug Cl in the elderly. If changes in the hydroxylation pathway as reflected in the debrisoquine "metabolic ratio" were insufficient to account for age-related differences in total Cl of antidepressants, the importance of alternative routes of drug metabolism or diminished renal Cl in older persons would be strengthened.

The above alterations in pharmacokinetics in the aged have generally been ascribed to reductions in the rate of hepatic biotransformation of drugs. Only relatively recently has the role of renal function in antidepressant metabolism been fully appreciated. It has long been recognized that lithium (Li) does not undergo metabolism but is nearly completely excreted through the kidneys. Thus, the normal decline in glomerular filtration rate (GFR) with age causes up to a 60% decrease in Li renal Cl, with lengthening of $t_{1/2}$ up to 36 hr in the elderly versus 24 hr in middle-aged adults.[41] Consequently, there was a significant, though modest, correlation between age and relative serum Li level in one outpatient study.[42]

Although unchanged TCAs are not subject to renal excretion, their metabolites are. With the demonstration of biological activity of tricyclic hydroxy metabolites[43] came realization of the clinical significance of changes in the elimination rate of these substances. It was originally shown by Alexanderson and Borgå[44] that the renal Cl of hydroxy NT was over 100 ml/min. A similarly high urinary Cl has been shown for unconjugated hydroxy DMI[45] with a decrease in elderly depressed patients ($r = -0.73$ between age and 2-OH-DMI renal Cl, Fig. 2) accompanied by an increase in the plasma OH-DMI/DMI ratio (Table I). This is interpreted to mean that the rate of DMI hydroxylation was not altered, since DMI plasma concentrations were the same in the elderly as in younger populations.[16] Even without truly elderly subjects[46] our Chinese and Caucasian healthy volunteers demonstrated after single-dose DMI a significant reduction in renal Cl of OH-DMI with increasing age.[47] A weaker, but still significant, association has been noted between age and plasma OH-NT/NT ratio in patients treated with NT or AT,[48-50] despite unchanged levels of the unmetabolized drug. Absolute plasma concentrations of 10-OH-NT were twice as high in elderly as in young adult patients during NT treatment in two studies.[49,50] Relating these findings to declining renal excretion of the metabolite, Young and associates[50] noted a positive, though limited ($r = 0.40$), correlation between the plasma OH-NT/NT ratio and serum creatinine concentrations in all subjects. A group of IMI-treated patients that included 21% slow hydroxylators[31] did not show this higher hydroxy metabolite ratio in the elderly. In 45

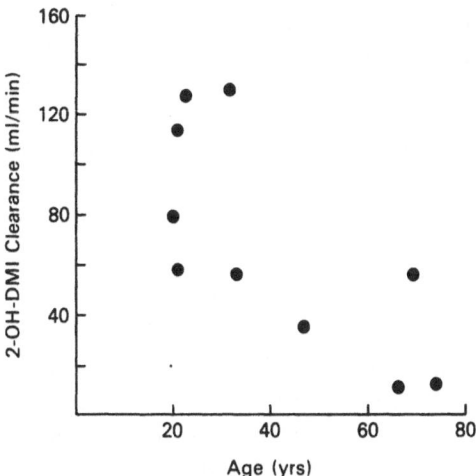

FIGURE 2   Relationship between age and renal clearance of the hydroxy metabolite during desipramine treatment ($r = -0.73$, $p < 0.05$).[45]

DMI-treated patients with a nearly fourfold age range[51] there was as well only a weak trend toward higher relative levels of hydroxy metabolite with increasing age. Therapeutic or toxic actions mediated by TCA hydroxy metabolites thus may be exaggerated in geriatric patients unless allowances are made in dosing. Furthermore, state-of-the-art pharmacokinetic monitoring during drug trials in the elderly now calls for routine measurement of active metabolite concentrations, especially OH-NT in patients treated with NT.

Although not frequently prescribed for older patients because of their potential toxicity, the monoamine oxidase inhibitors (MAOIs) may have great potential therapeutic value in this population. In theory this may relate to the increased MAO levels observed in platelets and in the brain with aging.[52] Pharmacokinetic data for MAOIs in geriatric populations are limited. As reviewed elsewhere,[11] the metabolism of MAOIs is complex and difficult to study, involving acetylation, formation of phenylethylamine metabolites and amphetaminelike compounds, and possibly self-inhibition of degradative processes. Since available MAOIs are irreversible inhibitors of the enzyme, which takes weeks to be resynthesized, pharmacokinetics will have less clinical relevance; effect will be related to total amount of drug and not to steady-state concentrations.

Robinson[53] retrospectively reviewed his group's experience with phenelzine in double-blind studies. He found a tendency for higher phenelzine plasma concentrations in older patients, with no effect of age on drug acetylactor subtype, and postulated a slower rate of phenelzine Cl in the elderly. An indirect bioassay of pharmacologically active drug concentration useful for a "hit-and-run" drug (irreversible inhibitor) such as an MAOI is measurement of the reduction of the norepinephrine metabolite 3-methoxy-4-hydroxyphenylglycol (MHPG). In our investigations with the type A MAOI clorgyline, the degree of MHPG reduction in the cerebrospinal fluid during treatment was not related to age.[54]

3.1. Drug Interactions

Polypharmacy is frequent among older persons. Other drugs often interact dynamically or kinetically with antidepressants. For the clinical investigator of antidepressant drugs, this cuts both ways. Ideally, clinical trials should be conducted in individuals not receiving any drugs other than the one under study. In addition, though, it is important to remember that actual clinical use of these compounds will include their combination with other substances. This, too, warrants systematic study.

The nature and effects of such kinetic interactions have been reviewed in detail[10,11] and can be briefly summarized here. Most commonly, drugs that stimulate hepatic oxidizing enzymes will enhance the degradation of TCAs and thereby lower their Css. Such agents include the barbiturates and, possibly, anticonvulsants,[55] though not benzodiazepines.[56] Alternatively, drugs that compete for the same liver enzyme system will slow TCA Cl. Thus Css of TCAs are higher when these antidepressants are combined with antipsychotic compounds,[57] the central stimulant methylphenidate,[58] or the histamine H_2-receptor antagonist cimetidine.[59]

The dependence of Li Cl on renal excretion means that concomitant drug treatment with agents that affect the kidney can produce interactions with Li. Such drugs include sodium-depleting diruetics, which lead to increased Li reabsorption and higher Li blood levels.[12] In addition, a variety of nonsteroidal antiinflammatory agents, apparently by inhibiting prostaglandin-dependent renal mechanisms, can decrease Li Cl with resultant elevation in Li plasma concentrations.[60]

3.2. Coexisting Organic Pathology

The presence of physical illness, not uncommon in the elderly, may aggravate age-related kinetic changes in antidepressant metabolism or introduce new abnormalities. Any step in the process may be affected. Thus, delayed gastric emptying may cause increased exposure of the weakly basic TCAs to gastric acidity, reducing speed of absorption, whereas rapid movement of gastric contents can increase the amount of TCAs presented to the small intestine for absorption.[61] Some intestinal disorders may directly limit absorption, though inflammatory bowel disease may be associated with increased vascularity or membrane damage causing increased amounts of drug to enter the blood. Nonetheless, since overall absorption of antidepressants is so high, peak concentrations and transient side effects only, and not Css, are likely to be affected.

As already noted, changes in plasma proteins can affect the highly bound TCAs. Hepatic and renal disease, as well as malnutrition, can reduce plasma proteins. On the other hand, a number of illnesses, including postoperative states, inflammatory disease,[23] especially rheumatoid arthritis and Crohn's disease, and cancer,[62] are associated with increased plasma protein binding and consequently higher total TCA concentrations. In monitoring drug concentrations and response in clinical trials, the investigator must note that in these altered states of binding the free, pharmacologically active concentration of drug generally remains constant,[10] rendering clinical effect unchanged but interpretation of (total) TCA blood levels tenuous. Acidosis from any cause may reduce binding affinity, thereby leading to higher concentrations of unbound drug; the opposite is true of alkalosis.

Despite the primary role of the liver as the site of oxidative metabolism of most antidepressants, reduced drug metabolism does not necessarily occur even in severe hepatic impairment. The cytochrome P-450 enzyme system generally functions in diseased as well as healthy liver tissue. A single case report of disproportionately high AT and NT Css in a man with biopsy-proven alcoholic cirrhosis[63] could only speculate that there was a causal relationship between the two phenomena and has not been replicated. Systematic studies of antidepressant metabolism by diseased livers in any age range are lacking.

Blood flow changes may directly or indirectly alter hepatic drug metabolism. Cl of highly extracted drugs such as the TCAs is particularly sensitive to changes in hepatic blood flow. Congestive heart failure has been suggested[64] to induce alterations in hepatic blood flow, with consequent reduction in demethylation of TCAs. Such cardiac patients, when treated with IMI, yielded higher plasma concentrations of IMI and DMI than healthy controls. Again, controlled studies are needed.

Lithium kinetics are susceptible to even minor changes in excretory function since this ion is cleared completely by renal elimination without biotransformation. The interaction between renal disease and Li retention is well established, with elevated Li serum concentrations in patients with diminished renal Cl.[65]

In contrast, throughout most of their history TCAs have been thought to depend little on renal elimination as a mode of termination of their activity, as it was believed these drugs were nearly completely metabolized to inactive by-products prior to reaching the kidney.[61] However, with the recognition of biological activity of unconjugated hydroxy TCA metabolites[43] came realization of the functional importance of renal Cl of these compounds.[47] Indeed, renal Cl of these metabolites, measured as high as 500 ml/min, greatly exceeds GFR, suggesting active secretion.[66] The clinical significance of reduced GFR and the resulting buildup of plasma concentrations of pharmacologically active TCA hydroxy metabolites has been reviewed here in the context of the physically healthy elderly; parallel changes follow from impairment of excretory function secondary to kidney disease or drugs. For instance, the elimination $t_{1/2}$ of the second-generation antidepressant nomifensine was prolonged in marked renal failure.[67]

Renal Cl of the hydroxy metabolite of the β-receptor antagonist metoprolol, which is metabolized in a similar fashion to the TCAs, appeared reduced in elderly healthy volunteers[68] and correlated directly with GFR in four patients with impaired renal function.[69] In an initial study of 20 patients with chronic renal failure who were administered a single dose of NT, Dawling et al.[70] found no changes in parent drug Cl or $t_{1/2}$ in patients compared to controls. These investigators speculated that hydroxy metabolites of NT might accumulate during renal insufficiency and proceeded to subsequently demonstrate this phenomenon, with elevated plasma concentrations of 10-OH-NT, in some of these same patients.[71]

In addition to these alterations in kinetics, it is necessary to identify determinants of the partition of antidepressant drugs into the central nervous system and to understand the effects of aging on each determinant. Factors other than peripheral drug metabolism that may influence brain drug levels include the drug's water–octanol partition coefficient and the brain lipid composition, the integrity of the blood–brain barrier, and cerebral blood flow. Changes in all these parameters occur in aging and require systematic clinical study.

4. PHARMACODYNAMIC FACTORS

Having accounted for the kinetic parameters, one still is left with a variety of desired as well as unwanted effects of antidepressant drugs in the elderly, related directly or indirectly to age.

4.1. INTRINSIC DRUG ACTIONS

Appropriately diagnosed depression is as responsive to pharmacotherapy in older as in younger patients.[3] Lower doses than usual may be effective in the elderly. We have already seen how pharmacokinetic factors may account for this finding in some aged individuals by producing a disproportionately high drug concentration from a relatively small daily dose. However, this is not always the case. Cutler et al.[16] demonstrated good clinical response in depressed elderly women despite low doses with commensurately "subtherapeutic" plasma drug concentrations. This suggests that the substrate for drug activity may be changed in the elderly.

All effective antidepressant drugs appear to influence the noradrenergic system,[72] even when the initial biochemical action is directed toward a different neurotransmitter system.[73] In turn, changes in the noradrenergic system are seen with aging. Plasma norepinephrine (NE) concentration is elevated in old as compared to young subjects,[74] with a weak, but significant, positive correlation between plasma norepinephrine and age. This relationship holds as well for individuals with depression and/or cardiac disease.[75] The main metabolite of norepinephrine, MPHG, is also increased in the urine and cerebrospinal fluid of drug-free older patients.[47] The most consistent functional finding with advancing age involves diminished responsiveness of β-adrenoceptor-mediated events. One of the observations leading to this conclusion was that larger doses of the β agonist isoproterenol were required to increase heart rate by a fixed amount, particularly in subjects over 70 years of age.[76] In addition, the effectiveness of any given free concentration of the nonspecific β blocker propranolol in blunting the isoproterenol-induced tachycardia was reduced with increasing age. Similarly, human lymphocyte response as measured by isoproterenol-stimulated cyclic AMP is decreased with aging. Although the actual number of β receptors is unchanged on lymphocytes from elderly as compared to young subjects, the proportion of β receptors in the "high" (versus "low") affinity state is diminished.[77] At this time it cannot be stated whether the central components of the noradrenergic system in humans change with aging, although in aged rodents the total number of β-receptors in the cortex is decreased.

Against this backdrop preliminary efforts to study antidepressant effects on noradrenergic function in the elderly have been conducted. Several aged women treated with DMI were compared with younger patients from other studies.[78] Despite an initially very high plasma NE concentration measured in the supine position, there was a significant mean rise of 28% with treatment. This DMI-induced increase of plasma NE concentrations in elderly depressed patients was seen to be consistent with and of the same order of magnitude as those observed in other settings with younger subjects, despite the latter's lower baseline NE levels. These comparison groups include young, healthy volunteers receiving chronic DMI,[79] enuretic prepubescent males treated with IMI or DMI,[80] and middle-aged patients with or without depression and heart disease treated with a variety of tertiary amine TCAs.[75] Cutler *et al.*[78] ruled out a pharmacokinetic effect by use of DMI and OH-DMI plasma concentrations. On the other hand, we have found a decrease in plasma NE induced by a single 100-mg dose of DMI in middle-aged, healthy volunteers (versus an increase in plasma NE in young subjects), with control for pharmacokinetic differences.[81] The implications of this finding for the elderly require further study with healthy aged subjects.

It is in terms of actual noradrenergic function that the age effect on TCA pharmacodynamics is most evident. In the face of the significant elevation of plasma NE concentrations during DMI treatment, the elderly women studied by Cutler *et al.*[78] showed no alteration of baseline heart rate and blood pressure. This is in contrast to the significant increases in supine heart rate and blood pressure in younger subject groups[79,80] or the rise in supine heart rate in the middle-aged patients of Veith *et al.*[75] These findings were interpreted to reflect a reduction of β-adrenoceptor function associated with aging. Concordant with these findings are those by Branconnier *et al.*,[82] wherein physically healthy depressives showed a significant negative correlation between heart rate and age during fixed-dose AT therapy. Controlled studies utilizing antidepressant drugs in healthy elderly volunteers could help clarify the role of depression- versus age-related noradrenergic dysfunction in determining pharmacological action in this population.

4.2. COEXISTING ORGANIC PATHOLOGY

The plethora of antidepressant drug actions may differ in diseased and normal tissue. In many cases physical illness serves to further exaggerate an apparently age-related pattern of differential prominent therapeutic or adverse effects. For instance, the quinidinelike slowing of intracardiac conduction by the TCAs is expressed as a silent electrocardiographic finding in physically healthy individuals[83] but as a frank antiarrhythmic effect in cardiac patients with premature ventricular contractions.[84]

As with pharmacokinetic factors, there is a great deal of interindividual variability in sensitivity to antidepressant effects in the physically ill. Thus, 24 depressed patients with well-documented chronic heart disease tolerated IMI or doxepin well, with no demonstrable adverse effects on hemodynamic function.[85] On the other hand, a 68-year-old depressed man with no history of cardiovascular symptoms but with left bundle branch block on his admission electrocardiogram developed congestive heart failure during treatment with nortriptyline.[86] It was speculated that the disproportionately high plasma concentration of 10-OH-NT found in this patient may have contributed to the acute cardiac decompensation, an example of the interaction between pharmacokinetic and -dynamic factors.

Blood pressure change with antidepressants may also reflect an underlying vulnerability in the patient as much as a direct drug action. Glassman and his associates[87] were surprised to find that the best predictor of orthostatic hypotension during IMI therapy was the degree of pretreatment orthostatic blood pressure change. In contrast, the variables of age, preexisting cardiac disease, and drug plasma concentration, although expected to have predictive power, had no bearing on the IMI-related blood pressure drop. Still, from a practical point of view, should orthostatic hypotension develop, the effect may be more serious in the older person who is more prone to fall and to suffer fractured bones. Similarly, the danger of a hypertensive crisis resulting from ingestion of pressor amines in food or other drugs in an elderly patient taking an MAOI may depend less on the person's age per se or even the specific pharmacological actions but rather on the status of the individual's vasculature,[12] with fragile, atherosclerotic cerebral vessels more likely to rupture. From a preventive point of view, impaired cognitive function, whether resulting from age, depression, or a combination of the two, must be viewed as a limiting factor to use of the MAOIs with their unyielding dietary/medication compliance requirements.

Anticholinergic side effects of current antidepressant medications may be particularly troublesome in older patients, who may already be experiencing dry mouth and constipation as depressive symptoms.[7] Again, the patient's underlying state of health may determine whether these adverse effects are nuisance symptoms or serious complications. For many aged individuals, severe dryness of the mouth may interfere with proper fitting of dentures. For the older man with prostatic hypertrophy, drug-induced anticholinergic activity may precipitate urinary retention. Glaucoma may also be aggravated. A central nervous system anticholinergic toxic syndrome may also be more common in elderly patients treated with TCAs.[12] This will present as an acute organic brain syndrome (delirium) with agitation, disorientation, confusion, memory loss, and visual and tactile hallucinations. Once more, in addition to the higher risk of such toxicity in the aged central nervous system, one must remain attentive to pharmacokinetic variability. A retrospective review of 125 TCA-treated inpatients[88] determined plasma drug level to be positively associated with documented tricyclic-induced delirium. In that report, delirium occurred in 35% of patients with a TCA (generally AT) plasma concentration greater than 300 ng/ml. Interestingly, Livingston and associates[88] identified both age and race as independent risk factors, with delirium during TCA treatment significantly more common in older and in black patients as well as in those with higher drug levels.

4.3. ADDITIVE DRUG EFFECTS

Most pharmacodynamic interactions of antidepressants with other drugs in older patients consist of additive adverse effects. For example, sedation from hypnotics or analgesics, anticholinergic activity from antispasmodic agents, or increased cardiac conduction time or even heart block may result from the combination of TCAs and medications used to treat the other physical illnesses common in the elderly. Further complicating matters, drugs prescribed for organic disease may produce or aggravate psychiatric symptomatology. The renal disease patients we spoke of in the context of reduced excretion of certain drugs and metabolites are often treated with concomitant medication, e.g., steroids and antihypertensives, which are known to precipitate depression in vulnerable individuals. The pharmacokinetic implications of polypharmacy have already been noted, including possible steroid-induced increase in antidepressant metabolism via induction of hepatic enzymes. Salzman[12] pointed out that increased drinking by an elderly patient suffering from TCA-induced dry mouth can lead to water intoxication in the presence of a diuretic, which can impair free-water excretion. Clearly, physical health and medication status must be carefully assessed in any antidepressant trial. Furthermore, although some of the newer antidepressants offer a reduced side effect profile, e.g., less anticholinergic activity with trazodone or bupropion,[35] an "ideal" antidepressant for the elderly does not exist at present.

In conclusion, the study of antidepressant drugs in the elderly is more similar to than different from that in other age groups. Working with aged subjects, the investigator is confronted with a number of additional sources of variance on both a pharmacokinetic and pharmacodynamic basis. These pose obstacles only if unaccounted for in the research design. With careful attention to, and controls for, the variables within subjects, including physical status and use of additional drugs, as well as proper monitoring of outcome, with measurement of active metabolites, valuable studies can be achieved with old and new antidepressant compounds. Of primary importance, the homogeneity of subject groups must be assured, whether these consist of healthy volunteers or depressed patients with coexisting organic illness. Through such clinical research, the goal of more effective and less toxic antidepressant agents for the elderly can be meaningfully sought.

REFERENCES

1. Fassler LB, Gaviria M: Depression in old age. *J Am Geriatr Soc* 1978; 26:471–479.
2. Gurland BJ: The comparative frequency of depression in various adult age groups. *J Gerontol* 1976; 31:283–292.
3. Ouslander JG, Small GW: Management of depression in the elderly patient with physical illness. *Geriatr Med Today* 1984; 3:90–96.
4. Feighner JP, Robins E, Guze SB, et al: Diagnostic criteria for use in psychiatric research. *Arch Gen Psychiatry* 1972; 25:57–63.
5. Spitzer RL, Endicott J, Robins E: *Research Diagnostic Criteria for a Selected Group of Functional Disorders*, ed 2. New York, Biometrics Research Division, New York State Psychiatric Institute, 1981.
6. American Psychiatric Association, Task Force on Nomenclature and Statistics: *Diagnostic and Statistical Manual of Mental Disorders*, ed 3. Washington, DC, American Psychiatric Association Press, 1980.
7. Lehmann HE: Affective disorders in the aged. *Psychiatr Clin North Am* 1982; 5:27–44.
8. Rudorfer MV, Clayton PJ: Pseudodementia: Use of the DST in diagnosis and treatment monitoring. *Psychosomatics* 1982; 23:429–431.
9. Bressler R: Treating geriatric depression: Current options. *Drug Ther* 1984; 14:129–144.
10. Potter WZ, Bertilsson L, Sjöqvist F: Clinical pharmacokinetics of psychotropic drugs: Fundamental and practical aspects, in van Praag HM, Rafaelson O, Lader M, Sachar E (eds): *The Handbook of Biological Psychiatry: Part VI, Practical Applications of Psychotropic Drugs and Other Biological Treatments*. New York, Marcel Dekker, Inc., 1981, pp 71–134.
11. Rudorfer MV, Potter WZ: Metabolism of drugs used in affective disorders, in Dewhurst WG, Baker GB (eds): *Pharmacotherapy of Affective Disorders: Theory and Practice*. London, Croom Helm Ltd, 1985, pp 382–448.
12. Salzman C: Primer on geriatric psychopharmacology. *Am J Psychiatry* 1982; 139:67–74.
13. Schulz P, Turner-Tamiyasu K, Smith G, et al: Amitriptyline disposition in young and elderly normal men. *Clin Pharmacol Ther* 1983; 33:360–366.
14. Abernethy DR, Kerzner L: Age effects on alpha-1-acid glycoprotein concentration and imipramine plasma protein binding. *J Am Geriatr Soc* 1984; 32:705–708.
15. Freilich DI, Giardina E-GV: Imipramine binding to alpha-1-acid glycoprotein in normal subjects and cardiac patients. *Clin Pharmacol Ther* 1984; 35:670–674.
16. Cutler NR, Zavadil AP III, Eisdorfer C, et al: Concentrations of desipramine in elderly women. *Am J Psychiatry* 1981; 138:1235–1237.
17. Dawling S, Crome P, Braithwaite R: Pharmacokinetics of single oral doses of nortriptyline in depressed elderly hospital patients and young healthy volunteers. *Clin Pharmacokin* 1980; 5:394–401.
18. Potter WZ, Zavadil AP III, Kopin IJ, et al: Single-dose kinetics predict steady-state concentrations of imipramine and desipramine. *Arch Gen Psychiatry* 1980; 37:314–320.
19. Hrinda PD, Rovei V, Henry JF, et al: Comparison of single-dose pharmacokinetics of imipramine and maprotiline in the elderly. *Psychopharmacology* 1980; 70:29–34.
20. Åsberg M, Price Evans DA, Sjöqvist F: Genetic control of nortriptyline kinetics in man. A study of propositi with high plasma concentrations. *J Med Genetics* 1971; 8:129–135.

21. Ziegler VE, Biggs JT: Tricyclic plasma levels: Effect of age, race, sex, and smoking. *JAMA* 1977; 238:2167–2169.
22. Linnoila M, George L, Guthrie S, et al: Effect of alcohol consumption and cigarette smoking on antidepressant levels of depressed patients. *Am J Psychiatry* 1981; 138:841–842.
23. Kragh-Sørensen P, Larsen N-E: Factors influencing nortriptyline steady-state kinetics: Plasma and saliva levels. *Clin Pharmacol Ther* 1980; 28:796–803.
24. Nies A, Robinson DS, Friedman MJ, et al: Relationship between age and tricyclic antidepressant plasma levels. *Am J Psychiatry* 1977; 134:790–793.
25. Rudorfer MV, Robins E: Amitriptyline overdose: Clinical effects on tricyclic antidepressant plasma levels. *J Clin Psychiatry* 1982; 43:457–460.
26. Dawling S, Ford S, Rangedara DC, et al: Amitriptyline dosage prediction in elderly patients from plasma concentration at 24 hours after a single 100 mg dose. *Clin Pharmacokin* 1984; 9:261–266.
27. Gram LF, Sondergaard I, Christiansen J, et al: Steady-state kinetics of imipramine in patients. *Psychopharmacology* 1977; 54:255–261.
28. Träskman L, Åsberg M, Bertilsson L, et al: Plasma levels of chlorimipramine and its desmethyl metabolite during treatment of depression. *Clin Pharmacol Ther* 1979; 26:600–610.
29. John VA, Luscombe DK, Kemp H: Effects of age, cigarette smoking and the oral contraceptive on the pharmacokinetics of clomipramine and its desmethyl metabolite during chronic dosing. *J Int Med Res* 1980; 8(suppl 3):88–95.
30. Bjerre M, Gram LF, Kragh-Sørensen P, et al: Dose-dependent kinetics of imipramine in elderly patients. *Psychopharmacology* 1981; 75:354–357.
31. Gram LF, Bjerre M, Kragh-Sørensen P, et al: Imipramine metabolites in blood of patients during therapy and after overdose. *Clin Pharmacol Ther* 1983; 33:335–342.
32. Potter WZ, Calil HM, Sutfin TA, et al: Active metabolites of imipramine and desipramine in man. *Clin Pharmacol Ther* 1982; 31:393–401.
33. Dugas JE, Bishop DS: Nonlinear desipramine pharmacokinetics: A case study. *J Clin Psychopharmacol* 1985; 5:43–45.
34. Cooke RG, Warsh JJ, Stancer HC, et al: The nonlinear kinetics of desipramine and 2-hydroxydesipramine in plasma. *Clin Pharmacol Ther* 1984; 36:343–349.
35. Rudorfer MV, Golden RN, Potter WZ: Second-generation antidepressants. *Psychiatr Clin North Am* 1984; 7:519–534.
36. Brogden RN, Heel RC, Speight TM, et al: Mianserin: A review of its pharmacological properties and therapeutic efficacy in depressive illness. *Drugs* 1978; 16:273–301.
37. Heel RC, Morley PA, Brogden RN, et al: Zimelidine: A review of its pharmacological properties and therapeutic efficacy in depressive illness. *Drugs* 1982; 24:169–206.
38. Greenblatt DJ, Divoll M, Abernethy DR, et al: Alprazolam kinetics in the elderly. *Arch Gen Psychiatry* 1983; 40:287–290.
39. Mellström B, Bertilsson L, Säwe J, et al: E- and Z-10-hydroxylation of nortriptyline: Relationship to polymorphic debrisoquine hydroxylation. *Clin Pharmacol Ther* 1981; 30:189–193.
40. Rudorfer MV, Lane EA, Potter WZ: Interethnic dissociation between debrisoquine and desipramine hydroxylation. *J Clin Psychopharmacol* 1985; 5:89–92.
41. Prien RF: Age-related changes in lithium pharmacokinetics, in Raskin A, Robinson DS, Levine J (eds): *Age and the Pharmacology of Psychoactive Drugs.* New York, Elsevier, 1981, pp 163–169.
42. Slater V, Milanes F, Talcott V, et al: Influence of age on lithium therapy. *South Med J* 1984; 77:153–158.

43. Potter WZ, Calil HM: Metabolites of tricyclic antidepressants: Biological activity and clinical implications, in Usdin E (ed): *Clinical Pharmacology in Psychiatry*. New York, Elsevier, 1981, pp 311–324.
44. Alexanderson B, Borgå O: Urinary excretion of nortriptyline and five of its metabolites in man after single and multiple doses. *Eur J Clin Pharmacol* 1973; 5:174–180.
45. Kitanaka I, Ross RJ, Cutler NR, et al: Altered hydroxydesipramine concentrations in elderly depressed patients. *Clin Pharmacol Ther* 1982; 31:51–55.
46. Rudorfer MV, Lane EA, Chang W-H, et al: Desipramine pharmacokinetics in Chinese and Caucasian volunteers. *Br J Clin Pharmacol* 1984; 17:433–440.
47. Potter WZ, Lane EA, Rudorfer MV: Hydroxy metabolite concentrations: Role of renal clearance, in Gram LF, Usdin E, Dahl SG, et al (eds): *Clinical Pharmacology in Psychiatry: Bridging the Experimental-Therapeutic Gap*. London, Macmillan Press, 1983, pp 203–216.
48. Sjöqvist F: General issues related to age and the pharmacology of psychoactive drugs, in Raskin A, Robinson DS, Levine J (eds): *Age and the Pharmacology of Psychoactive Drugs*. New York, Elsevier, 1981, pp 195–204.
49. Pottash ALC, Martin DM, Extein I, et al: The prediction of therapeutic nortriptyline dosage regimes and related plasma concentrations of hydroxylated metabolites in geriatric depressives. *Abstr Soc Neurosci* 1983; 9:428.
50. Young RC, Alexopoulos GS, Shamoian CA, et al: Plasma 10-hydroxynortriptyline in elderly depressed patients. *Clin Pharmacol Ther* 1984; 35:540–544.
51. Bock JL, Nelson JC, Gray S, et al: Desipramine hydroxylation: Variability and effect of antipsychotic drugs. *Clin Pharmacol Ther* 1983; 33:322–328.
52. Robinson DS, Sourkes JL, Nies A, et al: Monoamine metabolism in human brain. *Arch Gen Psychiatry* 1977; 34:89–92.
53. Robinson DS: Monoamine oxidase inhibitors and the elderly, in Raskin A, Robinson DS, Levine J (eds): *Age and the Pharmacology of Psychoactive Drugs*. New York, Elsevier, 1981, pp 157–162.
54. Potter WZ, Scheinin M, Golden RN, et al: Selective antidepressants and cerebrospinal fluid: Lack of specificity on norepinephrine and serotonin metabolites. *Arch Gen Psychiatry* 1985, 42:1171–1177.
55. Nawishy S, Hathway N, Turner P: Interactions of anticonvulsant drugs with mianserin and nomifensine. *Lancet* 1981; 2:871–872.
56. Ballinger BR, Presly A, Reid AH, et al: The effects of hypnotics on imipramine treatment. *Psychopharmacologia (Berlin)* 1974; 39:267–274.
57. Gram LF: Factors influencing the metabolism of tricyclic antidepressants. *Dan Med Bull* 1977; 24:81–89.
58. Wharton RN, Perel JM, Dayton PG, et al: A potential clinical use for methylphenidate with tricyclic antidepressants. *Am J Psychiatry* 1971; 127:1619–1625.
59. Miller DD, Macklin M: Cimetidine-imipramine interaction: Case report and comments. *Am J Psychiatry* 1984; 141:153.
60. Ragheb M, Ban TA, Buchanan D, et al: Interaction of indomethacin and ibuprofen with lithium in manic patients under a steady-state lithium level. *J Clin Psychiatry* 1980; 41:397–398.
61. Siris SG, Rifkin A: The problem of psychopharmacotherapy in the medically ill. *Psychiatr Clin North Am* 1981; 4:379–390.
62. Schulz P, Luttrell S: Increased plasma protein binding of imipramine in cancer patients. *J Clin Psychopharmacol* 1982; 2:417–420.
63. Giller EL Jr, Bialos DS, Docherty JP, et al: Chronic amitriptyline toxicity. *Am J Psychiatry* 1979; 136:458–459.

64. Glassman AH, Johnson LL, Giardina E-GV, et al: The use of imipramine in depressed patients with congestive heart failure. *JAMA* 1983; 250:1997–2001.
65. Amdisen A: Serum level monitoring and clinical pharmacokinetics of lithium. *Clin Pharmacokinet* 1977; 2:73–92.
66. Potter WZ, Rudorfer MV, Lane EA: Active metabolites of antidepressants: Pharmacodynamics and relevant pharmacokinetics, in Usdin E, Bertilsson L, Sjöqvist F (eds): *Frontiers in Biochemical and Pharmacological Research in Depression.* New York, Raven Press, 1984, pp 373–390.
67. Brogden RN, Heel RC, Speight TM, et al: Nomifensine: A review of its pharmacological properties and therapeutic efficacy in depressive illness. *Drugs* 1979; 18:1–24.
68. Lundborg P, Regardh CG, Landahl S: The pharmacokinetics of metoprolol in healthy elderly individuals. *Clin Pharmacol Ther* 1982; 31:246.
69. Hoffman K-J, Regardh CG, Aurell M, et al: The effect of impaired renal function on the plasma concentration and urinary excretion of metoprolol metabolites. *Clin Pharmacokin* 1980; 5:181–191.
70. Dawling S, Lynn K, Rosser R, et al: The pharmacokinetics of nortriptyline in patients with chronic renal failure. *Br J Clin Pharmacol* 1981; 12:39–45.
71. Braithwaite RA, Dawling S: The pharmacokinetics and metabolism of tricyclic antidepressant drugs in patients with chronic renal failure, in Usdin E, Dahl SG, Gram LF, et al (eds): *Clinical Pharmacology in Psychiatry: Neuroleptic and Antidepressant Research.* London, Macmillan, 1981, pp 285–295.
72. Linnoila M, Karoum F, Calil HM, et al: Alteration of norepinephrine metabolism with desipramine and zimelidine in depressed patients. *Arch Gen Psychiatry* 1982; 39:1025–1028.
73. Rudorfer MV, Scheinin M, Karoum F, et al: Reduction of norepinephrine turnover by serotonergic drug in man. *Biol Psychiatry* 1984; 19:179–193.
74. Lake CR, Ziegler MG, Coleman MD, et al: Age-adjusted plasma norepinephrine levels are similar in normotensive and hypertensive subjects. *N Engl J Med* 1977; 296:208–209.
75. Veith RC, Raskind MA, Barnes RF, et al: Tricyclic antidepressants and supine, standing, and exercise plasma norepinephrine levels. *Clin Pharmacol Ther* 1983; 33:763–769.
76. Vestal RE, Wood AJJ, Shand DG: Reduced beta-adrenoceptor sensitivity in the elderly. *Clin Pharmacol Ther* 1979; 26:181–186.
77. Feldman RD, Limbird LE, Nadeau J, et al: Dynamic regulation of leukocyte beta adrenergic receptor-agonist interactions by physiological changes in circulating catecholamines. *J Clin Invest* 1983; 72:164–170.
78. Cutler NR, Zavadil AP III, Linnoila M, et al: Effects of chronic desipramine on plasma norepinephrine concentrations and cardiovascular parameters in elderly depressed women: A preliminary report. *Biol Psychiatry* 1984; 19:549–556.
79. Ross RJ, Zavadil AP III, Calil HM, et al: Effects of desmethylimipramine on plasma norepinephrine, pulse, and blood pressure. *Clin Pharmacol Ther* 1983; 33:429–437.
80. Lake CR, Mikkelsen EJ, Rapoport JL, et al: Effect of imipramine on norepinephrine and blood pressure in enuretic boys. *Clin Pharmacol Ther* 1979; 26:647–653.
81. Lesieur P, Ross RJ, Rudorfer MV, et al: Different noradrenergic responses to norepinephrine reuptake inhibition in middle-aged vs young volunteers. *Clin Pharmacol Ther* 1984; 35:255.
82. Branconnier RJ, Harto-Truax NE, Cole JO: The effect of aging on the positive chronotropic response to amitriptyline. *Psychopharmacology* 1984; 82:256–257.
83. Rudorfer MV, Young RC: Desipramine: Cardiovascular effects and plasma levels. *Am J Psychiatry* 1980; 137:984–986.

84. Bigger JT Jr, Giardina E-GV, Perel JM, et al: Cardiac antiarrhythmic effect of imipramine hydrochloride. *N Engl J Med* 1977; 296:206–208.
85. Veith RC, Raskind MA, Caldwell JH, et al: Cardiovascular effects of tricyclic antidepressants in depressed patients with chronic heart disease. *N Engl J Med* 1982; 306:954–959.
86. Young RC, Alexopoulos GS, Shamoian CA, et al: Heart failure associated with high plasma 10-hydroxynortriptyline levels. *Am J Psychiatry* 1984; 141:432–433.
87. Glassman AH, Bigger JT Jr, Giardina EV, et al: Clinical characteristics of imipramine-induced orthostatic hypotension. *Lancet* 1979; 1:468–472.
88. Livingston RL, Zucker DK, Isenberg K, et al: Tricyclic antidepressants and delirium. *J Clin Psychiatry* 1983; 44:173–176.

CHAPTER 10

METHODOLOGICAL CONCERNS FOR CLINICAL TRIALS IN GERIATRICS
Benzodiazepines

DARRELL R. ABERNETHY

1. INTRODUCTION

Since the introduction of chlordiazepox de into clinical practice in the United States in 1960, benzodiazepine m...or tranquilizers and sedative hypnotics have replaced other drugs as the agents of choice in the treatment of anxiety and insomnia.[1] During this 25-year period a total of 13 benzodiazepine derivatives have been marketed in the United States. For this to occur a vast literature has accumulated reporting the preclinical and clinical testing of these agents. Since the mid-1970s many reports have indicated that elderly patients have alterations in both the pharmacokinetics and pharmacodynamics of many of the benzodiazepines.[2] However, even at the present time with the large research base available, many questions, both basic and clinical, remain concerning mechanisms of the kinetic and dynamic alterations and also concerning clinical use of benzodiazepines in a geriatric population. These gaps in knowledge are the result of several factors, including slow development of analytical

DARRELL R. ABERNETHY • Division of Clinical Pharmacology, Brown University, and Department of Medicine, Roger Williams General Hospital, Providence, Rhode Island 02908.

189

methodology for determination of benzodiazepine concentrations in body fluids for conduct of pharmacokinetic trials and lack of drug-sensitive measures to quantify benzodiazepine pharmacodynamics. Unfortunately, in many instances, these deficiencies have been compounded by deficiencies in clinical trial design, execution, and analysis, with the end result being publication of many studies that do not significantly contribute to the understanding of benzodiazepine pharmacokinetics and pharmacodynamics in the elderly.

2. PHARMACOKINETIC TRIALS

Proper design and conduct of pharmacokinetic trials for benzodiazepines in the geriatric population do not differ from those concepts developed for all other drug groups.[3]

2.1. SELECTION OF SUBJECT POPULATIONS

Subject populations must be selected carefully to isolate the variable being studied. In the case of a geriatric trial, longitudinal studies in which the subject serves as his or her own control are rarely practical, and though they are advocated for other kinds of geriatric clinical investigation as the most pure design, an important limitation is introduction of a time variable. Environmental factors implicated in alterations in drug biotransformation and clearance, such as diet and occupational exposure,[4] are difficult to control, and onset of subclinical disease processes (hepatic, cardiovascular, or renal dysfunction accelerated over the expected rate for the geriatric age group) may obscure the single variable (aging) selected for study. When longitudinal studies are impossible, cross-sectional (across age groups) study designs are appropriate. These must be rigorously controlled for environmental, genetic, and sociocultural factors. One effective approach is to use cohort populations matched as closely as possible for all variables except age.

Another important consideration is clear understanding of the specific hypothesis being tested. If isolated aging effects are to be studied, the subject population must be carefully screened to exclude coexisting illness, differences in environmental exposure, and differences in genetic makeup. In contrast, if the effects of aging as it relates to a specific disease process (i.e., hypertension, thyroid disease, renal disease, anxiety, depression, insomnia) are being evaluated, all other variables must be eliminated to the greatest extent possible. Patients of varying age must be evaluated carefully to exclude other concurrent illness and therapeutic modalities that may be used to treat the disease process being studied. Patients in the various groups must then be carefully matched for severity of illness, prior treatment history, and other potentially confounding factors such as cigarette smoking[5] and sex hormone status (pre- versus postmenopausal, oral contraceptive use, steroid use, estrogen replacement therapy).[6] When close matching is not possible, sample size must be increased and multiple factorial approaches implemented to obtain reliable data.[7] Benzodiazepine pharmacokinetic studies in geriatric populations must have subjects or patients sex-matched owing to the consistent finding of greater age-related pharmacokinetic changes in males than in females.[8] Finally, careful consideration of the age of individuals selected for study is warranted. Most geriatric pharmacokinetic studies have included a young (often less than 40 years) and "elderly" (greater than 60 or 65 years) group. However, several assumptions are made with such an approach. When correlational analysis of the data (kinetic parameters such as clearance or elimination half-life versus age) is used, the conclusion that age-related decline in drug biotransformation is linear with respect to age will inevitably ensue since only two relatively discrete age ranges have been studied. Though the literature is replete with such data, there is no evidence that age-related changes in benzodiazepine biotransformation are described by a linear function. Subject samples appropriately include individuals outside these age ranges in large enough numbers to address this issue. In this context, subjects from 40 to 60 years of age should be included, but even more important, subjects or patients greater than 70–75 years of age also require pharmacokinetic evaluation. This latter "very elderly" group is of special importance since their benzodiazepine exposure may be extensive, and they may be particularly at risk for benzodiazepine-induced adverse effects. Access to "very elderly" populations who can be appropriately matched to the younger groups can provide a major obstacle, however, one of such importance that researchers should confront the issue.

After the subject or patient population has been carefully selected to directly address the hypothesis being tested, and written informed consent is obtained, design of the pharmacokinetic trial requires careful attention.

2.2. DESIGN OF THE PHARMACOKINETIC STUDY

Only with the advent of specific analytic techniques of very high sensitivity, such as gas chromatography with electron capture and nitrogen–phosphorus detection or coupled with mass spectrometry, have meaningful benzodiazepine pharmacokinetic studies been possible.[9–12]

Because of the low drug doses and extensive distribution into peripheral tissues, drug concentrations are consistently in the nanogram/milliliter range after single-dose administration.

2.2.1. SINGLE-DOSE PHARMACOKINETIC STUDIES. As benzodiaze-pines administered to elderly individuals in single therapeutic doses have been considered quite safe, a large literature has accumulated that rep-etitiously documents that clearance of benzodiazepines which are bio-transformed by hepatic oxidative (cytochrome P-450) processes is de-creased in men and either remains unchanged or decreases less dramatically in women.[8,12-15] Since volume of distribution of benzodiazepines in-creases with age in both sexes,[13-15] these two independent pharmaco-kinetic variables combine [owing to the relationship shown in Eq. (1)]:

$$t_{1/2} = \frac{0.693 \times VD}{Clearance} \tag{1}$$

to result in marked prolongation of the dependent variable, elimination half-life. Decreased clearance in elderly men has also been described for high-clearance (owing to extensive first-pass hepatic extraction) benzo-diazepines, which are currently being emphasized in development.[16,17] In contrast, as for other drugs that undergo conjugative biotransfor-mation, little change in drug clearance has been described for elderly populations.[18-20]

The list of benzodiazepines in Table I is not all-inclusive; however, it does demonstrate several general points. Low-clearance benzodiaze-pines, therefore those with a long elimination half-life (40–100 hr in healthy young subjects), are uniformly biotransformed by hepatic oxi-dation. Further decreases in clearance, as demonstrated for many of these specific agents in elderly male and to a lesser extent female subjects, result in very slow rates of elimination. Each of these drugs has one or more pharmacologically active metabolites that contribute to observed clinical effects.

Intermediate clearance benzodiazepines (elimination half-life rang-ing from 5 to 20 hr) are biotransformed by conjugation (glucuronyl transferase mediated) to inactive glucuronides, oxidation (alprazolam), or nitroreduction (nitrazepam, flunitrazepam). Nitroreduction is im-paired in elderly males, similar to the oxidative pathways;[21] however, capacity for glucuronide conjugation of benzodiazepines is little affected by age.[18-20] In contrast to the low-clearance benzodiazepines, these drugs do not have pharmacologically active metabolites.

High-clearance benzodiazepines are biotransformed by oxidation; however, they undergo extensive first-pass extraction after oral admin-

TABLE I   ROUTE OF METABOLIC BIOTRANSFORMATION AND CLEARANCE RATE
FOR BENZODIAZEPINES

Drug[a]	Route of biotransformation	Active metabolites	Relative clearance rate
Chlordiazepoxide	Oxidation	Desmethylchlordiazepoxide Demoxepam Desmethyldiazepam	Low
Diazepam	Oxidation	Desmethyldiazepam	Low
[Flurazepam]	Oxidation	Flurazepam aldehyde Hydroxyethyl flurazepam Desalkylflurazepam	Low
[Clorazepate]	Oxidation	Desmethyldiazepam	Low
Clonazepam	Oxidation	—	Intermediate
[Prazepam]	Oxidation	Desmethyldiazepam	Low
Alprazolam	Oxidation	—	Low
Nitrazepam	Nitroreduction	—	Intermediate
Flunitrazepam	Nitroreduction	—	Intermediate
[Quazepam]	Oxidation	2-Oxoquazepam Desalkylflurazepam	Low
[Halazepam]	Oxidation	Desmethyldiazepam	Low
Clobazam	Oxidation	Desmethylclobazam	Low
[Ketazolam]	Oxidation	Desmethyldiazepam	Low
Clotiazepam	Oxidation	Hydroxyclotiazepam Desmethylchlotiazepam	Intermediate
Lorazepam	Conjugation	—	Intermediate
Oxazepam	Conjugation	—	Intermediate
Temazepam	Conjugation	—	Intermediate
Lormetazepam	Conjugation	—	Intermediate
Triazolam	Oxidation (high first-pass extraction)	—	High
Midazolam	Oxidation (high first-pass extraction)	—	High
Brotizolam	Oxidation (high first-pass extraction)	—	High

[a] [] Denotes prodrug, which is rapidly transformed to the pharmacologically active metabolite(s) listed.

istration, and variability in their clearance is predominantly related to factors that alter hepatic blood flow and extraction processes.[22] Pharmacokinetics of midazolam and triazolam in elderly subjects indicate decreased clearance, with consequent prolongation of elimination half-life, again greater in elderly males than females.[16,17] These data suggest that physiological decreases in hepatic blood flow or extraction processes associated with aging have pharmacokinetic consequences. The high-clearance benzodiazepines do not have pharmacologically active metabolites.

The mechanism for these pharmacokinetic changes associated with aging has not been well established in man; however, studies in aging animals indicate both decreased activity and quantity of cytochrome c (P-450) reductase and cytochrome P-450, the enzyme system that mediates benzodiazepine oxidation.[23] An extensive, yet inconclusive, literature, has developed describing gender differences in studies of benzodiazepine kinetics. Currently it is unknown why males exhibit a greater *in vivo* decrease in benzodiazepine-oxidizing capacity than do females; however, loss of estrogen effect in the aging female may preserve biotransformation capacity if inhibitory effect of estrogen-containing oral contraceptive steroids on drug oxidation can be used as a model.[24] Relative insensitivity of benzodiazepine glucuronidation to the aging process is consistent with lack of impairment in glucuronyl transferase activity. Relative resistance of drug glucuronidation has been noted for other diseases associated with impairment in hepatic function such as cirrhosis[25] and Gilbert's disease.[26] Impairment of hepatic extraction of high-clearance benzodiazepines in elderly subjects may be due to decreased hepatic blood flow associated with aging or impairment in the extraction process itself. In any case, these findings are consistent with the age-related impairment in extraction and clearance of other high-clearance drugs such as imipramine[27] and lidocaine.[28]

2.2.2. MULTIPLE-DOSE BENZODIAZEPINE PHARMACOKINETICS. Much less information is available describing steady-state kinetics of benzodiazepines in the elderly. However, the data available support what would be predicted from the single-dose data for a class of drugs with linear pharmacokinetics. Accumulation of desalkylflurazepam, the pharmacologically active metabolite of flurazepam that is biotransformed by hepatic oxidation, is increased in elderly patients.[29] Clobazam and its active metabolite desmethylclobazam both accumulated more extensively in elderly men than young men during chronic dosing; however, as in the single-dose studies, no age-related differences in steady-state kinetics were noted for women.

2.2.3. ANALYSIS OF SINGLE-DOSE AND STEADY-STATE BENZODIAZEPINE PHARMACOKINETIC STUDIES. Careful pharmacokinetic study of the benzodiazepines was not possible until highly sensitive analytical methodology became available. An additional contributing factor was lack of appreciation for the sometimes dramatic prolongation of oxidized benzodiazepine elimination half-life in elderly individuals that results from both impaired drug clearance and increased volume of distribution. Plasma sampling must continue for two or three elimination half-lives to appropriately characterize the terminal elimination rate constant. Some oxidized benzodiazepines, particularly desmethyldiazepam, may have an elimination half-life as prolonged as 200 hr in some elderly males.[30] Sampling for 400–600 hr dictates periodic blood samples must be obtained 3–4 weeks after a single dose in such individuals.

As noted in Table I, a number of clinically used benzodiazepines are precursors for low-clearance pharmacologically active benzodiazepine metabolites. Therefore, parent drug may have relatively high clearance; however, this clearance is due to biotransformation to a low-clearance oxidized benzodiazepine such as desmethyldiazepam, desmethylchlordiazepoxide, desalkylflurazepam, or desmethylclobazam. In such cases, to properly characterize the pharmacokinetic profile of the parent drug in the elderly, both parent drug and active metabolite pharmacokinetic characteristics must be established.[31] Similarly, when human pharmacokinetic evaluation of a new benzodiazepine is planned, the profile of metabolites identified from prior distribution and toxicity studies must be closely examined. When plasma samples from human pharmacokinetic studies are analyzed, analytical conditions capable of identifying active metabolites should be employed. For example, halazepam was initially described to have a rather rapid rate of elimination; however, prior toxicity studies indicated desmethyldiazepam was formed in significant amounts after administration of the parent drug. Predictably, in humans after halazepam administration, biotransformation to desmethyldiazepam in a nearly quantitative fashion results in this pharmacologically active metabolite, with very low clearance, providing the basis for persistence of drug in plasma for a prolonged time.[31] Therefore, during chronic halazepam administration, one can confidently predict accumulation of desmethyldiazepam in a manner similar to its accumulation after administration of any other of its precursors such as clorazepate, prazepam, and quazepam. This becomes an important consideration for pharmacokinetic studies in the elderly in the prediction of drug accumulation, which has been associated with increased incidence of adverse effect for some low-clearance benzodiazepines.

As a class benzodiazepines are extensively bound to plasma proteins, predominantly albumin.[32-34] Pharmacokinetic studies of these drugs in the elderly should include evaluation of plasma protein binding. Alteration in binding for diazepam, an extensively bound (99%) benzodiazepine, has been demonstrated to markedly alter drug distribution and clearance.[35] Therefore, the inference that changes in drug clearance are the result of changes in drug-metabolizing-enzyme capacity cannot be drawn unless both total clearance and clearance of drug not bound to plasma protein are determined. Evaluation of differences in distribution volume of a benzodiazepine in young versus elderly subjects can be distorted as well unless evaluation of distribution volume of drug not bound to plasma protein is also made.

Interpretation of pharmacokinetic parameters derived from studies of benzodiazepine pharmacokinetics in the elderly must be made with some caution. To date these studies have uniformly been cross-sectional rather than longitudinal in design. Advantages and pitfalls of these two dramatically different approaches are described above. With a cross-sectional design, by definition many environmental- and time-related variables besides age are by necessity present. Certainly, the lifetime environmental exposure of an 80-year-old subject, who has lived during times of war, economic depression, and when infectious diseases such as influenza and tuberculosis were widespread, can never be duplicated by a 25-year-old who has had childhood immunization, is likely to engage in national and international recreational travel, and lives in a time of increasing concern for personal physical fitness and well-being. To suggest such a pair of individuals can be matched for factors other than age is presumptuous indeed. However, these concerns must be tempered with the understanding that pharmacokinetic studies of benzodiazepines may not only be useful to further understanding of effects of the aging process on drug metabolism, but also are necessary to predict disposition of these drugs in populations who will have extensive exposure to benzodiazepines.

3. BENZODIAZEPINE PHARMACODYNAMIC STUDIES IN GERIATRICS

In contrast to the vast literature describing benzodiazepine pharmacokinetics in elderly individuals, only limited information about benzodiazepine pharmacodynamics is currently available. This may be attributed to the extraordinary difficulty in selecting pharmacodynamic measures, or end points, that are predictably influenced by benzodiazepine exposure. Since the pharmacodynamic end points of interest relate to psychomotor function, degrees of subjective alertness, and capacity to perform specific tasks, studies in the elderly are further complicated by changes in baseline psychomotor function as compared to a younger population. In addition, careful validation of the various pharmacodynamic measures is available only for younger patient populations.

3.1. Drug-Sensitive Pharmacodynamic Measures

Across all subject groups, certain measures of psychomotor performance have rather consistently been "benzodiazepine sensitive" while others are "benzodiazepine insensitive," as shown in Table II. Only selected pharmacodynamic measures have been reported in geriatric patients, and at present no measure has been systematically validated in the geriatric population. Therefore, comparison between age groups of a specific end point is hazardous and at best a suggestion of differential sensitivity to a specific "benzodiazepine-sensitive" function can be inferred.

3.2. Review of Benzodiazepine Pharmacodynamics in the Elderly

Early indication that benzodiazepines may have altered pharmacodynamics in geriatric patients was obtained from epidemiological study of adverse drug reactions in hospitalized patients. Reports from the Boston Collaborative Drug Surveillance Program indicated that age and dose of chlordiazepoxide, diazepam, flurazepam, and nitrazepam were positively correlated with incidence of adverse drug reactions as defined by any unwanted or unintended drug effect.[64-66] These epidemiological observations led to testing of the specific hypothesis that age may be an independent factor that explains a significant proportion of variance in benzodiazepine pharmacodynamic effects seen in a population.

Evaluation of nitrazepam using a letter cancellation task indicated increased psychomotor impairment in the elderly at similar plasma nitrazepam concentrations.[52] Using a quite different pharmacodynamic end point, response to painful stimulation, diazepam was noted to blunt this response to a greater extent in elderly patients undergoing electrical cardioversion than in a group of young patients.[41] Diazepam has also been reported to impair immediate and delayed recall and impaired performance on a reaction time test in elderly subjects; however, no comparison was made to a younger subject group.[45] In contrast, after single flurazepam doses, using self-rated sedation scales, no difference in perceived sedation or depth of sleep was noted in young versus elderly patients.[29] Evaluation of temazepam in young versus elderly subjects using postural sway, critical flicker fusion frequency, and choice reaction time as pharmacodynamic measures indicated the elderly had greater impairment in each of these psychomotor tasks than did young subjects.[63]

TABLE II   CLINICAL ASSESSMENT OF BENZODIAZEPINE PHARMACODYNAMICS

Performance task	Benzodiazepine sensitivity[a]	Patient group	Ref.
Simple reflexes			
Auditory reaction time	−	Controls	36
		Psychiatric inpatients	37
Visual reaction time	−	Psychiatric inpatients	38
Tapping speed	−	Controls	36, 39, 40
Pain response	+ +	Geriatric patients	41
Critical flicker fusion threshold			
Visual	+ +	Controls	42–44
		Geriatric patients	45
Auditory	+ +	Dental patients	46
Decision making			
Sorting	+ −	Controls	43, 44, 47
		Psychiatric inpatients	37
Choice reaction	−	Controls	36, 46
	+ −	Psychiatric inpatients	38
Memory			
Recall	+ +	Controls	39, 43, 44, 47
		Geriatric patients	45
Short-term	−	Controls	42, 47
Long-term	−	Controls	42, 47
Concentration task			
Digit symbol substitution	+ −	Controls	37, 39, 48–50
Letter cancellation	+	Controls	51
	+ +	Geriatric patients	52
Perceptual–motor skills			
Tracing tasks	+ +	Controls	37, 40, 43, 44, 53
Driving simulation	+	Controls	54–56
Pegboard performance	+	Controls	46, 48, 55
Pursuit rotor	−	Controls	40, 57, 58
Saccadic eye movement	+ +	Controls	59, 60
Time estimation	+ +	Controls	43, 44, 46, 54, 58
Self-rated sedation	+ +	Controls	49, 61
		Geriatric patients	29
Postural sway	+ +	Controls	62
		Geriatric patients	63

[a] − = no effect; + − = disagreement among studies of effect; + = consistent effect; + + = function is highly benzodiazepine sensitive.

Currently the mechanism for this increased sensitivity to psychomotor impairment induced by benzodiazepines is not well understood. Certainly, the pharmacokinetic consequences of aging may play a role during chronic treatment with an oxidized benzodiazepine since increased drug accumulation, at least in males, may be predicted. However, most of these observations were made after single-dose administration, and the differential drug effects were noted at similar plasma drug concentrations. Other pharmacokinetic factors that could influence the pharmacodynamic observations include increased rate and/or extent of drug absorption after oral administration, but with the exception of high-clearance benzodiazepines such as midazolam or triazolam, absorptive changes with aging have been documented not to occur with benzodiazepine. Finally, since only benzodiazepine not bound to plasma protein traverses the blood–brain barrier,[67] the decreased extent of protein binding reported in elderly subjects suggests that at a given total plasma benzodiazepine concentration, a slightly higher concentration of benzodiazepine may diffuse into the central nervous system in the elderly individual. With description of specific high-affinity binding sites for benzodiazepines in the central nervous system,[68,69] considerable enthusiasm for the concept of benzodiazepine receptor alterations with aging developed. Currently no definitive changes in benzodiazepine receptor number or affinity in aging man or the animal model have been identified.[70]

In summary, neither the description nor the mechanism of benzodiazepine pharmacodynamics in geriatric populations is well defined. This remains an important area for investigation; however, such studies must be approached with a clear understanding of the hypothesis being addressed to permit recruitment of the proper patient population, utilization of appropriate pharmacodynamic measures, and effective interpretation of study findings. Questions that must be answered include the following:

3.2.1. PATIENT POPULATION. Healthy volunteers, geriatric medical patients, psychogeriatric patients? Each of these groups may respond quite differently to pharmacodynamic measures owing to health or disease independent of age.

3.2.2. PHARMACODYNAMIC END POINTS. As noted previously, no dynamic measures, with the partial exception of self-rated sedation scales, have been validated in geriatric populations. To extrapolate proper validation in younger individuals to the elderly is not appropriate. Elderly individuals may have an "impaired" baseline for many of the psychomotor tests simply because of physical infirmity, such as arthritis, impaired vision, or altered attention span, that is completely unrelated to drug effect. Comparison of magnitude of effect caused by drug while starting from different baseline performance capacity is fraught with hazard. In some instances slight increases in sedation and lethargy over an abnormal baseline may have profound clinical consequences, whereas in the younger individual a quantitatively similar change may be of no clinical consequence.

3.2.3. DATA INTERPRETATION.   As implied here, great care must be taken to rigorously characterize the subject or patient population to be evaluated. If effects of physiological aging on benzodiazepine pharmacodynamics are to be isolated and studied, groups of subjects across the age range to be evaluated must be carefully screened for coexisting psychiatric or medical disease. Ideally they would come from a genetically and culturally homogeneous community. Subject groups would be stratified for gender, environmental exposure, and body habitus. Individuals physically incapable of performing the pharmacodynamic tasks would be excluded. Very important, and for the most part absent in the current literature, the pharmacodynamic measures themselves must then be characterized in the study population while in the control state. This can be done as an extension of the control repetitions that are necessarily used to abolish learning effects on the particular pharmacodynamic measure to be evaluated. After multiple repetitions, and after improvement in psychomotor function due to simple repetition (learning) disappears, a significant number of baseline performances in subgroups of the study population will establish baseline mean and variance in task performance for the different groups. This should be viewed simply as further characterization of the subject population to be studied, prior to initiation of blinded placebo (for experimental controls) versus active drug comparisons. To assess relative contribution of pharmacokinetic differences in the population to the observed pharmacodynamic effects, it is desirable to obtain plasma samples for determination of drug concentration (preferably both total and concentration not bound to plasma proteins) at the time of the pharmacodynamic testing. Concern about equilibration between the plasma compartment and the central nervous system is not warranted for benzodiazepines, which partition into the central nervous system and attain equilibrium with the plasma very rapidly. With acquisition of the data, obviously appropriate statistical treatment should be applied.

3.3. PHARMACODYNAMICS IN ELDERLY PATIENTS
     WITH UNDERLYING DISEASE

When the effect of age on drug response is studied for a certain disease process, patient selection must be at least as rigorous as that described for healthy subjects. Disease (i.e., anxiety, panic disorder, insomnia) must be defined as rigorously as is clinically possible. Then, a study population over the specified age range must be selected that has the isolated disease process of similar severity. The literature is replete with studies that compare healthy young subjects to aged patients with medical or psychiatric disease and then attempt to conclude that experimental findings were due to age or disease or both. Obviously, such study design provides little information about either question. Again, validation of the specific pharmacodynamic measures to be tested for disease-related age effect should be done *in the specific population studied*. Proper pharmacokinetic characterization of the benzodiazepine in the population is important to separate pharmacodynamic alterations that are simply due to changes in drug concentration in plasma or at the effector site and those which are due to intrinsic differences in sensitivity to the benzodiazepine-induced effect.

One limitation of such a study design (disease-related age effects) is the variability in expression of clinical disease as a function of age. For example, insomnia in the young patient may be secondary to a variety of factors quite different from those in the elderly patient. Clinical manifestation of depression may be quite different over an age spectrum, such that recruitment of comparable patient populations for study represents a challenging task. However, this should not deter clinical investigation in this extremely important area. In the animal model there is now clear-cut evidence that disease-related alteration in phenobarbital pharmacodynamics occurs with experimentally induced uremia.[71] This should stimulate clinical investigation of potential disease-induced (medical and psychiatric) alterations in pharmacodynamics of other central nervous system depressants such as benzodiazepines. Currently, no information is available other than observation in association with benzodiazepine pharmacokinetic trials in different disease states.

4. CONCLUSION

Study of benzodiazepine pharmacokinetics has given important insight into the effect of the aging process on drug biotransformation in humans. Furthermore, these studies have at least partly explained the early epidemiological observations of increased incidence of adverse drug effect in elderly patients who received benzodiazepine therapy. As new benzodiazepines become available for clinical testing, an understanding of their pharmacokinetic profile in elderly individuals is important.

Benzodiazepine pharmacodynamic evaluation in the elderly is at a much earlier stage of development. Little attention has been paid to differences in pharmacodynamic effects associated with acute, single-dose versus chronic drug administration. As starting places for potentially useful pharmacodynamic measures, extensive reviews of "drug-sensitive" measures in control populations are available.[72-74] Validation of specific pharmacodynamic measures in the geriatric patient is mandatory. Selection of study populations requires continuous refinement as more understanding of physiological changes associated with aging is gained, which may affect performance of tasks used for assessment of benzodiazepine pharmacodynamics. In addition, the effects of age on diagnosis and expression of clinical disease must continue to evolve to permit study of benzodiazepine pharmacodynamics as a function of age in patient populations who are candidates for benzodiazepine therapy. Careful and thoughtful experimental approaches to these most challenging problems may be expected to further insight into clinical effects of benzodiazepines in the elderly and ultimately improve clinical geriatric medicine.

ACKNOWLEDGMENT.   This work was supported in part by Grants AM-33479 and GM-34120 from the U.S. Public Health Service.

REFERENCES

1. Greenblatt DJ, Shader RI, Abernethy DR: Current status of benzodiazepines. *N. Engl J Med* 1983; 309:354–358, 410–416.
2. Meyer BR: Benzodiazepines in the elderly. *Med Clin North Am* 1982; 66:1017–1035.
3. Rowland M, Tozer TN: *Clinical Pharmacokinetics*. Philadelphia, Lea & Febiger, 1980.
4. Vesell ES, Penno MB: Assessment of methods to identify sources of interindividual pharmacokinetic variation. *Clin Pharmacokinet* 1983; 8:378–409.
5. Vestal RE, Norris AH, Tobin JD, et al: Antipyrine metabolism in man: influence of age, alcohol, caffeine, and smoking. *Clin. Pharmacol Ther* 1975; 18:425–433.
6. Goble FC: Sex as a factor in metabolism, toxicity, and efficacy of pharmacodynamic and chemotherapeutic agents. *Adv Pharmacol Chemother* 1975: 13:173–252.
7. Jusko WJ, Gardner M, Mangione A, et al: Factors affecting theophylline clearances: Age, tobacco, marihuana, cirrhosis, congestive heart failure, obesity, oral contraceptives, benzodiazepines, barbiturates, and ethanol. *J Pharm Sci* 1979; 68:1358–1366.
8. Greenblatt DJ, Divoll M, Abernethy DR, et al: Antipyrine kinetics in the elderly: prediction of age-related changes in benzodiazepine oxidizing capacity. *J Pharmacol Exp Ther* 1982; 220:120–126.
9. de Silva JAF, Bekersky I, Puglisi CV, et al: Determination of 1,4-benzodiazepines and -diazepine-2-ones in blood by electron-capture gas-liquid chromatography. *Anal Chem* 1976; 48:10–19.
10. Sadee W, van der Kleijn E.: Thermolysis of 1,4-benzodiazepines during gas chromatography and mass spectroscopy. *J Pharm Sci* 1971; 60:135–137.
11. Strojny N, Puglisi CV, de Silva JAF: Determination of chlordiazepoxide and its metabolites in plasma by high pressure liquid chromatography. *Anal Lett B* 1978; 11:135–160.

12. Greenblatt DJ, Divoll M, Moschitto LJ, et al: Electron-capture gas chromatographic analysis of triazolobenzodiazepines alprazolam and triazolam. *J Chromatogr* 1981; 225:202–207.

13. Klotz U, Avant GR, Hoyumpa A, et al: The effects of age and liver disease on the disposition and elimination of diazepam in adult man. *J Clin Invest* 1975; 55:347–359.

14. Macklon AF, Barton M, James O, et al: The effect of age on the pharmacokinetics of diazepam. *Clin Sci* 1980; 59:479–483.

15. Greenblatt DJ, Allen MD, Harmatz JS, et al: Diazepam disposition determinants. *Clin Pharmacol Ther* 1980; 27:301–312.

16. Greenblatt DJ, Divoll M, Abernethy DR, et al: Reduced clearance of triazolam in old age: relation to antipyrine oxidizing capacity. *Br J. Clin Pharmacol* 1983; 15:303–309.

17. Greenblatt DJ, Abernethy DR, Locniskar A, et al: Effect of age, gender and obesity on midazolam kinetics. *Anesthesiology* 1984; 61:27–35.

18. Greenblatt DJ, Divoll M, Harmatz JS, et al: Oxazepam kinetics: effects of age and sex. *J Pharmacol Exp Ther* 1980; 215:86–91.

19. Kraus JW, Desmond PV, Marshall JP, et al: Effects of aging and liver disease on disposition of lorazepam. *Clin Pharmacol Ther* 1978; 24:411–419.

20. Divoll M, Greenblatt DJ, Harmatz JS, et al: Effect of age and gender on disposition of temezepam. *J Pharm Sci* 1981; 70:1104–1107.

21. Greenblatt DJ, Aberncthy DR, Locniskar A, et al: Nitrazepam kinetics in old age and obesity. *Clin Pharmacol Ther* 1985; 37:199.

22. Wilkinson GR, Shand DG: A physiologic approach to hepatic drug clearance. *Clin Pharmacol Ther* 1975; 18:377–390.

23. Schmucker DL, Wang RK: Age-dependent changes in rat liver microsomal NADPH Cytochrome c (P-450) reductase: A kinetic analysis. *Exp Gerontol* 1983; 18:313–321.

24. Abernethy DR, Greenblatt DJ, Divoll M, et al: Impairment of diazepam metabolism by low-dose estrogen oral contraceptive steroids. *N Engl J Med* 1982; 306:791–792.

25. Shull HJ, Wilkinson GR, Johnson, et al: Normal disposition of oxazepam in acute viral hepatitis and cirrhosis. *Ann Intern Med* 1976; 84:420–425.

26. Shader RI, Divoll M, Greenblatt DJ: Kinetics of oxazepam and lorazepam in two subjects with Gilbert's syndrome. *J Clin Psychopharmacol* 1981; 1:400–402.

27. Abernethy DR, Greenblatt DJ, Shader RI: Imipramine and desipramine disposition in the elderly. *J Pharmacol Exp Ther* 1985; 232:183–188.

28. Abernethy DR, Greenblatt DJ: Impairment of lidocaine clearance in elderly male patients. *J Cardiovasc Pharmacol* 1983; 5:1093–1096.

29. Greenblatt DJ, Divoll M, Harmatz JS, et al: Kinetics and clinical effects of flurazepam in young and elderly noninsomniacs. *Clin Pharmacol Ther* 1981; 30:475–486.

30. Allen MD, Greenblatt DJ, Harmatz JS, et al: Desmethyldiazepam kinetics in the elderly after oral prazepam. *Clin Pharmacol Ther* 1980; 28:196–202.

31. Greenblatt DJ, Locniskar A, Shader RI: Halazepam, another precursor of desmethyl-diazepam. *Lancet* 1982; 1:1358–1359.

32. Johnson RF, Schenker S, Roberts RK, et al: Plasma binding of benzodiazepines in humans. *J Pharm Sci* 1979; 68:1320–1322.

33. Divoll M, Greenblatt DJ: Effect of age and sex on lorazepam protein binding. *J Pharm Pharmacol* 1982; 34:122–123.

34. Moschitto LJ, Greenblatt DJ: Concentration-independent plasma protein binding of benzodiazepines. *J Pharm Pharmacol* 1983; 35:179–180.

35. Ochs HR, Greenblatt DJ, Kaschell HJ, et al: Diazepam kinetics in patients with renal insufficiency or hyperthyroidism. *Br J Clin Pharmacol* 1981; 12:829–832.

36. Bernheim J, Michiels W: Effects psychophysiques du diazepam (Valium) et d'une faible dose d'alcool ches l'homme. *Schweiz Med Wochenschr* 1973; 103:863–870.
37. Tansella M, Zimmermann-Tansella C, Lader M: The residual effects of N-desmethyl-diazepam in patients. *Psychopharmacologia* 1974; 38:81–90.
38. Savage PPE, Wilkinson V: Reaction time in psychiatric patients: A pilot study. *NZ Med J* 1971; 73:285–288.
39. Jaattela A, Mannisto R, Paatero H, Tuomisto J: The effects of diazepam or diphen-hydramine on healthy human subjects. *Psychopharmacologia* 1971; 21:202–211.
40. Milner G, Landauer AA: Haloperidol and diazepam alone and together with alcohol, in relation to driving safety. *Blut Alkohol* 1973; 10:247–254.
41. Reidenberg MM, Levy M, Warner H, et al: Relationship between diazepam dose, plasma level, age, and central nervous system depression. *Clin Pharmacol Ther* 1978; 23:371–374.
42. Grove-White IG, Kelman FR: Critical flicker frequency after small doses of methohex-itone, diazepam and sodium 4-hydroxybutyrate. *Br J Anaesth* 1971; 43:110–112.
43. Haffner JFW, Morland J, Setekleiv J, et al: Mental and psychomotor effects of di-azepam and alcohol. *Acta Pharmacol Toxicol* 1973; 32:161–178.
44. Morland J, Setekleiv J, Haffner JFW, et al: Combined effects of diazepam and ethanol on mental and psychomotor functions. *Acta Pharmacol Toxicol* 1974; 34:5–15.
45. Pomara N, Stanley B, Block R, et al: Diazepam impairs performance in normal elderly subjects. *Psychopharmacol Bull* 1984; 20:137–139.
46. Healy TEJ, Lautch H, Hall N, et al: Interdisciplinary study of diazepam sedation of outpatient dentistry. *Br Med J* 1970; 3:13–17.
47. Clark PRF, Eccersley PS, Frisby JP, et al: The amnesic effect of diazepam (Valium). *Br J Anaesth* 1970; 42:690–697.
48. Lawton MP, Cahn B: The effects of diazepam (Valium) and alcohol on psychomotor performance. *J Nerv Ment Dis* 1963; 136:550–554.
49. Greenblatt DJ, Abernethy DR, Morse DS, et al: Clinical importance of the interaction of diazepam and cimetidine. *N Engl J Med.* 1984; 310:1639–1643.
50. Wittenborn JR, Flaherty CF, McGough WE, et al: Psychomotor changes during the initial day of benzodiazepine medication. *Br J Clin Pharmacol* 1979; 7:695–765.
51. Malpas A, Joyce CRB: Effects of nitrazepam, amylobarbitone, and placebo on some perceptual, motor, and cognitive tasks in normal subjects. *Psychopharmacologia* 1969; 14:167–177.
52. Castelden CM, George CF, Marcer D, et al: Increased sensitivity to nitrazepam in old age. *Br Med J* 1977; 1:10–12.
53. Dixon RA, Thornton JA: Tests of recovery from anaesthesis and sedation: Intravenous diazepam in dentistry. *Br J Anaesth* 1973; 45:207–215.
54. Linnoila M, Haskkinen S: Effects of diazepam and codeine alone and in combination with alcohol on simulated driving. *Clin Pharmacol Ther* 1974; 15:368–373.
55. Hindmarch I, Gudgeon AC: The effects of clobazam and lorazepam on aspects of psychomotor performance and car handling ability. *Br J Clin Pharmacol* 1980; 10:145–150.
56. de Gier JJ, Hart BJ't, Nelemans FA, et al: Psychomotor performance and real driving performance of outpatients receiving diazepam. *Psychopharmacology* 1981; 73:340–344.
57. Hughes FW, Forney RB, Richards AB: Comparative effects in human subjects of chlordiazepoxide, diazepam, and placebo on mental and physical performance. *Clin Pharmacol Ther* 1965; 6:139–145.
58. Wittenborn JR, Flaherty CF, Hamilton LW, et al: The effect of minor tranquilizers on psychomotor performance. *Psychopharmacology* 1976; 74:281–286.

59. Gentles W, Thomas EL: Effect of benzodiazepines upon saccadic eye movements in man. *Clin Pharmacol Ther* 1971; 12:563–574.
60. Rodnight E, Gooch RN: A new method for the determination of individual differences in susceptibility to a depressant drug, in Eysenck HR (ed): *Experiments with Drugs*. London, Permagon Press, 1963, pp 169–193.
61. Greenblatt DJ, Shader RI, Harmatz JS, et al: Self-rated sedation and plasma concentrations of desmethyldiazepam following single doses of clorazepate. *Psychopharmacology* 1979; 66:289–290.
62. Ellinwood EH, Linnoila M, Angle HV: Use of simple tasks to test for impairment of complex skills by a sedative. *Psychopharmacology* 1981; 73:350–354.
63. Crook J: Aging and drug disposition—pharmacodynamics. *J Chron Dis* 1983; 36:85–90.
64. Boston Collaborative Drug Surveillance Programs: Clinical depression of the central nervous system due to diazepam and chlordiazepoxide in relation to cigarette smoking and age. *N Engl J Med* 1973; 288:277–280.
65. Greenblatt DJ, Allen MD, Shader RI: Toxicity of high-dose flurazepam in the elderly. *Clin Pharmacol Ther* 1977; 21:355 –361.
66. Greenblatt DJ, Allen MD: Toxicity of nitrazepam in the elderly: A report from the Boston Collaborative Drug Surveillance Program. *Br J Clin Pharmacol* 1978; 5:407–413.
67. Arendt RM, Greenblatt DJ, de Jong RH, et al: *In vitro* correlates of benzodiazepine cerebrospinal fluid uptake, pharmacodynamic action, and peripheral distribution. *J Pharmacol Exp Ther* 1983; 227:98–106.
68. Braestrup C, Albrechtsen R, Squires RF: High densities of benzodiazepine receptors in human cortical areas. *Nature* 1977; 269:702–704.
69. Mohler H, Okada T: Benzodiazepine receptor: demonstration in the central nervous sytem. *Science* 1977; 198:849–851.
70. Tsang CC, Speeg KV, Wilkinson GR: Aging and benzodiazepine binding in rat cerebral cortex. *Life Sci* 1982; 30:343–346.
71. Dahof M, Hisaoka M, Levy G: Kinetics of drug action in disease states, II. Effect of experimental renal dysfunction on phenobarbital concentrations in rats at onset of loss of righting reflex. *J Pharmacol Exp Ther* 1984; 230:627–631.
72. Kleinknecht RA, Donaldson D: A review of the effects of diazepam on cognitive and psychomotor performance. *J Nerv Ment Dis* 1975; 161:399–411.·
73. Wittenborn JR: Effects of benzodiazepines on psychomotor performance. *Br J Clin Pharmacol* 1979; 7:615–675.
74. Johnson LC, Chernik DA: Sedative-hypnotics and human performance. *Psychopharmacology* 1982; 76:101–113.

CHAPTER 11

EFFECT OF AGE ON THE CLINICAL PHARMACOKINETICS OF ANTIARRHYTHMIC DRUGS

J. W. MASSARELLA AND K.-C. KHOO

1. INTRODUCTION

An accurate description of the pharmacokinetics of antiarrhythmic drugs is clearly warranted since plasma concentrations of many agents in this class have been correlated with therapeutic effects. It is thus important to understand the influence of various factors such as age, disease states, and concurrent therapy on the drug's kinetics, as these factors often alter drug pharmacokinetics (and therefore therapeutic efficacy) and are present in a large percentage of cardiac patients. This chapter provides a general description of the pharmacokinetics of the most commonly used antiarrhythmics as well as those which have recently entered the marketplace or will do so in the near future. Clinical studies investigating the effect of age on the pharmacokinetics of antiarrhythmic drugs are clearly lacking in the literature. Although the drugs in this class have been studied in the patient population for which the drug is intended and these patients are typically older, few studies have been conducted in which age alone was the influencing factor. Thus, much of the in-

J. W. MASSARELLA AND K.-C. KHOO • Department of Drug Metabolism, Hoffmann-La Roche, Inc., Nutley, New Jersey 07110.

formation presented is based on results of studies conducted in patients with arrhythmias or impaired function of another major organ system.

2. MAJOR ANTIARRHYTHMIC AGENTS

2.1. DISOPYRAMIDE

Disopyramide is a class I antiarrhythmic agent with properties similar to those of quinidine. It has been marketed as the base and as the phosphate salt. The usual oral dosage is 100–200 mg every 6 hr. Intravenous therapy begins with an i.v. bolus of about 1–2 mg/kg followed by a lower maintenance dose over a longer period of time.

Maximum disopyramide plasma concentrations after single 3- and 6-mg/kg oral doses to healthy volunteers are 1.5 and 3.0 μg/ml, respectively, and occur between 0.5 and 3 hr.[1] Absorption of the phosphate salt appears to be slower but more complete than that of the base. The absolute bioavailability of the phosphate salt and the free base is 82 and 67%, respectively, determined from the plasma concentration–time profiles, and 95 and 88%, respectively, determined from the urinary excretion data for unchanged drug obtained in the same study.[2] This suggests a moderate degree of presystemic metabolism or the existence of route-dependent alterations in clearance. Support for the latter hypothesis has been provided by Cunningham et al.[3]

Unlike other antiarrhythmic agents, the unbound fraction of disopyramide varies at therapeutic plasma concentrations. The plasma protein binding of disopyramide (primarily to α_1-acid glycoprotein) is saturable and ranges from 70 to 30% in normal volunteers at plasma concentrations between 0.4 and 4 μg/ml and from 80 to 50% in patients at average total drug concentrations of 2–8 μg/ml.[4-7]

The pharmacokinetics of disopyramide have been studied in older arrhythmia patients following single and multiple doses given both intravenously and orally.[2,8,9] Relevant pharmacokinetic parameters are given in Table I. Overall, the pharmacokinetics reported in these patients do not differ from those found in healthy younger volunteers.[2,10] In general, the reported elimination half-life is 6–9 hr, the total clearance is 0.7–1.5 ml/min per kg, and the volume of distribution is 0.5–1 liter/kg. Two investigators, however, have reported a much longer half-life. Rangno et al.[11] found a mean half-life of 12 hr, and Hulting and Rosenhamer[12] reported a widely variable half-life (range 9.0–34 hr) with a mean of 18 hr. Neither of these authors characterized their patient population adequately to discuss factors responsible for the prolongation.

TABLE I   PHARMACOKINETIC PARAMETERS OF DISOPYRAMIDE

Subjects	Mean age (range)	Dose[a] (mg)	Half-life (hr)	Clearance (ml/min per kg)	Distribution volume (liters/kg)	Ref.
Single dose						
8 healthy subjects	25 (21–31)	2 mg/kg i.v.	7.8 ± 1.6	0.8 ± 0.3[b]	0.8 ± 0.3[b]	2
8 healthy subjects	25 (21–31)	300 p.o. P	8.9 ± 1.6	—	—	2
		300 p.o. B	8.0 ± 1.4			
6 healthy subjects	(30–45)	150 p.o.	6.3 ± 1.0	1.2 ± 0.2[c]	0.5 ± 0.1	10
8 healthy subjects	(20–30)	1.5 mg/kg i.v.	6.0 ± 1.0	1.2 ± 0.4	0.6 ± 0.1[d]	5
				5.4 ± 2.8[c]	1.7 ± 0.8[c]	
4 MI patients	(45–65)	100 i.v. or p.o.	7.0	0.7[c]	0.5[c]	9
8 arrhythmia patients	61 (43–71)	200 p.o.	18.6 ± 9.4	—	—	12
13 arrhythmia patients	62 (36–77)	2 mg/kg i.v.	11.8	0.8[c]	0.4[c]	11
Multiple dose						
10 arrhythmia patients	67 (51–85)	100 p.o. q6h	7.3 ± 2.5	1.2 ± 0.6	0.9 ± 0.3	8
10 arrhythmia patients	62 (43–78)	200 p.o. q6h	6.4 ± 2.5	1.5 ± 0.5	1.1 ± 0.4	8

[a] P = phosphate salt; B = base.
[b] Normalized to mean weight.
[c] Normalized to 70 kg.
[d] Volume of distribution—steady state.
[e] Determined from free (unbound) fraction of drug in plasma.

Disopyramide is eliminated by both renal excretion and hepatic metabolism.[1] The primary metabolite is mono-N-dealkylated disopyramide (NMD). The relative antiarrhythmic activity of the metabolite has not been established. About 35–44% of the dose is recovered in urine as unchanged drug and an additional 26–47% as NMD after i.v. or oral administration.[1,3,13] Although previous studies demonstrated that the renal clearance is dependent on the unbound fraction in plasma, the renal clearance of unbound drug appears to depend on time after dose and route of administration. As a result of the concentration-dependent binding of disopyramide to plasma proteins, the elimination kinetics of total plasma drug concentration are nonlinear. In healthy volunteers after i.v. administration, the mean renal clearance values of total and unbound drug were 0.6 and 2.8 ml/min per kg over the urine collection period of 0 and 12 hr, and increased at later times.

Plasma and serum concentrations of 3–8 μg/ml have been well correlated to therapeutic response in both healthy, younger subjects and older arrhythmia patients following both single and multiple doses.[1,2,8,11,12] Bryson et al.[2] studied the time course of pharmacological response following both single and multiple doses. The absence of hysteresis indicated that the kinetics of the pharmacological effect are closely related to those of the drug in plasma. Similar results were observed by Ranney et al.[14] following both intravenous and oral administrations.

2.2. LIDOCAINE

Lidocaine is a widely used local anesthetic and antiarrhythmic agent. Its major use in cardiac patients is in the prophylaxis of ventricular fibrillation after myocardial infarction. Many approaches have been suggested for rapidly achieving and maintaining therapeutic concentrations of the drug. One method is a rapid, high-dose infusion of 50–100 mg over 20–60 min followed by a slower maintenance infusion of 1–2 mg/min.

The absolute bioavailability of orally administered lidocaine is only 35% due to extensive first-pass metabolism.[15] Attempts to give higher doses to overcome the first-pass effect and give concentrations similar to intravenous doses are complicated by side effects, probably due to the formation of toxic metabolites. For this reason, lidocaine is normally administered parenterally.

Lidocaine is metabolized predominantly in the liver. Its two major metabolites, monoethylglycine xylidine and glycinexylidine, possess antiarrhythmic activity and are found in significant concentrations during lidocaine therapy.[16] Accumulation of these metabolites during chronic dosing may account for toxicity occasionally observed in patients even though lidocaine concentrations are within the therapeutic range.

The large hepatic extraction ratio has led to the suggestion that lidocaine clearance is determined primarily by liver blood flow. The results of animal experiments using drugs known to alter liver perfusion and human studies conducted in patients with disease states known to alter hepatic blood flow have also given support to the perfusion-limited clearance concept.[17-19] In addition, coadministration of beta blockers, which alter cardiac output and, therefore, liver blood flow, has been shown to moderately reduce lidocaine clearance.[20,21]

Lidocaine is rapidly distributed following intravenous injection. Plasma concentration–time profiles decline biphasically with an initial half-life of 5–10 min and a slower terminal elimination phase of 1–2 hr.[22] The disposition parameters of lidocaine show marked intersubject variability, which may be expected of a compound that is subject to such a high degree of hepatic extraction. In general, however, these parameters are only moderately affected by age (Table II). Nation et al.[16] studied the pharmocokinetics of lidocaine in six healthy elderly subjects (aged 61–71 years) and in four healthy younger subjects (aged 22–26 years). Total clearance did not differ between groups. Compared to the younger subjects, the elderly group exhibited a longer elimination half-life (2.3 versus 1.3 hr) and a larger volume of distribution (1.6 versus 0.9 liter/kg).

Thomson et al.[22] studied the pharmacokinetics of lidocaine in patients with heart failure, liver disease, and renal failure. Eight heart failure patients (age range 44–73 years) exhibited a half-life similar to that of a younger control group ranging in age from 24 to 57 years. The volume of distribution and plasma clearance, however, were significantly reduced in these patients, probably due to decreased tissue perfusion. Patients with renal failure did not show changes in any pharmacokinetic parameter probably due to the fact that renal clearance accounts for only a small fraction of the total clearance of lidocaine. The eight liver disease patients (mean age 52 years) displayed a threefold increase in half-life, a significant increase in volume of distribution, and a decrease in plasma clearance. These alterations are to be expected, given that lidocaine undergoes a high degree of hepatic metabolism and that hepatic disease results in alterations in organ perfusion and plasma or tissue binding. These changes observed in these patients could not be correlated to severity of the disease state. These results are in good agreement with those of another study in which the elimination of lidocaine was determined in 23 patients of widely varying age with different types of liver disease.[23] Although the half-life showed marked interindividual variability (range 1.8–19 hr), no age relationship could be delineated. More recently, Cusson and others[18] studied age-dependent lidocaine disposition in patients with acute myocardial infarction. Lidocaine clearance decreased with age, partly as a result of a tendency

TABLE II PHARMACOKINETIC PARAMETERS OF LIDOCAINE

Subjects	Mean age (range)	Dose (mg)	Half-life (hr)	Clearance (ml/min per kg)	Distribution volume (liters/kg)	Ref.
10 healthy subjects	(24–57)	50–170 i.v.	1.8	10.0	1.3[a]	22
6 renal disease pts	(24–38)	50–170 i.v.	1.3	13.7	1.2[a]	22
8 liver disease pts	52 (45–65)	50–170 i.v.	4.9	6.0	2.3[a]	22
8 heart failure pts	61 (44–73)	50–170 i.v.	1.9	6.3	0.9[a]	22
4 healthy subjects	24 (22–26)	50 i.v.	1.3 ± 0.2	7.6 ± 1.6	0.9	16
6 elderly subjects	65 (61–71)	50 i.v.	2.3 ± 1.1	8.1 ± 1.9	1.6	16
13 cardiac patients	64 (35–81)	Variable[b]	6.0 ± 6.0	—	—	16
24 MI patients[c]	60 (35–85)	3–4 g infused over 18–59 hr	3.2 ± 1.2	10.5 ± 3.6	2.8 ± 1.3	18
11 MI patients[d]	73 (44–91)	3–4 g infused over 18–59 hr	5.2 ± 2.4	6.7 ± 2.2	2.9 ± 1.5	18
23 liver disease pts	50 (27–68)	400 p.o.	6.6 ± 1.1	—	—	23

[a] Volume of distribution—steady state.
[b] Infusion rate varied according to patient response.
[c] Without cardiac failure.
[d] With cardiac failure.

for body weight to decrease with age. The terminal half-life and volume of distribution were not altered by aging. The half-life, however, increased significantly from a mean of 3.2 hr in patients without heart failure (mean age 60) to a mean of 5.2 hr in patients with heart failure (mean age 73).

2.3. PROCAINAMIDE

Procainamide is an effective agent in the treatment of ventricular and supraventricular arrhythmias. Although many investigators have studied the pharmacokinetics of procainamide in the past decade, few of these studies have focused on the elderly population.

Procainamide is a weak base (pKa 9.3) and is thus almost completely ionized in the stomach. The small intestine appears to be the major site of absorption, and considerable variability exists in both the rate and extent of absorption from this site.[24] Absorption may be significantly delayed in elderly patients, in patients given drug soon after myocardial infarction, and in patients who are receiving drugs known to alter splanchnic blood flow or intestinal pH.[25,26]

The plasma concentration–time curve following an intravenous dose of procainamide is characterized by an initial rapid decline with a distribution half-life of approximately 5 min. This rapid distribution is due, in part, to the relatively low degree of plasma protein binding (15–20%).[27,28] The large unbound fraction results in a large volume of distribution (2–3 liters/kg). The apparent mean elimination half-life of procainamide in healthy subjects is 2–6 hr[29–31] (Table III) depending, in large part, on the rate at which an individual acetylates the drug.

Procainamide is cleared from the body via both renal and nonrenal mechanisms. Approximately one-half of the drug that enters the systemic circulation is found unchanged in the urine. Acidification or alkalinization of urine does not affect procainamide excretion, so it is unlikely that passive diffusion according to the pH partition hypothesis is an important mechanism of renal elimination at normal plasma concentrations.[32] Active tubular secretion appears to be the primary mechanism for renal elimination at therapeutic concentrations. At higher concentrations, this process may become saturated, leaving glomerular filtration as the major pathway of renal elimination. The aging process appears to decrease the renal tubular secretion of procainamide and, as a result, increases its half-life and steady-state drug concentrations.[33,34] Thus, dosage adjustments may be necessary in the elderly.

Wyman et al.[35] conducted a study designed to identify factors influencing the total body clearance of procainamide in arrhythmia patients

TABLE III PHARMACOKINETIC PARAMETERS OF PROCAINAMIDE

Subjects	Mean age (range)	Dose (mg)	Half-life (hr)	Clearance (ml/min per kg)	Distribution volume (liters/kg)	Ref.
3 healthy subjects	37 (36–38)	—	2.5	8.2 ± 7.5	1.4[a]	29
9 healthy subjects	(25–40)	1000 p.o.	3.0 ± 0.1	10.0 ± 0.7	2.6 ± 0.1	30
5 healthy subjects	34 (28–40)	1000 p.o.	2.9 ± 0.5	—	—	31
1 MI patient	67	7000 p.o. accidental	10.5	0.90	0.76	39
29 arrhythmia patients	61 (38–82)	500–1500 i.v.	5.2 ± 1.8[b] 2.7 ± 1.6[c]	5.7 ± 1.5[b] 9.7 ± 2.2[c]	2.6 ± 1.0[b] 2.1 ± 0.9[c]	26
15 MI patients	58 (31–78)	4–5 mg/min for 24 hr.	—	5.9	—	35
39 arrhythmia patients	60	—	5.2	4.3	1.93	34
5 cardiac patients	57 (42–65)	500–750 p.o. q6h	5.5 ± 0.9	—	—	31

[a] Volume of distribution—steady state.
[b] Slow acetylators
[c] Fast acetylators

ranging in age from 31 to 78 years. Results indicated that, in patients with good renal and hepatic function, body weight correlated better with total body clearance than did age and was the most important parameter in selecting an initial procainamide infusion rate. Other studies showed that changes in renal function may profoundly affect the elimination half-life of procainamide.[28,36] Patients with severe renal impairment may have half-lives greater than 20 hr.

The liver represents the major nonrenal site of procainamide elimination. Hepatic biotransformation results in the production of the major metabolite, N-acetylprocainamide (NAPA), and two metabolites of minor importance. Metabolism to NAPA is related to acetylator phenotype, with slow and fast acetylators converting 20% and 40% of the procainamide dose, respectively, to NAPA.[37,38] There is no evidence that drug acetylation decreases with age. Thus, one's phenotype remains the same throughout life and, in the case of lidocaine, may be used to predict the risk of developing clinically important adverse reactions. It has also been shown that decreased liver function caused by hepatic cirrhosis, chronic liver disease, or hepatic impairment secondary to cardiac disease or congestive heart failure may significantly decrease the elimination of procainamide.[39,40]

NAPA has been shown to have antiarrhythmic activity comparable to that of procainamide. Following procainamide administration, NAPA levels may be comparable to or even exceed those of the parent drug.[41] Reidenberg et al.[37] found that the plasma concentration ratio of NAPA to procainamide may be an indicator of the acetylator phenotype of the patient. In patients with normal renal function, slow and fast acetylators were found to have ratios of 0.61 and 1.8, respectively. Dutcher and co-workers[29] simultaneously studied the kinetics of coadministered procainamide and [^{13}C]NAPA in three healthy subjects who were fast acetylators. The mean elimination half-life of NAPA was more than twice that of the parent. Mean renal clearance of both compounds exceeded glomerular filtration, indicating that the metabolite is also eliminated in part by tubular secretion.

2.4. QUINIDINE

Quinidine is a type I antiarrhythmic agent, effective against both supraventricular and ventricular arrhythmias. It is a natural product, the dextrorotary isomer of quinidine, and is prepared by isolation from *Cinchona* bark or by isomerization of quinidine. The most common marketed forms of quinidine for oral administration are the sulfate and gluconate salts, with the gluconate salt representing most sustained-release forms. The gluconate and lactate salts are available for parenteral administration. Although plasma or serum quinidine concentrations have been found to be related to therapeutic effect, interindividual variability in the absorption, distribution, metabolism, and elimination of quinidine results in a poor dose–concentration relationship and therefore a poor dose–response relationship.[42,43]

Quinidine is a weak base with pKa's of 4.3 and 8.4 and is thus completely ionized in the acid environment of the stomach. In relative bioavailability comparisons conducted by Covinsky et al.[44] and Greenblatt et al.,[45] single oral doses of quinidine sulfate or quinidine gluconate, as solid oral dosage forms, were given to healthy volunteers between 20 and 54 years of age. Maximum plasma concentrations following the sulfate salt and the gluconate salt occurred in 1–2 hr and 4–5 hr, respectively. In another study,[46] the gluconate salt was administered as an oral solution in healthy older subjects, aged 40–78 years. In this case, quinidine absorption was rapid, with drug detected in the plasma within 15 min after ingestion. Maximum plasma concentrations occurred at times similar to those observed for the sulfate salt in previous studies. Thus, it appears that quinidine is rapidly absorbed from the GI tract, regardless of the large ionized fraction present in the stomach, and that this rapid absorption is independent of age.

The pharmacokinetics of quinidine have been described in numerous studies conducted in healthy subjects of various ages. A summary of the disposition parameters from some of these studies is given in Table IV. In general, the pharmacokinetic parameters are independent of age, and differences in the reported parameters are often a function of study design. Although the drug is rapidly absorbed in healthy older subjects, Ueda et al.[47] found that the absolute bioavailability of orally administered quinidine was 44–89%, with the reduced bioavailability due primarily to first-pass hepatic drug removal. The volume of distribution is on the order of 2–3 liters/kg and is similar in both young and old. Quinidine is eliminated via both renal and hepatic mechanisms. Less than 20% of an oral or intravenous dose appears in the urine as unchanged drug. Ochs et al.[48] observed a decrease in total clearance from 4.0 ml/min per kg in young subjects to 2.6 ml/min per kg in elderly subjects, due to reductions in both renal and nonrenal elimination pathways. The terminal elimination half-life in this study was 7.2 ± 0.7 hr in the 14 young subjects and 9.7 ± 0.9 hr in the eight elderly (age 60–69 years) subjects. In another study, however, Ueda et al.[49] found a terminal half-life of 6.3 hr in 12 arrhythmia patients with normal hepatic and renal function (age 44–78 years). The discrepancy in these studies appears to be due to the different methods used to analyze the data. The

TABLE IV PHARMACOKINETIC PARAMETERS OF QUINIDINE

Subjects	Mean age (range)	Dose[a] (mg)	Half-life (hr)	Clearance (ml/min per kg)	Distribution volume (liters/kg)	Ref.
11 healthy subjects	28 (23–31)	330 B i.v.	7.3 ± 0.8	3.8 ± 0.3	2.3 ± 0.2	45
14 healthy subjects	29 (23–34)	180–300 B i.v.	7.2 ± 0.7	4.0 ± 0.3	2.4 ± 0.2	48
8 healthy subjects	66 (60–69)	180–300 B i.v.	9.7 ± 0.9	2.6 ± 0.2	2.2 ± 0.2	48
20 healthy subjects	(21–54)	200 S	6.4 ± 1.7	6.4[b]	3.5[b]	44
20 healthy subjects	(21–54)	330 G tab	7.7 ± 1.0	10.1[b]	6.7[b]	44
12 arrhythmia patients	54 (44–78)	2.6–5.2 mg/kg B i.v.	6.3 ± 0.5	4.7 ± 0.5	3.0 ± 0.3[c]	49
8 healthy subjects	57 (40–78)	400 G i.v., p.o.	6.2 ± 1.8	5.0 ± 1.4	2.7 ± 1.2	46,47
9 CHF patients	56 (42–79)	400 G i.v.	6.8 ± 1.6	3.2 ± 1.1	1.8 ± 0.5	46
9 arrhythmia patients	49 (16–76)	2–6 mg/kg B i.v.	7.8 ± 0.7	4.8 ± 0.8	3.0 ± 0.5	51
		4–10 mg/kg B p.o.	11.0 ± 1.5	—	—	
8 healthy subjects	(43–67)	200 S p.o.	6.0 ± 0.5	5.0 ± 0.4	2.5 ± 0.1	52
8 cirrhotic patients	49 (44–54)	200 S p.o.	9.0 ± 1.0	5.2 ± 0.6	3.8 ± 0.4	52

[a] B = base; G = gluconate; S = sulfate.
[b] Normalized to 70 kg.
[c] Volume of distribution—steady state.

former study included plasma concentrations observed at 24 and 30 hr after drug administration, whereas the latter study used data only until 12 hr postadministration and excluded a 24-hr data point that fell above the terminal log–linear regression line in all cases. Thus, the half-life observed in the latter study probably underestimates the true half-life in the elderly population.

The pharmacokinetics of quinidine have also been reported in patients with renal,[50] cardiac,[46,51] and hepatic[52] disease (Table IV). In a study conducted in nine arrhythmia patients of various ages (range 16–76 years), pharmacokinetic parameters were independent of age and were similar to those obtained in other studies conducted in healthy subjects.[51] Ueda and Dzindzio[46] studied the pharmacokinetics of quinidine in nine patients aged 42–79 years who exhibited moderate to severe congestive heart failure (CHF). Mean plasma concentrations were higher in this group of subjects when compared to a similarly aged group of healthy subjects. The elimination half-life was primarily associated with nonrenal elimination pathways in each group. The authors concluded that the decrease in clearance in the CHF patients, coupled with a similar elimination half-life, suggested that the apparent volume of distribution was decreased in the presence of CHF. Similar findings were reported by Crouthamel[53] and by Kessler et al.[54] Crouthamel[53] also suggested that the bioavailability of quinidine is reduced in CHF as a result of a slower absorption rate due to reduced splanchnic blood flow, which allows more drug to be metabolized in the first pass through the liver.

Patients with cirrhosis had a significantly longer half-life (9 hr) when compared to a similarly aged control group (6 hr).[52] This was not related to reduced quinidine clearance but to an increased volume of distribution. This was probably due to the increased free (unbound) fraction of quinidine observed in cirrhotics compared to control patients. Therapeutically, the overall effect is unpredictable since the increased volume of distribution will result in lower plasma concentrations, but the decrease in binding results in an increase in free (active) drug concentration.

Studies conducted by Bellet et al.[50] and Ochs et al.[48] have shown that plasma or serum quinidine concentrations increase in patients with renal insufficiency owing to a decrease in glomerular filtration rate. These findings are in contrast to an earlier report showing no impairment of quinidine elimination in patients with poor renal function.[54] Excretion of quinidine was found to vary inversely with urine pH. Clearance was decreased by 50% and serum concentrations increased during alkalinization.[55]

Numerous investigators have studied the effects of other drugs on the pharmacokinetics of quinidine. Patients receiving propranolol together with quinidine had a half-life similar to those receiving quinidine alone.[52] However, plasma concentrations were significantly higher in propranolol patients due to decreased clearance and increased volume of distribution. It was therefore recommended that the dosage be decreased and the frequency adjusted in patients receiving this combination.

The administration of quinidine to patients receiving digoxin results in increased digoxin concentrations due to a decrease in total body clearance.[56-58] The mechanism by which quinidine increases digoxin plasma concentrations is thought to be a decrease in the renal tubular secretion and/or extrarenal clearance of digoxin or a displacement of tissue-bound digoxin.

3. New Antiarrhythmic Agents

3.1. Amiodarone

Amiodarone is a class III agent that has been used as an effective antianginal and antiarrhythmic agent in Europe and South America since 1967 and has recently received approval for marketing in the United States. Amiodarone is given intravenously as a dose of 5 mg/kg over 1 min and orally as maintenance doses of 200–800 mg daily.[59-61]

The absorption of amiodarone is slow and erratic in both young and old.[62,63] The absolute bioavailability varies from 20 to 100% in healthy younger subjects[64] and may be decreased somewhat in older subjects.[65,66]

The pharmacokinetics of amiodarone following single oral and intravenous doses have been reported by many investigators in both young and elderly patients. The disposition parameters reported in some of these studies are summarized in Table V. The large amount of variability seen in those parameters precludes any definitive comments regarding kinetic alterations in the elderly. The wide range of observed half-lives (17–595 hr) is probably explained by factors in addition to intersubject variability. The most likely explanation is that some investigators used more sensitive assay techniques or obtained blood samples longer after drug administration and were thus able to observe slower terminal phases of the plasma concentration–time profile.

Amiodarone is primarily eliminated from the body via hepatic metabolism. Several studies have shown that no unchanged drug appears in the urine following oral administration. The major metabolite has been identified as N-desethyl amiodarone.[62] During long-term oral dos-

TABLE V PHARMACOKINETIC PARAMETERS OF AMIODARONE

Subjects	Mean age (range)	Dose (mg)	Half-life	Clearance (ml/min per kg)	Distribution volume (liters/kg)	Ref.
Single dose			(hours)			
6 healthy subjects	27	400 i.v.	595	1.9	65.8	71
3 healthy subjects	26 (25–27)	150 i.v.	17.4 ± 4.6	8.5 ± 1.2	12.9 ± 4.0	65
5 healthy subjects	27 (22–32)	200 p.o.	17.1 ± 6.0	4.9 ± 2.4	8.7 ± 7.3	66
6 arrhythmia patients	59 (35–73)	1400–1800 p.o.	7.2 ± 5.0	—	—	62
8 arrhythmia patients	52 (23–70)	150 i.v.	4.3 ± 1.0	3.5 ± 1.2	1.3 ± 0.4	66
7 arrhythmia patients	58 (49–67)	400 i.v.	11	—	—	73
3 arrhythmia patients	57 (43–71)	260–275 i.v.	16.2 ± 4.1	8.4 ± 5.1	11.7	65
3 arrhythmia patients	57 (50–69)	400 p.o.	9.2 ± 3.2	—	—	65
8 arrhythmia patients	62 (50–70)	200–400 p.o.	17.1 ± 5.2	3.1 ± 1.0	4.8 ± 2.0	66
8 arrhythmia patients	56 (22–93)	800 p.o.	4.6	—	—	60
Multiple dose			(days)			
4 arrhythmia patients	67 (60–72)	600–1000/day p.o.	29 ± 19	—	—	62
8 arrhythmia patients	53	200–600/day p.o.	52.6	—	—	71
7 arrhythmia patients	53 (32–76)	200–600/day p.o.	40.7 ± 18.8	3.0 ± 1.3	—	72

ing, the serum concentration of the metabolite may rise to as much as 50% of that of amiodarone.[67] Although the relative activity of the metabolite has not been studied, the findings from several chronic dosing studies suggest that the antiarrhythmic action of amiodarone may be due, in part, to the formation of an active metabolite.[59,62]

The apparent elimination half-life of amiodarone has been shown to increase dramatically in patients during chronic dosing. Kannan et al.[62] found that the half-life of 7.2 hr following a single oral dose increased to 29 days after multiple dosing in older arrhythmia patients. It was hypothesized in this and in other studies that the increase in half-life was due to the accumulation of drug in adipose tissue and other organs that have strong affinity for the drug.[68] Other studies have suggested that the true half-life may be even longer.[69-72] The significantly longer half-life following chronic dosing suggests that the pharmacokinetics of amiodarone are nonlinear. It is possible, however, that the long half-life is due to accumulation of plasma concentrations into the measurable range. Indeed, Riva et al.[61] have recently shown that the plasma concentration–time data obtained following a single dose and a multiple dose regimen could be described by a single equation without the need to invoke nonlinearity into the pharmacokinetics of amiodarone. The long apparent half-life and extensive accumulation following multiple oral dosing suggest that a chronic oral dosing regimen must be continuously adjusted for individual patients. The presence of a deep tissue for sequestering drug during chronic dosing suggests that elderly patients may handle the drug differently, since the composition of the body changes significantly with advancing age. However, studies have not been conducted that report on the pharmacokinetics of amiodarone as a function of age.

The relationship between serum amiodarone concentration and the drug's antiarrhythmic effect is complicated by its long apparent half-life and extensive tissue accumulation following multiple oral dosing. Despite high serum concentrations of amiodarone during the initial chronic oral dosing, there is a delayed onset of antiarrhythmic effect.[69] During long-term dosing, however, serum concentrations of 0.6–3.0 μg/ml have shown therapeutic effect.[62,73,74]

Arboix et al.[75] showed that the effective acenocoumarol daily dose could be reduced in patients undergoing simultaneous amiodarone therapy. The effect was shown to be age-independent in patients over 40. In another study, Serlin et al.[76] observed a gradual rise in warfarin plasma concentrations after beginning amiodarone treatment. Whether or not the adjustment was necessary because of a protein binding interaction was not determined. Lalloz et al.[77] have shown that, although amiodarone is approximately 96% bound to plasma proteins, it does not displace warfarin because the two drugs bind to different sites.

3.2. CIBENZOLINE

Cibenzoline is a class I agent with a unique chemical structure being evaluated clinically for both ventricular and supraventricular arrhythmias. The usual dosage is 130–160 mg twice daily. An intravenous dosing regimen has been suggested consisting of a 0.25 mg/kg infusion over 1 min followed by a 1–1.5 mg/kg per hr for 1 hr and 0.2–0.4 mg/kg per hr maintenance infusion.[78]

Cibenzoline is well absorbed following oral administration with an absolute bioavailability of approximately 85% in healthy young subjects and in older patients with renal failure.[79,80] Plasma concentrations increase proportionally with doses ranging from 65 to 260 mg.[81] Plasma concentrations decline biexponentially following oral administration with an apparent elimination half-life between 7 and 15 hr (Table VI). Cibenzoline is eliminated primarily by renal excretion of unchanged drug, with as much as 60% of the dose excreted in the urine. The renal clearance of unchanged drug in healthy volunteers greatly exceeds the glomerular filtration rate, indicating that renal tubular secretion contributes to its urinary excretion.

The pharmacokinetics of cibenzoline were studied in 36 healthy subjects between 20 and 80 years of age (six per decade).[82] The mean apparent elimination half-life was found to increase with age from approximately 7 hr in the 20–30 age group to 10.5 hr in the 70–80 age group. The increased half-life was due to decreases in both renal and nonrenal clearance.

The relationship between cibenzoline plasma concentration and antiarrhythmic effect was evaluated in 25 patients ranging in age from 34 to 76 years (mean 61 years) after ascending multiple doses of the drug.[83] Although there was considerable interpatient variability in antiarrhythmic response, plasma concentrations of greater than 300 ng/ml were associated with some decrease in PVC frequency in virtually all patients.

3.3. ENCAINIDE

Encainide is a class I agent being evaluated for the suppression of ventricular arrhythmias. The drug is administered intravenously at doses of up to 1 mg/kg over 15–30 min and orally at daily doses of up to 325 mg.[84–86]

TABLE VI   PHARMACOKINETIC PARAMETERS OF CIBENZOLINE

Subjects	Mean age (range)	Dose (mg)	Half-life (hr)	Clearance (ml/min per kg)	Distribution volume (liters/kg)	Ref.
6 healthy subjects	24 (20–29)	160 p.o.	7.0	14.5 ± 5.3	8.5 ± 3.1	82
6 healthy subjects	35 (30–39)	160 p.o.	10.5	13.6 ± 1.0	8.0 ± 0.6	82
6 healthy subjects	43 (40–49)	160 p.o.	8.7	10.1 ± 5.0	6.0 ± 2.9	82
6 healthy subjects	54 (50–59)	160 p.o.	9.1	9.6 ± 0.9	5.7 ± 0.5	82
6 healthy subjects	65 (60–69)	160 p.o.	9.9	8.7 ± 1.7	4.9 ± 1.3	82
6 healthy subjects	74 (70–79)	160 p.o.	10.5	6.3 ± 1.1	3.8 ± 0.6	82
6 healthy subjects	64 (51–78)	100 i.v.	9.8	6.9 ± 1.5	5.9 ± 0.9	78

Absorption of encainide is quite variable, possibly related to saturable presystemic metabolism. Absolute bioavailability ranges from 7 to 82%.[85,86] Following single 75-mg oral doses to older arrhythmia patients (age range 49–77), maximum plasma encainide concentrations ranging from 36 to 587 ng/ml occur at 1.5–3 hr. These concentrations are in contrast to those following a 20-min 75-mg intravenous infusion in which maximum concentrations ranged from 640 to 1556 ng/ml.[86]

The pharmacokinetic parameters of encainide are given in Table VII. A definitive study comparing the pharmacokinetics in the young and elderly has not been reported. However, several studies have been conducted in older arrhythmia patients. Encainide has a large volume of distribution in older arrhythmia patients (mean 3.5 ± 1.5 liters/kg), which does not appear to be influenced greatly by body weight over the range of 70–120 kg. Thus, body weight need not be a major dosing consideration in the elderly. The drug has a large and variable systemic clearance, ranging from 4 to 22 ml/min per kg, and a half-life ranging from 2 to 7 hr following single doses.[86,87]

Encainide is rapidly eliminated from the body via hepatic biotransformation. Three metabolites have been identified: O-desmethyl encainide (ODE), 3-methoxy-0-desmethyl encainide (MODE), and N-desethylencainide. The results of studies by Kates et al.[88] indicate that MODE is equipotent to encainide and ODE is significantly more potent. The mean half-life of ODE following multiple oral dosing to patients was 11 hr, ranging from 5 to 37 hr. The elimination of MODE was somewhat slower, although its half-life was not reported. There appears to be a subset of the population (5–10%) who are slow metabolizers of encainide and exhibit markedly different encainide pharmacokinetics compared to the rest of the population.[88,89] These subjects do not form two of the metabolites.

A slightly longer elimination half-life has been observed during multiple oral dosing.[85] In addition, a nonlinear relationship between steady-state encainide concentrations and dosage has been reported.[85] An increase in dose from 25 to 75 mg q6h resulted in a 12-fold increase in mean trough plasma concentrations suggesting saturable presystemic elimination. During chronic dosing of 100–250 mg/day to patients, the mean steady-state concentrations of encainide, ODE, and MODE were 56, 215, and 185 ng/ml, respectively.[90] The accumulation of ODE and MODE during chronic dosing may contribute substantially to its efficacy during long-term oral therapy.

Encainide clearance was reduced following i.v. or oral administration in a study conducted in patients with liver disease.[91] However, there was no significant difference in encainide half-life, volume of distribu-

TABLE VII PHARMACOKINETIC PARAMETERS OF ENCAINIDE

Subjects	Mean age (range)	Dose (mg)	Half-life (hr)	Clearance (ml/min per kg)	Distribution volume (liters/kg)	Ref.
Single dose						
9 arrhythmia patients	60 (49–77)	75 i.v.	3.4 ± 1.7	13.2 ± 5.6	3.5 ± 1.5	86
9 arrhythmia patients	60 (49–77)	75 p.o.	2.5 ± 0.8	—	—	86
11 arrhythmia patients	(32–67)	25 p.o.	2.7 ± 0.2	—	—	85
Multiple dose						
11 arrhythmia patients	(32–67)	100–300/day p.o.	3.4 ± 0.3	—	—	85
13 arrhythmia patients	59 (43–76)	100–250/day	1.2 ± 0.5	—	—	90

tion, protein binding, or bioavailability compared to healthy subjects. Plasma ODE and MODE concentration–time profiles did not differ between healthy subjects and liver patients. The results of this study suggest that dosage modifications in liver disease are not necessary.

The relationship between plasma encainide concentration and antiarrhythmic effect is complicated by the formation of active metabolites. Following i.v. administration to arrhythmia patients, minimal effective plasma concentrations were higher (range 3.5–170 ng/ml) than after oral administration (range 1.5–4.8 ng/ml), possibly due to formation of the active metabolites after oral administration.[86]

3.4. FLECAINIDE

Flecainide acetate is a class I agent with possible class III antiarrhythmic activity which has been recently approved in the United States for oral use and is being evaluated for intravenous administration. Flecainide is given intravenously at doses up to 2 mg/kg over 5 min and orally at doses of 100–250 mg every 12 hr.[92,93]

The pharmacokinetics of flecainide have been reviewed previously.[92] Following single dose oral administration of 60–240 mg flecainide to healthy young subjects, maximum plasma concentrations occur at 1.5–3 hr. The absolute bioavailability is approximately 95%, and plasma concentrations are proportional to dose. The bioavailability is not reduced by the presence of food or antacid, and the absorption rate is comparable in older patients with arrhythmias, renal impairment, and congestive heart failure. Flecainide is approximately 40% bound to plasma proteins over the concentration range of 15–3400 ng/ml.[94] Because of the relatively low degree of protein binding, it is unlikely that drug–drug interactions based on binding displacement phenomena or distribution volume alterations due to plasma protein concentration changes in the elderly would occur. Indeed, studies have shown that no binding interaction is seen between flecainide and a variety of other antiarrhythmics and cardioactive agents.

Following intravenous administration, flecainide plasma concentrations show an initial rapid decline during the first 15–30 min followed by a relatively slow elimination phase with a terminal half-life of 7–23 hr (Table VIII). The mean plasma clearance of flecainide following i.v. administration is approximately 7.6 ml/min per kg (range 4.6–12.1 ml/min per kg), and the volume of distribution is 8.7 liters/kg (range 5.0–13.4 liters/kg).[92] Similar values are seen following oral administration.[92,95] The pharmacokinetics of flecainide do not appear to change significantly during multiple dosing, with single-dose kinetics being predictive of

TABLE VIII Pharmacokinetic Parameters of Flecainide

Subjects	Mean age (Range)	Dose (mg)	Half-life (hr)	Clearance (ml/min per kg)	Distribution volume (liters/kg)	Ref.
Single dose						
9 healthy subjects	NR[a]	200 p.o.	13.8 ± 2.9	10.2 ± 3.8	8.4	95
4 healthy subjects	48 (36–56)	200 p.o. (^{14}C)	16 (8.5–21.4)[b]	—	—	98
8 healthy subjects	NR	0.6–1.7 mg/kg i.v.	14 (7–19)	7.6 (4.6–12.1)	8.7 (5.0–13.4)	92
10 CHF patients	NR	200 p.o.	19.4 ± 5.2	8.1 ± 3.5	9.4	95
10 renal patients	NR	200 p.o.	17 (12–26)	6.7 (2.2–13.9)	—	92
Multiple dose						
11 arrhythmia patients	51 (32–65)	200–600/day p.o.	20.3 ± 4.4	6.2 (3.1–12.6)	—	93
9 arrhythmia patients	54 (28–72)	100–250/day p.o.	18.8 ± 3.8	—	—	96
11 arrhythmia patients	57 (27–68)	200–600/day p.o.	19.7 ± 4.6	—	—	97

[a] NR = not reported.
[b] Values in parentheses indicate range.

steady state.[92,93,96,97] Conard and Ober[92] examined the effect of age on the half-life and clearance of flecainide. Over the range of 28–69 years, the half-life tended to be longer and the plasma clearance smaller for older patients compared to younger patients, although the correlation was not statistically significant.

Following an oral dose of [^{14}C]flecainide to healthy volunteers (mean age 48), 81–90% of the administered radioactivity was recovered in the urine and 4–6% in the feces.[98] Thirty-five to fifty percent of the administered dose was excreted unchanged in the urine within the first 72 hr. Alkalinization of the urine to pH above neutral resulted in a significantly longer elimination half-life and a greater fraction of drug appearing unchanged in the urine.[99] Thus, dietary factors should be taken into consideration in dosing elderly patients.

Studies have shown that flecainide may affect steady-state digoxin and propranolol plasma concentrations.[100,101] These studies were conducted in healthy young subjects, and the effect was not considered to be of therapeutic importance. A definitive study has not been conducted in elderly subjects, although similar results may be expected.

The minimal effective concentration range was 245–980 ng/ml in 11 arrhythmia patients (mean age 51) after receiving oral doses of 100–300 mg every 12 hr.[93] In another study in which a multiple oral dose regimen of 50–200 mg b.i.d. was given to eight patients (mean age 53) over 12 days, the minimum plasma flecainide concentrations for therapeutic effect ranged from 200 to 400 ng/ml.[92]

3.5. LORCAINIDE

Lorcainide is a class I agent that can be administered both intravenously and orally. The usual i.v. dose is 100 mg or 2 mg/kg given over 5 min, and the oral doses are 100–200 mg given every 6, 8, or 12 hr.[102–105]

Following oral administration of lorcainide, maximum plasma concentrations occur within 1–4 hr.[106,107] Bioavailability appears to increase with dose, suggesting the presence of extensive but saturable presystemic metabolism.[107,108] Jahnchen et al.[106] reported an absolute bioavailability of 0–78% following a 150-mg dose and 12–100% following a 300-mg dose to arrhythmia patients ranging in age from 22 to 69. The fraction absorbed at each dose was highly variable and not related to age. The plasma protein binding of lorcainide averages approximately 79% in arrhythmia patients and is not affected by age.[106,107]

There is considerable intersubject variability in the pharmacokinetics of lorcainide. The mean elimination half-life following i.v. administration is slightly longer in patients than healthy volunteers, but the total plasma clearance and apparent volumes of distribution are approximately the same (Table IX). Clearance of lorcainide is within the normal physiological range of liver blood flow, predicting the low bioavailability observed at lower doses.

Kates et al.[109] studied the pharmacokinetics of lorcainide in 14 arrhythmia patients varying in age from 37 to 74 years. There did not appear to be age-related differences in any pharmacokinetic parameter. These results are in agreement with those of other investigators who have collectively studied a total of 40 patients ranging in age from 22 to 78 years.[106,107,110]

Lorcainide is predominantly eliminated from the body via biotransformation to an active dealkylated metabolite, norlorcainide.[106,107,109] Less than 5% of an administered dose is recovered unchanged in the urine. Kates et al.[109] found that, following oral dosing of 100 mg given to eight patients every 12 hr for one week, the half-life of norlorcainide ranged from 19 to 38 hr as compared to 6 to 14 hr for lorcainide.

Jahnchen et al.[106] studied the disposition of lorcainide in patients (mean age 50) during long-term i.v. and oral treatment. Low plasma concentrations of norlorcainide were present during i.v. treatment. When oral dosing was initiated, however, norlorcainide concentrations exceeded those of unchanged lorcainide after the third dose. Estimates of bioavailability during chronic dosing were considerably higher than those following single oral doses, suggesting that the clearance of lorcainide decreased during long-term therapy. Age did not appear to influence the bioavailability during chronic oral dosing. Similar results were reported by Klotz et al.[107] and Kates et al.[109] After several weeks of multiple oral dosing of 100 mg of lorcainide every 12 hr to eight patients (mean age 52), steady-state plasma concentrations of norlorcainide ranged from 102 to 678 ng/ml. These concentrations were two- to threefold higher than those of the parent drug, reflecting significant accumulation of the active metabolite. The accumulation of norlorcainide during chronic dosing probably contributes to a 10-fold range of intersubject variation in plasma lorcainide concentrations and the overall antiarrhythmic efficacy of lorcainide.[104,111]

After a single 100-mg i.v. dose of lorcainide to arrhythmia patients, there is a linear relationship between plasma drug concentrations and prolongation of the QRS complex.[104] Lower plasma lorcainide concentrations are required to produce the same QRS widening following oral administration than following intravenous administration, probably due to the presence of the active metabolite, norlorcainide. The relationship between lorcainide plasma concentrations and antiarrhythmic effect during chronic dosing is complicated by the accumulation of norlor-

TABLE IX PHARMACOKINETIC PARAMETERS OF LORCAINIDE

Subjects	Mean age (range)	Dose (mg)	Half-life (hr)	Clearance (ml/min per kg)	Distribution volume (liters/kg)	Ref.
Single dose						
5 healthy subjects	(22–26)	100 i.v.	5.1 ± 0.6	17.1	8.6 ± 2.4	107
6 arrhythmia patients	63 (49–78)	100 i.v.	7.6 ± 2.2	15.6	10.7 ± 4.2	107
10 arrhythmia patients	42 (22–64)	2 mg/kg i.v.	7.7 ± 4.0	24.1 ± 13.1	12.9 ± 3.9	106
6 arrhythmia patients	57 (49–72)	100–180 i.v.	7.8 ± 2.5	12.7 ± 5.4	7.9 ± 2.3	103
10 arrhythmia patients	66 (59–70)	182 ± 26 i.v.	13.1 ± 5.0[a]	15.6 ± 4.4	17.0 ± 7.9	110
14 arrhythmia patients	54 (37–74)	100–200 i.v.	7.8 ± 2.2	14.4 ± 3.3	6.3 ± 2.2	109
Multiple dose						
4 arrhythmia patients	63 (50–75)	100 p.o. b.i.d.	5.3 ± 0.3	—	—	107
6 arrhythmia patients	64 (45–80)	200–400/day p.o. or i.v.	—	11.3 ± 4.5	—	103
8 arrhythmia patients	52 (37–74)	200/day	9.6 ± 2.8	13.3 ± 2.0	6.4 ± 2.5	109

[a] Authors suggested that age may be responsible for prolonged half-life compared to other studies. However, no correlation between age and half-life was seen, and a later study in similarly aged subjects[73] failed to demonstrate an age-dependent increase in half-life.

cainide. Keefe *et al.*[112] reported an 82% suppression of PVCs in 10 patients receiving 100 mg twice daily. Effective lorcainide plasma concentrations ranged from 10 to 170 ng/ml and norlorcainide from 49 to 257 ng/ml.

3.6. MEXILETINE

Mexiletine is a class I antiarrhythmic agent that is structurally similar to lidocaine. It is a weak base with a pKa of 8.75. The usual regimen for intravenous therapy is an initial bolus of 150–250 mg over 5–10 min followed by a maintenance infusion of 20–40 mg/hr.[113] An oral loading dose of 400–600 mg followed by maintenance doses of 100–400 mg mexiletine every 6–8 hr has been reported.[113]

Mexiletine is well absorbed orally. Maximum plasma mexiletine concentrations of 0.6–1.5 μg/ml occur within 2–4 hr after administration of a 400-mg oral dose.[114] Mean systemic availability of mexiletine is approximately 90% in healthy subjects and may be higher in patients or older subjects. Delayed and incomplete absorption of mexiletine has been observed in patients with myocardial infarction and in patients taking narcotic analgesics.[115]

The moderate degree of plasma protein binding of mexiletine (70%) results in extensive distribution in the body and a relatively large distribution volume of 5–10 liters/kg (Table X). Reported mean half-lives in healthy volunteers range from 5 to 10 hr.[114–116] The mean total body clearance is 5–11 ml/min per kg following i.v. and oral administration. Clearance may be decreased in patients with acute myocardial infarction.[114,117–119]

Mexiletine is eliminated predominantly by hepatic metabolism. The major metabolites are parahydroxymexiletine, hydroxymethylmexiletine, and their corresponding alcohols.[120] These metabolites are not pharmacologically active and do not accumulate to a significant extent in the body.

At the usual pH of urine, 3–16% of an oral or i.v. dose of mexiletine is recovered intact in the urine. Mitchell *et al.*[121] found that urine pH may have a dramatic effect on the renal clearance of mexiletine. In healthy subjects, renal clearance of mexiletine was 4 ml/min at pH 5 compared to 168 ml/min at pH 8. The mean half-life was 6.7 hr under acidic conditions and 9.7 hr under alkaline conditions. Predicted steady-state mexiletine concentrations during infusion increased 5–95% with change in urine pH from 5 to 8. Thus, factors likely to alter urinary pH should be avoided during treatment.

Steady-state plasma concentrations are directly proportional to the

TABLE X PHARMACOKINETIC PARAMETERS OF MEXILETINE

Subjects	Mean age (range)	Dose (mg)	Half-life (hr)	Clearance (ml/min per kg)	Distribution volume (liters/kg)	Ref.
Single dose
6 healthy subjects	33 (27–47)	200 i.v.	6.3 ± 1.5	10.3 ± 2.3	5.5 ± 0.5	116
5 healthy subjects	(25–28)	200 i.v.	11.8 ± 1.5	6.1[a]	6.0[a]	115
4 healthy subjects	(28–42)	100 i.v.	10.4 ± 2.8	10.7 ± 5.9[b]	9.5 ± 3.4[b]	114
6 healthy subjects	(28–42)	3 mg/kg p.o.	9.3 ± 1.0	9.7 ± 2.5[b]	5.4[b]	114
7 MI patients	58 (50–60)	400 p.o.	15.0 ± 0.6	—	—	118
12 CCU patients	NR[c]	100 i.v.	16.7 ± 5.1	5.9 ± 1.0[b]	10.8 ± 7.4[b]	114
6 CCU patients	NR	3 mg/kg p.o.	12.1 ± 4.0	5.1 ± 1.3[b,d]	—	114
10 arrhythmia patients	NR	200 i.v.	13.2 ± 1.7	6.5 ± 1.2	6.6 ± 0.9	117
Multiple dose
9 healthy subjects	65 (50–77)	50 q8h	10.4 ± 3.2	—	—	122
15 renal failure pts	63 (31–73)	50 q8h	14.8	—	—	122
58 arrhythmia patients	66	200–250 q8h	9.6	6.3	5.3	121

[a] Normalized to mean body weight.
[b] Normalized to 70 kg.
[c] NR = not reported.
[d] Assumes 90% oral bioavailability.

dose but show large intersubject variability (five- to eightfold). The elimination rate appears to be independent of age and dose.[122] El Allaf *et al.*[123] administered multiple doses of oral mexiletine to three groups of older patients (mean age 63) with various degrees of renal insufficiency. The mean elimination half-lives were 10, 15, and 16 hr, respectively, in patients with creatinine clearance of >75, 10–30, and 0–10 ml/min. Although the patients in this study covered a wide range of ages (31–77 years), the degree of renal impairment appeared to be a more important variable in predicting the elimination half-life. Another study showed that peritoneal dialysis does not remove significant amounts of mexiletine from the body.[124]

The disposition of mexiletine is altered by several drugs commonly used in older patients. The absorption of mexiletine is delayed by aluminum hydroxide and atropine. Phenytoin and rifampicin have been shown to enhance its metabolism.[101] Leahey *et al.*[58] reported that serum digoxin concentrations were not increased significantly by the coadministration of mexiletine.

3.7. PIRMENOL

Pirmenol is a new antiarrhythmic agent being investigated for the suppression of ventricular arrhythmias. Intravenous doses of 70–150 mg infused over 30 min or single oral doses of 150–250 mg have been shown to be safe and effective.[125] A three-stage i.v. infusion regimen consisting of an initial bolus of 50 mg over 2 min followed by a loading infusion of 2.5 mg/min for 1 hr and a maintenance infusion of 0.25 mg/min satisfactorily achieves and maintains therapeutic plasma concentrations with no toxicity.[126]

Although several clinical studies have assessed the pharmacokinetics of pirmenol in patients covering a wide age range, none have provided sufficient information to make definitive statements regarding the effect of aging on the drug's kinetics. Table XI summarizes several studies conducted in older arrhythmia patients.

Pirmenol is rapidly absorbed, resulting in plasma concentrations that are proportional to dose. Maximum plasma pirmenol concentrations range from 0.7 to 2.0 µg/ml following single 150–250 mg doses to arrhythmia patients.[127] The mean absolute bioavailability is 83% (range 50–100%). Pirmenol is 82–89% bound to plasma proteins with the ratio of bound to free drug correlating well with the concentration of α_1-acid glycoprotein.[128]

Pirmenol plasma concentration–time data display a biexponential decline with a mean elimination half-life of approximately 8 hr. Ap-

TABLE XI PHARMACOKINETIC PARAMETERS OF PIRMENOL

Subjects	Mean age (range)	Dose (mg)	Half-life (hr)	Clearance (ml/min per kg)	Distribution volume (liters/kg)	Ref.
12 arrhythmia patients	60 (27–75)	150 i.v.	7.6	2.3[a]	1.4 ± 0.4	129
8 arrhythmia patients	56	50–250 p.o.	9.3	—	—	127
10 arrhythmia patients	54	0.5–150 i.v.	8.4	2.9[a]	2.0	128

[a] Normalized to 70 kg.

proximately one-third of an i.v. dose of pirmenol appears unchanged in the urine within 24 hr, with renal clearance ranging from 14 to 78 ml/min.[128,129] Similar values were reported for the nonrenal clearance of pirmenol.

Efficacy of pirmenol has been associated with plasma concentrations of 1–4 µg/ml, with arrhythmias reappearing at plasma concentrations of 0.8–1.4 µg/ml.[128] Hammill et al.[127] found greater than 90% reduction of PVCs with concentrations of 0.7–2.0 µg/ml following oral doses of 150–250 mg of pirmenol.

3.8. PROPAFENONE

Propafenone is being evaluated for treatment of ventricular and supraventricular arrhythmias. Oral doses of 450–1200 mg/day have been studied, usually given 3–4 times a day.[130]

Propafenone undergoes dose-dependent presystemic metabolism following oral administration.[131] Estimates of absolute bioavailability following oral administration of 150, 300, and 450 mg to healthy subjects were 13, 32, and 55%, respectively.

The pharmacokinetic parameters of propafenone are given in Table XII. The disposition of propafenone was studied in arrhythmia patients ranging in age from 19 to 76 following 1 mg/kg intravenous[132] and oral[133] doses. Large interpatient variability was observed in the half-life (range 1.9–15.3 hr), clearance (4.0–22.2 ml/min per kg), and volume of distribution (1.1–10.0 liters/kg). These parameters appeared independent of age, disease state, and concomitant medication. The pharmacokinetics of propafenone during multiple dosing have been studied in both healthy young subjects[131] and arrhythmia patients of various ages.[132,134] Results of these studies indicate that considerable intersubject variability is present in the disposition parameters of both populations. In general, decreases in total clearance occur during chronic dosing, resulting in small increases in the apparent elimination half-life.

3.9. TOCAINIDE

Tocainide is a primary amine analog of lidocaine, a weak base (pKa 7.7) with low water solubility. It does not undergo the extensive first-pass metabolism and thus, unlike lidocaine, can be given orally. The usual i.v. dose is 0.5–0.75 mg/kg per min infused over 15–20 min, and the usual oral dose is 400–600 mg given every 8 hr.[135–137]

Tocainide absorption is rapid following oral administration with a systemic bioavailability of approximately 100%. Following single oral

TABLE XII PHARMACOKINETIC PARAMETERS OF PROPAFENONE.

Subjects	Mean age (range)	Dose (mg)	Half-life (hr)	Clearance (ml/min per kg)	Distribution volume (liters/kg)	Ref.
Single dose						
8 healthy subjects	30 (19–45)	70 i.v.	2.8	13.6 ± 3.7	3.1 ± 0.6	131
8 healthy subjects	36 (25–47)	300 p.o. soln.	4.4 ± 2.4	15.1 ± 2.3[a]	5.6 ± 3.0[a]	131
8 healthy subjects	36 (25–47)	300 p.o. tab.	4.6 ± 2.5	15.1 ± 2.3[a]	6.1 ± 3.5[a]	131
15 arrhythmia patients	58 (25–73)	1 mg/kg i.v.	5.0 ± 3.6	11.2 ± 4.8	3.6 ± 2.1	132
7 arrhythmia patients	(19–76)	900 p.o.	3.6 ± 0.2	—	—	133
Multiple dose						
9 arrhythmia patients	60 (39–71)	150–300 q8h p.o.	6.2 ± 3.3	—	—	132
13 arrhythmia patients	(44–73)	300–900/day	6 (2.4–11.8)[c]	—	—	134
8 healthy subjects	30 (19–45)	70 IV[b]	3.9	12.0 ± 5.4	3.1 ± 0.7	131

[a] Corrected for absolute bioavailability.
[b] Following 12 days of oral propafenone (150 b.i.d.).
[c] Parentheses indicate range.

doses ranging from 10 to 1000 mg to healthy volunteers and arrhythmia patients, both maximum blood tocainide concentrations and AUCs increase in proportion to the administered dose.[138] The presence of a large meal in the stomach markedly suppresses C_{max}, but has little influence on AUC.[139] Tocainide is approximately 15–20% bound to plasma proteins in healthy volunteers, and binding is independent of serum drug concentration within the range of 4–12 µg/ml.[140]

No definitive studies have been conducted to assess the effect of age on the pharmacokinetics of tocainide. The results of studies performed in young healthy volunteers[135,141] and older arrhythmia patients,[138,142] however, suggest that the kinetics of this agent are minimally affected by age alone (Table XIII). Tocainide plasma concentration–time profiles are biphasic with an initial distribution half-life of approximately 10 min and a terminal elimination half-life of 10–15 hr.[135,139] The apparent volume of distribution is 2–3 liters/kg, and the total body clearance is 2–3 ml/min per kg.

Tocainide is eliminated from the body via metabolism and renal excretion. Approximately 30–50% of an administered dose is excreted unchanged in the urine with a renal clearance ranging from 0.6–1.3 ml/min per kg.[135,139] Two metabolites, tocainide carbamoyl O-β-D-glucuronide and lactoxylidide, have been identified but have failed to demonstrate activity.

The pharmacokinetics of tocainide have been studied in older patients with various disease states.[135,138,142–145] In patients with acute myocardial infarction (age range 51–75) there were no differences in pharmacokinetic parameters when compared to healthy younger subjects.[135] Mohiuddin et al.[144] found no significant differences in pharmacokinetic parameters between arrhythmia patients with and without congestive heart failure. In studies conducted in older patients with renal failure,[145,146] the plasma half-life was 17–43 hr. Creatinine clearance was shown to be a poor predictor of drug clearance, suggesting the existence of extrarenal elimination pathways. Four hours of hemodialysis resulted in the removal of 7–53% of the drug.

The dose–effect and concentration–effect relationship between tocainide plasma concentrations and premature ventricular contractions (PVCs) has been investigated in several studies conducted in patients varying in age from 37 to 86 who received doses of 400–600 mg three times daily.[136,137,147–149] In general, plasma concentrations of 4–10 µg/ml were associated with significant reductions in PVCs, and efficacy was not found to be age-related. The effectiveness of tocainide was at best comparable to that of quinidine but produced a greater incidence of side effects.

TABLE XIII PHARMACOKINETIC PARAMETERS OF TOCAINIDE

Subjects	Mean age (range)	Dose (mg)	Half-life (hr)	Clearance (ml/min per kg)	Distribution volume (liters/kg)	Ref.
Single dose						
6 healthy subjects	32 (25–41)	600 p.o.	12.4	2.9 ± 0.7	2.9	141
6 healthy subjects	23	100 i.v.	13.5 ± 2.3	2.6 ± 0.4	2.9 ± 0.2	135
5 AMI patients	67 (51–75)	0.5–0.75 mg/kg per min i.v.	14.3 ± 2.2	2.6 ± 0.6	3.2 ± 0.4	135
11 cardiac patients	48 (32–63)	0.5–0.75 mg/kg per min i.v.	12.2 ± 2.3	2.1 ± 0.3	—	142
8 arrhythmia patients	51 (38–56)	10–1200 p.o.	14.7 ± 1.7	2.4[a]	3.0[a]	138
6 patients with hepatic and renal dysfunction	57 (41–68)	100 i.v.	27.4 ± 15.4	1.8 ± 0.8	3.8 ± 1.1	143
Multiple dose						
22 cardiac patients	59 (44–86)	400 q8h	13.6	1.9 ± 0.6	2.2 ± 0.4	147
15 arrhythmia patients	56 (37–68)	400–600 q8h	13.5 ± 2.0	—	—	137
14 CHF patients	60 (23–82)	1200–2400/day	10.1 ± 2.4	1.8 ± 0.4	1.5 ± 0.5	144
9 renal failure pts	66 (57–77)	1200 q8h	29.7 ± 9.2	1.1 ± 0.4	2.6 ± 0.5	145

[a] Normalized to 70 kg.

4. COMMENTARY

Age-related alterations in drug disposition are often responsible for the exaggerated pharmacological responses observed in the elderly. Since the pharmacological response is largely dependent on the concentration of drug at the site of action, any factor that alters this concentration (absorption, distribution, metabolism, elimination) may affect the individual's response. During the aging process, these factors may be altered, resulting in changes in drug disposition and pharmacological effect.

The preceding discussion has focused on the pharmacokinetics of the most widely used agents employed in the treatment of cardiac arrhythmias. The disposition of virtually all the drugs discussed in this chapter were altered in older subjects when compared to healthy younger subjects. It is apparent, however, that very few studies have been conducted that assess solely the influence of age on the disposition of these agents. In most cases, the populations studied were patients with various cardiac disorders, renal impairment (or failure), or liver dysfunction. The population has often been poorly defined in these studies in terms of severity or subtype of the disease state evaluated. Authors should supply a demographic summary that gives the reader an adequate description of the patients studied. Minimally, this should include the age and weight of each patient and underlying disease states. Clearly, age-dependent changes in body composition may contribute significantly to alterations in such pharmacokinetic parameters as clearance and volume of distribution. Similarly, the pharmacokinetics of an antiarrhythmic agent may be expected to differ in patients with arrhythmias in the presence of edema due to congestive heart failure and in those with arrhythmias secondary to myocardial infarction.

The majority of the drugs discussed in this chapter are eliminated by renal excretion of parent drug or metabolites. It is well known that renal function tends to decline with age even in healthy subjects. Whereas young subjects usually have creatinine clearances of 100–120 ml/min, 60-year-old and 80-year-old subjects normally have creatinine clearances of 70–80 and 45–55 ml/min, respectively. Thus, it is to be expected that renally excreted drugs would display age-related increases in half-life. It should also be noted that patients with cardiac arrhythmias may also show decreased creatinine clearance due to decreases in renal blood flow secondary to decreased cardiac output. As a result, it may be difficult to separate the pharmacokinetic changes due to arrhythmias from those due solely to the aging process. If the objective of a study is to determine the effect of age on the pharmacokinetics of an antiarrhythmic (or any drug), healthy elderly subjects should be used and compared to a younger control group. If the objective is to determine the effect of cardiac arrhythmia on the pharmacokinetics, a group of similarly-aged healthy subjects should be included as a control. Alternatively, investigators may choose to compare data gathered from older arrhythmia patients to historical data obtained from a similar population of older healthy subjects. Historical control data must be used with caution, however, since studies may be conducted under different conditions and pharmacokinetic analyses may differ from one investigator to another.

Several of the drugs described in this chapter are eliminated primarily by hepatic biotransformation. As such, the pharmacokinetics of these drugs may be expected to be altered as a function of age-dependent decreases in liver function. Reidenberg suggested that drug metabolism is decreased in the elderly by one-third to one-half.[150] Such a statement is difficult to substantiate, however, given the considerable interindividual variation in liver function seen in healthy young subjects and as a result of chronological versus physiological aging. Sjoqvist and Alvan[38] have also pointed out that the effect of age is not easily discerned since a number of other factors that affect drug metabolism, such as dietary and smoking habits and drug interactions, also change with age. Therefore, in order to gain the most useful information from a study in liver patients, it is important to provide a good estimate of liver function for each patient. A study involving patients with liver dysfunction without cardiac disease should include the diagnosis and the degree of impairment (i.e., ICG perfusion). For those drugs which are metabolized to active compounds (e.g., procainamide, encainide, lorcainide), studies conducted in the elderly should adequately assess the pharmacokinetics (disposition, accumulation) of both parent drug and metabolite.

Although attempts have been made to correlate plasma concentrations with therapeutic effect for most of the drugs presented in this chapter, few attempts have been made to determine the concentration–effect relationship as a function of age. As stated previously, it is generally found that elderly patients are more sensitive to drugs than are the young. In addition to the presence of drug at the site of action, differences in sensitivity are hypothesized to be due to age-related alterations in the quality or quantity of drug receptors. It may therefore be expected that concentration–effect curves will appear significantly different in the elderly and in the young. Pharmacodynamic studies should utilize patients over as narrow an age range as feasible or, if possible, separate patients into several groups according to age.

For many of the drugs discussed in this chapter, individualized dosing may be indicated based on age, weight, and presence of major organ dysfunction. The therapeutic implications, however, should depend on the therapeutic index of the drug. In the case of quinidine, a well-defined and narrow therapeutic range has been established, and age-related alterations in kinetics indicate the need for dosage adjustment. For other drugs, particularly the newer agents, insufficient information is available to define the therapeutic range. Patients should be monitored closely for signs of toxicity or lack of efficacy. It should be emphasized again that the pharmacokinetic changes only partially explain the changes in drug sensitivity associated with aging. Drug therapy in the elderly must always be undertaken with caution, particularly with antiarrhythmics.

REFERENCES

1. Karim A: The pharmacokinetics of Norpace. *Angiology* 1975; 26(suppl 1):85–98.
2. Bryson SM, Whiting B, Lawrence JR: Disopyramide serum and pharmacologic effect kinetics applied to the assessment of bioavailability. *Br J Clin Pharmacol* 1978; 6:409–419.
3. Cunningham JL, Shen DD, Shudo I, et al: The effect of nonlinear disposition kinetics on the systemic availability of disopyramide. *Br J Clin Pharmacol* 1978; 5:343–346.
4. Cunningham JL, Shen DD, Shudo I, et al: The effect of urine pH and plasma protein binding on the renal clearance of disopyramide. *Clin Pharmacokin* 1977; 2:373–383.
5. Giacomini KM, Blaschke TF: Effect of concentration-dependent binding to plasma proteins on the pharmacokinetics and pharmacodynamics of disopyramide. *Clin Pharmacokin* 1984; 9(suppl 1):42–48.
6. Lima JJ, Boudoulas H, Blanford M.: Concentration-dependence of disopyramide binding to plasma protein and its influence on kinetics and dynamics. *J Pharmacol Exp Ther* 1981; 219:741–747.
7. Meffin PJ, Robert BW, Winkle RA, et al: Role of concentration-dependent plasma protein binding in disopyramide disposition. *J Pharmacokin Biopharm* 1979; 7:29–45.
8. Ueda CT, Dzindzio BS, Vosik WM: Serum disopyramide concentrations and suppression of ventricular premature contractions. *Clin Pharmacol Ther* 1984; 36:326–336.
9. Ward JW, Kinghorn, GR: The pharmacokinetics of disopyramide following myocardial infarction with special reference to oral and intravenous dose regimens. *J Int Med Res* 1976; 4(suppl 1):49–53.
10. Olsen H, Bredesen JE, Lunde PKM: Effect of ethanol intake on disopyramide elimination by healthy volunteers. *Eur J Clin Pharmacol* 1983; 25:103–105.
11. Rangno RE, Warnica W, Ogilvie RI, et al: Correlation of disopyramide pharmacokinetics with efficacy in ventricular tachyarrhythmia. *J Int Med Res* 1976; 4(suppl 1):54–58.
12. Hulting J, Rosenhamer G: Anti-arrhythmic and haemodynamic effects of intravenous and oral disopyramide in patients with ventricular arrhythmia. *J Int Med Res* 1976; 4(suppl 1):90–95.
13. Hinderling PH, Garrett ER: Pharmacokinetics of the antiarrhythmic disopyramide in healthy humans. *J Pharmacokin Biopharm* 1976; 4:199–230.
14. Ranney RE, Dean RR, Karim A, et al: Disopyramide phosphate: pharmacokinetic and pharmacologic relationships of a new antiarrhythmic agent. *Arch Int. Pharmacodyn* 1971; 191:162–188.

15. Boyes RN, Scott DB, Jebson PJ, et al: Pharmacokinetics of lidocaine in man. *Clin Pharmacol Ther* 1971; 12:105–116.
16. Nation RL, Triggs EJ, Selig M: Lignocaine kinetics in cardiac patients and aged subjects. *Br J Clin Pharmacol* 1977; 4:439–448.
17. Benowitz N, Forsyth RP, Melmon KL, et al: Lidocaine disposition kinetics in monkey and man. I. Prediction by a perfusion model. *Clin Pharmacol Ther* 1974; 16:87–98.
18. Cusson J, Nattel S, Matthews S, et al: Age-dependent lidocaine disposition in patients with acute myocardial infarction. *Clin Pharmacol Ther* 1985; 37:381–386.
19. Prescott LF, Adjepon-Yamoah KK, Talbot RG: Impaired lignocaine metabolism in patients with myocardial infarction and cardiac failure. *Br Med J* 1976; 1:939–941.
20. Conrad KA, Byers JM, Finley PR, et al: Lidocaine elimination: effects of metoprolol and of propranolol. *Clin Pharmacol Ther* 1983; 33:133–138.
21. Jordo L, Johnsson G, Lundborg P, et al: Pharmacokinetics of lidocaine in healthy individuals pretreated with multiple doses of metoprolol. *Int J Clin Pharmacol Ther Toxicol* 1984; 22:312–315.
22. Thomson PD, Melmon KL, Richardson JA, et al: Lidocaine pharmacokinetics in advanced heart failure, liver disease, and renal failure in humans. *Ann Intern Med* 1973; 78:499–508.
23. Forrest JA, Finlayson ND, Adjepon-Yamoah KK, et al: Antipyrine, paracetamol, and lignocaine elimination in chronic liver disease. *Br Med J* 1977; 1:1384–1387.
24. Manion CV, Lalka D, Baer DT, et al: Absorption kinetics of procainamide in humans. *J Pharm Sci* 1977; 66:981–984.
25. Koch-Weser J: Pharmacokinetics of procainamide in man. *Ann NY Acad Sci* 1971; 179:370–382.
26. Lima JJ, Conti DR, Goldfarb AL, et al: Clinical pharmacokinetics of procainamide infusion in relation to acetylator phenotype. *J Pharmacokin Biopharm* 1979; 7:69–85.
27. Galeazzi RL, Benet LZ, Sheiner LB: Relationship between the pharmacokinetics and pharmacodynamics of procainamide. *Clin Pharmacol Ther* 1976; 20:278–289.
28. Gibson TP, Atkinson AJ, Matusik E, et al: Kinetics of procainamide and N-acetyl-procainamide in renal failure. *Kidney Int* 1977; 12:422–429.
29. Dutcher JS, Strong JM, Lucas SV, et al: Procainamide and N-acetylprocainamide kinetics investigated simultaneously with stable isotope methodology. *Clin Pharmacol Ther* 1977; 22:447–457.
30. Christian CD, Meredith GCG, Speeg KV: Cimetidine inhibits renal procainamide clearance. *Clin Pharmacol Ther* 1984; 36:221–227.
31. Giardina EGV, Dreyfuss J, Briggs JT, et al: Metabolism of procainamide in normal and cardiac subjects. *Clin Pharmacol Ther* 1976; 19:339–351.
32. Galeazzi RL, Sheiner LB, Lockwood TL, et al: The renal elimination of procainamide. *Clin Pharmacol Ther* 1976; 19:55–62.
33. Reidenberg MM, Camacho M, Kluger J, et al: Aging and renal clearance of procainamide and acetylprocainamide. *Clin Pharmacol Ther* 1980; 28:732–735.
34. Grasela TH, Sheiner LB: Population pharmacokinetics of procainamide from routine clinical data. *Clin Pharmacokin* 1984; 9:545–554.
35. Wyman MG, Goldreyer BN, Cannom DS, et al: Factors influencing procainamide total body clearance in the immediate post myocardial infarction period. *J Clin Pharmacol* 1981; 21:20–25.
36. Gibson TP, Lowenthal DT, Nelson HA, et al: Elimination of procainamide in end stage renal failure. *Clin Pharmacol Ther* 1975; 17:321–329.
37. Reidenberg MM, Drayer DE, Levy M, et al: Polymorphic acetylation of procainamide in man. *Clin Pharmacol Ther* 1975; 17:722–730.

38. Sjoqvist F, Alvan G: Aging and drug disposition-metabolism. *J Chron Dis* 1983; 36:31–37.
39. Atkinson AJ, Krumlovsky Fa, Huang CM, et al: Hemodialysis for severe procain-amide toxicity: Clinical and pharmacokinetic observations. *Clin Pharmacol Ther* 1976; 20:585–592.
40. du Souich P, Erill S: Metabolism of procainamide and *p*-aminobenzoic acid in patients with chronic liver disease. *Clin Pharmacol Ther* 1977; 22:588–595.
41. Graffner C: Elimination rate of N-acetylprocainamide after a single intravenous dose of procainamide hydrochloride in man. *J Pharmacokin Biopharm* 1975; 3:69–76.
42. Ditlefsen EML, Knutsen B: Quinidine treatment in chronic auricular fibrillation. I. Conversion to sinus rhythm, related to quinidine serum concentration. *Acta Med Scand* 1956; 156:1–14.
43. Holford NHG, Coates PE, Guenter TW, et al: The effect of quinidine and its me-tabolites on the electrocardiogram and systolic time intervals: Concentration-effect relationships. *Br J Clin Pharmacol* 1981; 11:187–195.
44. Covinsky JO, Russo J, Kelly KL, et al: Relative bioavailability of quinidine gluconate and quinidine sulfate in healthy volunteers. *J Clin Pharmacol* 1979; 19:261–269.
45. Greenblatt DJ, Pfeifer HJ, Ochs HR, et al: Pharmacokinetics of quinidine in humans after intravenous, intramuscular and oral administration. *J Pharmacol Exp Ther* 1977; 202:365–378.
46. Ueda CT, Dzindzio BS: Quinidine kinetics in congestive heart failure. *Clin Pharmacol Ther* 1978; 23:158–164.
47. Ueda CT, Williamson BJ and Dzindzio BS: Absolute quinidine bioavailability. *Clin Pharmacol Ther* 1976; 20:260–265.
48. Ochs YW, Lambert HJ, Woo E, et al: Reduced quinidine clearance in elderly persons. *Am J Cardiol* 1978; 42:481–485.
49. Ueda CT, Hirschfield DS, Scheinmann MM, et al: Disposition kinetics of quinidine. *Clin Pharmacol Ther* 1976; 19:30–36.
50. Bellet S, Roman LR, Boza A: Relation between serum quinidine levels and renal function. *Am J Cardiol* 1971; 27:368.
51. Conrad KA, Molk BL, Chidsey CA: Pharmacokinetic studies of quinidine in patients with arrhythmias. *Circulation* 1977; 55:1–7.
52. Kessler KM, Humphries WC, Black M, et al: Quinidine pharmacokinetics in patients with cirrhosis or receiving propranolol. *Am Heart J* 1978; 96:627–635.
53. Crouthamel WG: The effect of congestive heart failure on quinidine pharmacoki-netics. *Am Heart J* 1975; 90:335–339.
54. Kessler KM, Lowenthal DT, Warner H, et al: Quinidine elimination in patients with congestive heart failure or poor renal function. *N Engl J Med* 1974; 290:706–709.
55. Gerhardt RE, Knouss RF, Thyrum PT, et al: Quinidine excretion in aciduria and alkaluria. *Ann Intern Med* 1969; 71:927–933.
56. Fichtl B, Doering W: The quinidine-digoxin interaction in perspective. *Clin Phar-macokin* 1983; 8:137–154.
57. Leahey EB, Bigger JT, Butler VP, et al: Quinidine-digoxin interaction: Time course and pharmacokinetics. *Am J Cardiol* 1981; 48:1141–1146.
58. Leahey EB, Reiffel JA, Giardina EV, et al: The effect of quinidine and other oral antiarrhythmic drugs on serum digoxin. A prospective study. *Ann Intern Med* 1980; 92:605–608.
59. Canada AT, Lesko LJ, Haffajee CI: Disposition of amiodarone in patients with tachyarrhythmias. *Cur Ther Res* 1981; 30:968–974.
60. Haffajee CI, Love JC, Canada AT, et al: Clinical pharmacokinetics and efficacy of amiodarone for refractory tachyarrhythmias. *Circulation* 1983; 67:1347–1355.

61. Riva E, Aarons L, Latini R, et al: Amiodarone kinetics after single IV bolus and multiple dosing in healthy volunteers. *Eur J Clin Pharmacol* 1984; 27:491–494.
62. Kannan R, Nademanee K, Hendrickson JA, et al: Amiodarone kinetics after oral doses. *Clin Pharmacol Ther* 1982; 31:438–444.
63. Tucker GT, Jackson PR, Storey CGA, et al: Bioavailability of amiodarone. *Eur J Clin Pharmacol* 1984; 26:533–534.
64. Pourbaix S, Berger Y, Desager J, et al: Absolute bioavailability of amiodarone in normal subjects. *Clin Pharmacol Ther* 1985; 37:118–123.
65. Riva E, Gerna M, Latini et al: Pharmacokinetics of amiodarone in man. *J Cardiovas Pharmacol* 1982; 4:264–269.
66. Anastasiou-Nana M, Levis GM, Moulopoulos S: Pharmacokinetics of amiodarone after intravenous and oral administration. *Int J Clin Pharmacol Ther Toxicol* 1982; 20:524–529.
67. Marchiset D, Bruno R, Djiane P, et al: Amiodarone and desethylamiodarone elimination kinetics following withdrawal of long-term amiodarone maintenance therapy. *Biopharm Drug Disp* 1985; 6:209–215.
68. Plomp TA, van Rossum JM, de Medina R, et al: Pharmacokinetics and body distribution of admiodarone in man. *Arzneim-Forsch/Drug Res* 1984; 34:513–520.
69. Siddoway LA, McAllister CB, Wilkinson GR, et al: Amiodarone dosing: A proposal based on its pharmacokinetics, *Am Heart J* 1983; 106:951–956.
70. Zipes DP, Prystowsky EN, Heger JJ: Amiodarone: Electrophysiologic actions, pharmacokinetics and clinical effects. *J Am Coll Cardiol* 1984; 3:1059–1071.
71. Holt DW, Tucker GT, Jackson PR, et al: Amiodarone pharmacokinetics. *Am Heart J* 1983; 106:840–848.
72. Staubli M, Bircher J, Galeazzi RL, et al: Serum concentrations of amiodarone during long-term therapy. Relation to dose, efficacy and toxicity. *Eur J Clin Pharmacol* 1983; 24:485–494.
73. Andreasen F, Agerbaek H, Bjerregaard P, et al: Pharmacokinetics of amiodarone after intravenous and oral administration. *Eur J Clin Pharmacol* 1981; 19:293–299.
74. Canada AT, Lesko LJ, Haffajee CI, et al: Amiodarone for tachyarrhythmias: pharmacology, kinetics and efficacy. *Drug Intell Clin Pharm* 1983; 17:100–104.
75. Arboix M, Frati ME, Laporte JR: The potentiation of acenocoumarol anticoagulant effect by amiodarone. *Br J Clin Pharmacol* 1984; 18:355–360.
76. Serlin MJ, Sibeon RG, Green GJ: Dangers of amiodarone and anticoagulant treatment. *Br Med J* 1981; 283:58.
77. Lalloz MRA, Byfield PGH, Greenwood RM, et al: Binding of amiodarone by serum proteins and the effect of drugs, hormones and other interacting ligands. *J Pharm Pharmacol* 1984; 36:366–372.
78. Brazzell RK, Khoo K-C, Szuna AJ, et al: Pharmacokinetics and pharmacodynamics of intravenous cibenzoline in normal volunteers. *J Clin Pharmacol* 1985; 25:418–423.
79. Canal M, Flouvat B, Aubert P, et al: Pharmacokinetics of cibenzoline in patients with renal impairment. *J Clin Pharmacol* 1985; 25:197–203.
80. Aronoff GR, Mayer ML, Barbalas M, et al: Bioavailability and elimination kinetics of cibenzoline in healthy volunteers and patients with renal failure. *Clin Pharmacol Ther* 1986; (in press).
81. Khoo K-C, Szuna AJ, Colburn WA, et al: Single dose pharmacokinetics and dose proportionality of oral cibenzoline. *J Clin Pharmacol* 1984; 24:283–288.
82. Brazzell RK, Rees MMC, Khoo K-C, et al: Age and cibenzoline disposition. *Clin Pharmacol Ther* 1984; 36:613–619.
83. Brazzell RK, Aogaichi K, Heger J, et al: Cibenzoline plasma concentration and antiarrhythmic effect. *Clin Pharmacol Ther* 1984; 35:307–316.

84. Keefe DL, Kates RE, Harrison DC: New antiarrhythmic drugs: Their place in therapy. *Drugs* 1981; 22:363–400.
85. Roden DM, Reele SB, Higgins SB, et al: Total suppression of ventricular arrhythmias by encainide. *N Engl J Med* 1980; 302:877–882.
86. Winkle RA, Peters F, Kates RE, et al: Clinical pharmacology and antiarrhythmic efficacy of encainide in patients with chronic ventricular arrhythmias. *Circulation* 1981; 64:290–296.
87. Roden DM, Duff HJ, Altenbern D, et al: Antiarrhythmic activity of the o-demethyl metabolite of encainide. *J Pharmacol Exp Ther* 1982; 221:552–557.
88. Kates RE, Woosley RL, Harrison DG: Clinical importance of metabolites of antiarrhythmic drugs. *Am J Cardiol* 1984; 53:248–251.
89. Woosley RL, Roden DM, Duff HJ, et al: Co-inheritance of deficient oxidative metabolism of encainide and debrisoquine. *Clin Res* 1981; 29:501A.
90. Kates RE, Harrison DC, Winkle RA: Metabolite cumulation during long-term oral encainide administration. *Clin Pharmacol Ther* 1982; 31:427–432.
91. Bergstrand RH, Wang T, Roden DM, et al: Effect of liver disease on encainide disposition. *Proceedings of the World Conference on Clinical Pharmacological Therapy,* Washington DC, American Society for Pharmacological Experimental Therapy, 1983, p A:11.
92. Conard GJ, Ober RE: Metabolism of flecainide. *Am J Cardiol* 1984; 53:41B–51B.
93. Duff HJ, Roden DM, Maffucci RJ, et al: Suppression of resistant ventricular arrhythmias by twice daily dosing with flecainide. *Am J Cardiol* 1981; 48:1133–1140.
94. Johnston A, Muhiddin KA, Hamer J: Serum protein binding of flecainide. *Br J Clin Pharmacol* 1982; 13:606.
95. Franciosa JA, Wilen M, Weeks CE, et al: Pharmacokinetics and hemodynamic effects of flecainide in patients with chronic low output heart failure. *J Am Coll Cardiol* 1983; 1:699.
96. Anderson JL, Stewart JR, Perry BA, et al: Oral flecainide acetate for the treatment of ventricular arrhythmias. *N Engl J Med* 1981; 305:473–477.
97. Hodges M, Haugland JM, Granrud G, et al: Suppression of ventricular ectopic depolarizations by flecainide acetate, a new antiarrhythmic agent. *Circulation* 1982; 65:879–883.
98. McQuinn RL, Quarfoth GJ, Johnson JD, et al: Biotransformation and elimination of ^{14}C-flecainide acetate in humans. *Drug Metab Disp* 1984; 12:414–420.
99. Muhiddin KA, Johnston A, Turner P: The influence of urinary pH on flecainide excretion and its serum pharmacokinetics. *Br J Clin Pharmacol* 1984; 17:447–451.
100. Lewis GP, Holtzman JL: Interaction of flecainide with digoxin and propranolol. *Am J Cardiol* 1984; 53:52B–57B.
101. Gillis AM, Kates RE: Clinical pharmacokinetics of the newer antiarrhythmic agents. *Clin Pharmacokin* 1984; 9:375–403.
102. Klotz U, Muller-Seydlitz PM, Heimburg P: Lorcainide infusion in the treatment of ventricular premature beats (VPB). *Eur J Clin Pharmacol* 1979; 16:1–6.
103. Klotz U, Muller-Seydlitz PM, Heimburg P: Disposition and antiarrhythmic effect of lorcainide. *Int J Clin Pharmacol Biopharm* 1979; 17:152–158.
104. Meinertz T, Kasper W, Kersting F, et al: Lorcainide II. Plasma concentration-effect relationship. *Clin Pharmacol Ther* 1979; 26:196–204.
105. Meinertz T, Kersting F, Kasper W, et al: Haemodynamic effects of a single intravenous dose of lorcainide in patients with heart disease. *Eur J Clin Pharmacol* 1980; 18:461–465.
106. Jahnchen E, Bechtold H, Kasper W, et al: Lorcainide I. Saturable presystemic elimination. *Clin Pharmacol Ther* 1979; 26:187–195.

107. Klotz U, Muller-Seydlitz PM, Heimburg P: Pharmacokinetics of lorcainide in man: A new antiarrhythmic agent. *Clin Pharmacokin* 1978; 3:407–418.
108. Amery WK, Heykants J, Bruyneel K, et al: Bioavailability and saturation of the presystemic metabolism of oral lorcainide therapy initiated in three different dose regimens. *Eur J Clin Pharmacol* 1983; 24:517–519.
109. Kates RE, Keefe DL, Winkle RA: Lorcainide disposition kinetics in arrhythmia patients. *Clin Pharmacol Ther* 1983; 33:28–34.
110. Somani P: Pharmacokinetics of lorcainide, a new antiarrhythmic drug, in patients with cardiac rhythm disorders. *Am J Cardiol* 1981; 48:157–163.
111. Winkle RA, Keefe DL, Rodriguez I, et al: Pharmacodynamics of the initiation of antiarrhythmic therapy with lorcainide. *Am J Cardiol* 1984; 53:544–551.
112. Keefe DL, Peters F, Winkle RA: Randomized double-blind placebo controlled crossover trial documenting oral lorcainide efficacy in suppression of symptomatic ventricular arrhythmias. *Am Heart J* 1982; 103:511–518.
113. Fenster PE, Dahl C: Mexiletine in the treatment of post-MI arrhythmias. *Drug Ther* 1981; 11(July):63–69.
114. Prescott LF, Pottage A, Clements JA: Absorption, distribution and elimination of mexiletine. *Postgrad Med J* 1977; 53(suppl 1):50–55.
115. Campbell NPS, Kelly JG, Adgey AAJ, et al: Mexiletine in normal volunteers. *Br J Clin Pharmacol* 1978; 6:372–373.
116. Haselbarth V, Doevendans JE, Wolf M: Kinetics and bioavailability of mexiletine in healthy subjects. *Clin Pharmacol Ther* 1981; 29:729–736.
117. Campbell NPS, Kelly JG, Adgey AAJ, et al: The clinical pharmacology of mexiletine. *Br J Clin Pharmacol* 1978; 6:103–108.
118. Pentikainen PJ, Halinen MO, Helin MJ: Pharmacokinetics of oral mexiletine in patients with acute myocardial infarction. *Eur J Clin Pharmacol* 1983; 25:773–777.
119. Pentikainen PJ, Halinen MO, Helin MJ: Pharmacokinetics of intravenous mexiletine in patients with acute myocardial infarction. *J Cardiovasc Pharmacol* 1984; 6:1–6.
120. Beckett AH, Chidomere EC: The distribution, metabolism and excretion of mexiletine in man. *Postgrad Med J* 1977; 53(suppl 1):60–66.
121. Mitchell BG, Clements JA, Pottage A, et al: Mexiletine disposition: individual variation in response to urine acidification and alkalinisation. *Br J Clin Pharmacol* 1983; 16:281–284.
122. Vozeh V, Katz G, Steiner V, et al: Population pharmacokinetic parameters in patients treated with oral mexiletine. *Eur J Clin Pharmacol* 1982; 23:445–451.
123. El Allaf D, Henrard L, Crochelet L, et al: Pharmacokinetics of mexiletine in renal insufficiency. *Br J Clin Pharmacol* 1982; 14:431–435.
124. Jones TE, Reece PA, Fisher GC: Mexiletine removal by peritoneal dialysis. *Eur J Clin Pharmacol* 1983; 25:839–840.
125. Reiter MJ, Hammill SC, Shand DG, et al: Efficacy, safety and pharmacokinetics of a concentration-maintaining regimen of intravenous pirmenol. *Am J Cardiol* 1983; 52:83–87.
126. Phillips HR, Stack RS, Davis DD, et al: Hemodynamic effects of the anitarrhythmic drug pirmenol. *Clin Pharmacol Ther* 1982; 32:235–239.
127. Hammill SC, Shand DG, Harrell FE, et al: Pirmenol kinetics and effective oral dose. *Clin Pharmacol Ther* 1982; 32:686–691.
128. Hammill SC, Shand DG, Routledge PA, et al: Pirmenol, a new antiarrhythmic agent: Initial study of efficacy, safety and pharmacokinetics. *Circulation* 1982; 65:369–375.
129. Sanders SW, Nappi JM, Foltz RL, et al: Disposition of intravenous pirmenol. *J Clin Pharmacol* 1983; 23:113–122.

130. Connolly SJ, Kates RE, Lebsack CS, et al: Clinical efficacy and electrophysiology of oral propafenone for ventricular tachycardia. *Am J Cardiol* 1983; 52:1208–1213.
131. Hollmann M, Brode E, Hotz D, et al: Investigations on the pharmacokinetics of propafenone in man. *Arzneim-Forsch* 1983; 33:763–770.
132. Connolly SJ, Lebsack CS, Winkle RA, et al: Propafenone disposition kinetics in cardiac arrhythmia. *Clin Pharmacol Ther* 1984; 36:163–168.
133. Keller K, Meyer-Estorf G, Beck OA, et al: Correlation between serum concentration and pharmacological effect on atrioventricular conduction time of the antiarrhythmic drug propafenone. *Eur J Clin Pharmacol* 1978; 13:17–20.
134. Connolly SJ, Kates RE, Lebsack CS, et al: Clinical pharmacology of propafenone. *Circulation* 1983; 68:589–596.
135. Graffner C, Conradson T-B, Hofvendahl S, et al: Tocainide kinetics after intravenous and oral administration in healthy subjects and in patients with acute myocardial infarction. *Clin Pharmacol Ther* 1980; 27:64–71.
136. Meffin PJ, Winkle RA, Blaschke TF, et al: Response optimization of drug dosage: antiarrhythmic studies with tocainide. *Clin Pharmacol Ther* 1977; 22:42–57.
137. Winkle RA, Meffin PJ, Fitzgerald JW, et al: Clinical efficacy and pharmacokinetics of a new orally effective antiarrhythmic, tocainide. *Circulation* 1976; 54:884–889.
138. McDevitt DG, Nies AS, Wilkinson GR, et al: Antiarrhythmic effects of a lidocaine congener, tocainide 2-amino-2',6-propionoxylidide, in man. *Clin Pharmacol Ther* 1976; 19:396–402.
139. Lalka D, Meyer MB, Duce BR, et al: Kinetics of the oral antiarrhythmic lidocaine congener tocainide. *Clin Pharmacol Ther* 1976; 19:757–766.
140. Elvin AT, Axelson JE, Lalka D: Tocainide protein binding in normal volunteers and trauma patients. *Br J Clin Pharmacol* 13:872–874.
141. Elvin AT, Lalka D, Stoeckel K, et al: Tocainide kinetics and metabolism: Effects of phenobarbital and substrates for gluceronyl transferase. *Clin Pharmacol Ther* 1980; 28:652–658.
142. Schwartz M, Covino B, Duce B, et al: Acute hemodynamic effects of tocainide in patients undergoing cardiac catheterization. *J Clin Pharmacol* 1979; 19:100–107.
143. Oltmanns D, Pottage A, Endell W: Pharmacokinetics of tocainide in patients with combined hepatic and renal dysfunction. *Eur J Clin Pharmacol* 1983; 25:787–790.
144. Mohiuddin SM, Esterbrooks D, Hilleman DE, et al: Tocainide kinetics in congestive heart failure. *Clin Pharmacol Ther* 1983; 34:596–603.
145. Wiegers U, Hanrath P, Kuck DH, et al: Pharmacokinetics of tocainide in patients with renal dysfunction and during haemodialysis. *Eur J Clin Pharmacol* 1983; 24:503–507.
146. Braun J, Sörgel F, Engelmaier F, et al: Pharmacokinetics of tocainide in patients with severe renal failure. *Eur J Clin Pharmacol* 1985; 28:665–670.
147. Sonnhag C: Efficacy and tolerance of tocainide during acute and long-term treatment of chronic ventricular arrhythmias. *Eur J Clin Pharmacol* 1980; 18:301–310.
148. Ronfeld RA, Wolshin EM, Block AJ: On the kinetics and dynamics of tocainide and its metabolites. *Clin Pharmacol Ther* 1982; 31:384–392.
149. Wasenmiller JE, Aronow WS: Effect of tocainide and quinidine on premature ventricular contractions. *Clin Pharmacol Ther* 1980; 28:431–435.
150. Reidenberg M: Drugs in the elderly. *Bull NY Acad Med* 1980; 56:703.

CHAPTER 12

BETA BLOCKERS IN THE ELDERLY

DAVID S. ROFFMAN AND ALAN FORREST

1. INTRODUCTION

Adrenergic blocking drugs have become one of the most widely prescribed groups of agents in the current armamentarium of drugs for
the treatment of a variety of cardiovascular and noncardiovascular disorders. Because diseases such as hypertension, angina, myocardial infarction, and arrhythmias are so prevalent among the elderly, beta blockers are commonly used in this population. There is, however, a significant
amount of controversy involving the need to treat some of these problems in the elderly and additional doubt as to whether geriatric patients
are less sensitive to the therapeutic effects of these drugs or are at greater
risk of toxicity from this class of agents. The purpose of this chapter is
to review the evidence documenting the efficacy of β-adrenergic blocking
agents in the elderly as well as the data that quantitate the altered pharmacokinetics of beta blockers in the geriatric population. In addition,
we will discuss methodology that is most appropriate to study β-blocking
drugs in the elderly.

DAVID S. ROFFMAN AND ALAN FORREST • Department of Clinical Pharmacy, School
of Pharmacy, University of Maryland at Baltimore, Baltimore, Maryland 21201.

2. Pharmacokinetics of Beta Blockers in the Elderly

The aging process has the potential to alter many of the factors that affect the pharmacokinetics of β-adrenergic blocking agents. Changes in the gastrointestinal mucosa and gastrointestinal motility may effect drug absorption. Altered hepatic and renal blood flow as well as excretory function can affect drug clearance. Changes in intravascular and total-body water, fat content, and serum protein concentration may alter drug binding and distribution as well as drug clearance. Furthermore, because of differences in physical and chemical characteristics among the various beta blockers, such as their lipid solubility, degree of first-pass hepatic metabolism, and extent of renal elimination, one cannot generalize and assume that the effects due to aging will be uniform throughout the entire group of drugs. It is necessary, therefore, to evaluate the data describing pharmacokinetic changes associated with aging, for each of the β-blocking agents used in clinical practice. As can be easily observed by the sparsity of such studies, much of this information is still lacking. A review of available pharmacokinetic studies should enable practitioners to make more appropriate use of β-blocking agents in the elderly. In addition, identification of information lacking and design flaws in the existing studies in this area should encourage more appropriate pharmacokinetic trials.

3. Systemic Bioavailability

A number of anatomical and functional changes occur in the elderly that have the potential to alter beta-blocker absorption. These changes include a decrease in gastric acid secretion, a reduction in gastric motility, as well as a decrease in splanchnic blood flow and intestinal surface area. A review of available data suggests that despite these physiological alterations in the elderly, the gastrointestinal absorption of the beta blockers remains unaltered. There is, however, evidence that bioavailability of some of these drugs is altered in the elderly. The definition of bioavailability that will be used here is the fraction of the administered dose that transfers to the systemic circulation in its active form.

Castleden and George[1] studied the pharmacokinetics of propranolol in seven young patients (mean ± S.E. age 29 ± 2 years) and eight older patients (age 78 ± 3 years). Subjects were given a single oral dose (40 mg) and an i.v. dose (0.15 mg/kg) and had blood sampled for 8 hr after each dose. Four of the young subjects (30 ± 4 years) and five of the older (83 ± 3 years) also received oral propranolol, 40 mg four times daily for 2 days. Blood was sampled on day 3 at predose and 2, 4, and 6 hr postdose. The single-dose first-pass extraction was 69.9 ± 4.5% in the young and 45.4 ± 8.0% in the older subjects ($p < 0.05$). The clearances were significantly ($p < 0.02$) less in the older subjects (7.8 ± 1.3 ml/min per kg versus 13.2 ± 1.4 ml/min per kg). These parameters are not specified for the subset of subjects who also received repeated oral doses. Concentrations achieved by repeated doses are not stated in the text but are displayed on a graph of mean (± S.E.) concentrations versus time. Estimating from this graph, the mean predose concentrations were approximately 120 ng/ml (older) versus 30 ng/ml (young), and the 2-hr concentrations were approximately 200 ng/ml (older) and 65 ng/ml (young). The authors suggest that the "substantially higher" concentrations seen in the older subjects after repeated oral doses were probably due to an age-related decrease in both hepatic blood flow, and thus systemic clearance, and first-pass extraction.

Barber et al.[2] gave oral propranolol, 80 mg daily for 8 days, to six young (23–33 years) and five older (66–72 years) subjects. Blood was sampled for 8 hr on day 8. The AUC(0 → 12) (area under the concentration–time curve from predose to 12 hr postdose) on day 8 was 380 ± 84 (mean ± S.E.) in the young and 760 ± 150 ng hr/ml in the older subjects. Serum protein binding was similar between groups. The authors observe that this "increase in propranolol bioavailability in the elderly subjects" is less than that noted by Castleden and George.[1] This study design does not allow one to differentiate bioavailability from clearance between groups. The magnitude of differences in AUC observed, combined with the small sample sizes, does not achieve significance at the 5% level.

Quarterman et al.[3] gave metoprolol tartrate, 100 mg orally twice daily, to eight young (18–25 years) and seven older (63–74 years) subjects. Blood samples were collected for 24 hr on days 1 and 8. The mean (± S.E.) AUC(0 → ∞) after dose 1, the AUC(0 → 12) on day 8, and the mean percent accumulation AUC(0 → ∞) on day 1 compared to AUC(0 → 12) on day 8 was:

	AUC(0 → ∞)	AUC(0 →12)	Accumulation %
Young	836 ± 210	1306 ± 289	62
Older	450 ± 97	851 ± 182	120

The differences in AUC, for young versus older subjects, did not achieve statistical significance. If bioavailability and serum clearance remained constant, the AUC(0 → ∞) after the first dose should have been equal (within subjects) to the AUC(0 → 12) after repeated dosing. In both the young and older subjects the difference in AUC within groups (day 1 versus day 8) was statistically significant ($p < 0.02$). The accumulation, beyond that which was expected, by day 8 appears greater in the older subjects (statistics not provided). Half-lives within groups were not different (day 1 versus day 8). Although subjects could have had a decrease in serum clearance, manifesting primarily as a decrease in apparent distribution volumes, the most probable explanation for this accumulation is a change in bioavailability, due to increased saturation of metabolizing enzymes, resulting from decreased first-pass metabolism. In contrast to Castleden's findings for oral propranolol, there was a (statistically nonsignificant) trend toward *lower* concentrations in the elderly when compared to younger patients.

Regardh and colleagues[4] gave six young (22–28 years) subjects 20 mg of [3H]-metoprolol i.v. and then 50 mg orally twice daily for 5 days. Blood was sampled for 7–8 hr after the i.v. dose and after the first and last oral doses. Patients were studied only once, but the data were used for comparison to populations in three other papers,[5–7] and all four references must be consulted to extract the pertinent methods and results. Though not explicitly stated, it is evident that one young subject was dropped from consideration in their evaluation of the effects of aging on metoprolol's disposition. In this study, 10 older subjects (71–74 years) were given 25 mg of metoprolol, orally, twice daily, for 3 days. On day 4, they received a 25-mg oral dose simultaneously with 0.16 mg of [3H]-metoprolol i.v. Serum clearances calculated from the i.v. doses were similar in the young (1.06 ± 0.08 liter/min) and older (0.91 ± 0.08 liter/min) subjects. Bioavailability (F) at steady state was estimated as:

$$F = \frac{\text{AUC}(0 \rightarrow \infty)\text{p.o.}}{\text{AUC}(0 \rightarrow \infty)\text{i.v.}} \frac{\text{DOSEi.v.}}{\text{DOSEp.o.}}$$

So determined, the bioavailability in the young subjects was $50 \pm 7\%$ and in the older subjects $39 \pm 4\%$ (not statistically significant). Unfortunately, this formula is incorrect. The steady-state AUC after repeated oral dosing (AUCp.o.) should have been that for one dose interval, 12 hr in this case. By extrapolating to infinity, the authors introduced a systematic error that resulted in all determinations of F being falsely elevated. As the terminal half-lives after repeated dosing were similar (young: 3.14 ± 0.68 hr versus older: 3.76 ± 0.31 hr), the percent error may have been similar. Also, interestingly, the young subjects did not demonstrate any trend toward accumulation as seen by Quarterman *et al.* One possible explanation for this is that the subjects of Quarterman and his co-workers received twice the dose of the young controls in Regardh's study. For a drug with potentially saturable first-pass metabolism, comparing bioavailability between populations that are given markedly different oral doses might also introduce ambiguity.

Kelly et al.[8] studied 10 subjects between 28 and 75 years of age and found labetalol bioavailability to be extremely variable, ranging from 8.9 to 68.4%. Linear regression of bioavailability versus age revealed a significant ($p < 0.05$) trend toward decreased bioavailability with age ($F = 0.81$ age $+ 0.78$, $r^2 = 0.49$).

Rubin et al.[9] gave seven young (23–32 years) and seven older (66–78 years) males atenolol, 100 mg orally and 10 mg intravenously, in random order, on two occasions at least 1 week apart. Serial blood samples were collected over 24 h. The mean (range) findings included

	t 1/2 i.v. (min)	t 1/2 p.o. (min)	Bioavailability
Young	200 (135–350)	282 (171–417)	0.56 (0.1–0.9)
Older	211 (161–275)	350 (259–488)	0.55 (0.35–0.9)

Clearly, there is no difference discernible between groups in the values determined for F. The substantial differences in half-life, following the i.v. versus oral doses, are troublesome. F is determined as the dose-normalized ratios of AUCp.o. to AUCi.v., both extrapolated to infinity, based on half-life. If the clearances during the two study periods were not equal, or if the half-life used for extrapolation is not "terminal," the values for F can not be so determined.

4. DISTRIBUTION

The distribution of drugs in the elderly can be affected by changes in binding to circulating proteins (albumin) and to extravascular tissues. There are few data concerning tissue uptake of drugs in the elderly with the exception of the general knowledge that the percentage of lean body tissue decreases and is replaced by adipose tissue. Thus, drugs that have high lipid solubility, such as diazepam, may have an altered distribution in the elderly. Of the beta blockers, acebutolol and sotalol have both been shown to have a decreased volume of distribution in the elderly, probably owing to variation in adipose tissue in relation to total body weight and total body water.[10,11] Changes in volume of distribution with the more lipid-soluble β-blocking agents such as propranolol and metoprolol have not been demonstrated.

β-Adrenergic blockers differ widely in their degree of binding to plasma proteins. Although most of the currently available agents are not very highly protein bound, some, for example propanolol, are highly bound to serum proteins. The binding of beta blockers involves serum proteins other than albumin. Propranolol, bound to total serum proteins by 93%, is only bound to albumin by 62%. Piafsky[12] demonstrated that the acute phase reactant, α-1-acid glycoprotein, is important to the plasma binding of propranolol. In the absence of disease, α-1-acid glycoprotein concentration is increased in the elderly and is increased even more dramatically in the presence of chronic inflammatory diseases such as arthritis. In addition, because the elderly often have a decrease in serum albumin concentration, two important determinants of propranolol binding are frequently altered in the geriatric patient. The net effect for propranolol protein binding in the elderly, then, is dependent on the relative balance between the decrease in serum albumin and the degree of increase in α-1-acid glycoprotein. In the study of Barber et al.[2] the percent of propranalol binding in young (88 ± 1) and older subjects (90 ± 2) did not differ.

5. CLEARANCE

Of all the pharmacokinetic parameters of beta blockers studied in the elderly, drug clearance has been the most extensively evaluated. Beta blocker elimination from the serum is dependent on serum protein binding, hepatic perfusion and intrinsic clearance, renal excretion (both glomerular filtration and active secretion), and, for some drugs, more than one of these. Others in this class, such as metoprolol, which are metabolized in the liver, are converted to active metabolites and are then renally eliminated.

As opposed to the well-documented age-related changes in renal physiology and the associated decrease in function, changes in hepatic histology are not as well correlated with changes in the liver's ability to metabolize drugs such as the lipid-soluble β-blocking agents. Hepatic blood flow in the elderly decreases by 0.3–1.5% yearly,[13] and the additional decrease in hepatic blood flow associated with decreasing cardiac output, even in the absence of clinical heart failure, can result in decreased drug clearance. Both first-pass hepatic clearance and intrinsic hepatic metabolism of beta blockers are thereby potentially affected by these age-related changes in hepatic function.

Multiple pharmacokinetic studies of the fate of propranolol in the elderly have produced conflicting results. In the study of Castleden and

George,[1] systemic clearance after the i.v. dose (0.15mg/kg infused at 1 mg/min) was estimated using suboptimal methods. Instead of computer-fitting an appropriate model to the data, or estimating clearance as DOSEi.v./AUC($0 \rightarrow \infty$), the authors determined an elimination rate constant via linear least-squares regression of the logarithm of the terminal concentration versus time plots. They then extrapolated this terminal exponential to "zero time" (end of infusion?) to determine "Co" and calculated an apparent distribution volume as "amount of drug in the body" (dose?) divided by Co. This volume (commonly called V extrapolated), times 0.693, divided by $t_{1/2}$ was designated clearance. For a drug that demonstrates polyexponential decline, this method would systematically overestimate clearance. The percent error in clearance may have been greater in the older subjects as the authors stated ". . . the distribution phase of the concentration–time curve appeared prolonged in the elderly; on average it lasted 75 min in the young compared to 108 min in the old." The eight older subjects who received an i.v. dose had significantly ($p < 0.02$) lower clearance (mean $\pm$ S.E.: 7.8 $\pm$ 1.3 ml/min per kg) than the seven younger subjects (13.2 $\pm$ 1.4 ml/min per kg). Half-lives averaged 254 $\pm$ 51.9 min in the elderly and 152 $\pm$ 10.3 min in the young subjects ($p < 0.05$). The same 15 subjects plus one additional young subject were also given a single 40-mg oral dose of propranalol. Terminal half-lives after this dose were 217 $\pm$ 13 min in the old versus 215 $\pm$ 20 min in the young. Concentrations achieved in the older subjects were as much as 2.3-fold higher than those in the young. Based, in part, on the observation that albumin concentrations were similar between groups, the authors propose that the plasma protein binding was probably similar and that the lower clearance seen in older subjects after the i.v. dose was probably due to an age-associated decrease in hepatic blood flow. The differences noted in first-pass extraction were indicative of an age-related decline in intrinsic hepatic clearing capacity. One elderly subject smoked, a factor implicated in altered hepatic clearance. None took other medications or had evidence of hepatic, cardiac, renal or pulmonary disease. However, an unspecified number of the older subjects had osteoarthritis. Rheumatoid arthritis and other inflammatory diseases have been shown to be associated with increased concentrations of α-1-acid glycoprotein and increased propranolol binding.

In Barber's six young and five older subjects, studied at steady state on propranalol, 80 mg orally per day, the mean ($\pm$ S.E.) AUC($0 \rightarrow 12$) was 339 $\pm$ 79 ng hr/ml in the young and 606 $\pm$ 125 ng hr/ml in the older subjects. Total body clearances (ml/min) were reported at 4428 $\pm$ 940 and 2126 $\pm$ 557, for the young and older subjects, respectively. Methods used to determine these parameters are not specified, but the study design dictates that the clearances must have been *apparent* total body clearances or actual clearance divided by bioavailability. Although the mean AUCs and clearances differed by more than 100%, neither of these parameters achieved statistical significance ($p > 0.05$). No subject data besides age and creatinine clearance are provided.

In studying the effect of a variable, such as age, on a drug's pharmacokinetics, most investigators try to control for other potentially confounding variables, such as other drugs or diseases. Among other reasons, this is done to minimize intragroup variability and, thus, improve the power of the analysis—an especially useful maneuver in studies comparing groups of small sample size. In so doing it is possible that important examples of covariance may be missed. An example is the interrelationship between the effects of age and smoking status on the clearance of a number of compounds including propranolol. A number of these studies have been performed and reviewed by Vestal and Wood.[14] Their findings (mean ± S.E.) for propranalol are summarized in Table I.

As can be seen in Table I, there is no change in HBFap associated with smoking, although there is a significant ($p < 0.025$) decrease with age. Also, smoking was associated with a significant ($p < 0.005$) increase in CLint in the younger subjects and a smaller (nonsignificant) increase in the older subjects. Although, in comparison to younger smokers, older smokers had a significantly ($p < 0.025$) lower CLint, there was no age-related difference between nonsmokers. Despite the younger subjects having a mean CLint approximately 50% greater than the older, this

TABLE I   EFFECTS OF AGE AND SMOKING ON PROPANOLOL CLEARANCE[a,d]

	CLtot	CLint	HBFap
Age: 21–37 years			
Nonsmokers ($n = 6$)	10.6 ± 1.31	24.4 ± 4.20[b]	19.1 ± 1.82
Smokers ($n = 7$)	14.9 ± 14.7[c]	65.5 ± 10.05[c]	20.0 ± 2.11
Total ($n = 13$)	12.9 ± 1.14[c]	46.5 ± 8.10	19.6 ± 1.36[c]
Age: 46–73 years			
Nonsmokers ($n = 6$)	9.0 ± .091	26.8 ± 5.59	14.7 ± 1.68
Smokers ($n = 8$)	10.4 ± 0.72[c]	35.3 ± 3.61[c]	15.4 ± 1.40
Total ($n = 14$)	9.8 ± 0.58[c]	31.6 ± 3.23	15.1 ± 1.04[c]

[a] Clearances and flows are in ml/min per kg, CLtot is total systemic clearance, CLint is intrinsic clearance, and HBFap is apparent hepatic blood flow.
[b] Significant differences between smokers and nonsmokers.
[c] Significant differences between young and older subjects.
[d] Data from Vestal and Wood.[14]

difference did not achieve significance. The systemic clearance, which is dependent on HBF, CLint, and, perhaps, serum protein binding, was not different between nonsmokers, but was significantly different ($p < 0.005$), between young and old smokers and between young and old subjects overall ($p < 0.025$). As has been demonstrated with antipyrine clearance, age seems to blunt the induction of propranolol clearance in smokers. Using indocyanine green, it has been shown that there is an age-related decrease in hepatic blood flow that can also affect the clearance of relatively flow-rate-sensitive compounds. As in many of these studies, the number of subjects evaluated was small. In addition, in the Vestal study the difference in age of the "younger" (up to 37 years) and "older" (as low as 46 years) subjects was small.

Hitzenberger et al.[15] evaluated the effects of age and smoking on the pharmacokinetics of propranalol and pindolol. Smokers were defined as using more than 10 cigarettes/day, nonsmokers as those who smoked less than 10/day, or who smoked none at all for 5 or more years. Sixteen young (20–30 years, eight smokers) and 16 older (60 or more years, eight smokers) subjects were given propranolol, 80 mg orally three times daily for 3 days. For propranalol on day 3, the mean ($\pm$ S.E.) AUC($0 \rightarrow 8$), in ng hr/ml, achieved was

	Smokers	Nonsmokers
Young	388 ± 59	327 ± 49
Older	835 ± 115	974 ± 210

Analysis of variance, examining the effects on AUC of age, smoking status, and their interaction (age × smoking status), showed age to be the only significant ($p < 0.001$) factor in this population. In comparing their findings to those of Vestal and Wood,[14] the authors postulate that their criteria for smoking status (less than versus greater than 10 cigarettes/day) and/or differences in environmental factors that may also affect liver enzyme status may explain the divergent findings. Other studies,[16,17] evaluating the effects of aging on propranolol disposition, have also had small numbers of subjects and showed age-associated differences in apparent clearance that did not achieve significance.

In addition to the many examples of suggestive differences that fail to achieve statistical significance (possible β or type II errors), metoprolol represents a drug for which another of the potential pitfalls of small sample size may be demonstrated. Although only recently recognized, there is a subset of patients (9% of people in the United Kingdom, 1% of Arabs, 3% of Hong Kong Chinese) with a genetically determined deficiency in hepatic hydroxylation capacity.[18] One or two of these outliers, in a 5- to 10-subject population, can easily introduce sufficient variance to obscure other important trends and relationships. Regardh and colleagues[4–7] found the following terminal half-lives ($t_{1/2}$ in hr) and systemic clearances (CLtot in liters/min) in six young subjects given i.v. metoprolol:

Subject:	KS	CGH	RK	LS	LA	ML
$t1/2$:	2.9	9.5	2.8	3.6	3.9	2.6
CLtot:	0.98	0.34	0.72	0.87	0.78	1.1

The second subject (CGH) was deficient in his ability to hydroxylate metoprolol. When the data from these young subjects were combined with data from six others and compared to those from 10 older subjects, CGH had apparently been dropped without comment or explanation. The mean ($\pm$ S.E.) clearances seen were 1.06 ± 0.08 liter/min in the remaining ($n = 11$) young, versus 0.91 ± 0.08 liter/min in the 10 older subjects ($p < 0.05$). One older subject, who was not excluded, had a clearance of approximately 0.5 liter/min and appeared as much an outlier as CGH.

In Quarterman et al.,[3] evaluation of the effects of age on orally administered metoprolol concentrations, mean AUC($0 \rightarrow \infty$), after dose 1 was 86% *higher* in the young compared to older subjects, and AUC($0 \rightarrow 12$) at steady state was 53% higher in the young. These differences did not achieve statistical significance. These concentrations are, of course, an index of both clearance and bioavailability. In addition, the renally eliminated α-OH metabolite of metoprolol, which is also pharmacologically active, showed substantial differences between young and older subjects, after both the first dose and at steady state. Mean ($\pm$ S.E.) AUCs as ng hr/ml were

	Metoprolol AUC		α-OH-Metoprolol AUC	
	Dose 1($0 \rightarrow \infty$)	Day 8($0 \rightarrow 12$)	Dose 1($0 \rightarrow \infty$)	Day 8($0 \rightarrow 12$)
Young	836 ± 210	1306 ± 289	757 ± 90	571 ± 40
Older	450 ± 97	851 ± 182	1276 ± 177	958 ± 116

As can be seen, compared to young controls, the older subjects had higher ($p < 0.01$) metabolite concentrations on both study days. The change in ratio of metoprolol to its major metabolite, on day 1 compared to day 8, seen in both groups, gives further support to the theory of decreased first-pass metabolism ocurring simultaneously with accumulation to steady state. Individual subjects' data, sufficient to comment on the presence or absence of outliers that would contribute to within-group variance, are not provided.

Briant et al.[19] gave eight young (27–39 years) and seven older (67–94 years) subjects 100 mg of oral metoprolol. Mean (± S.E.) AUC(0 → ∞) as ng hr/ml was 1550 ± 564 in the young and 2012 ± 678 in the older subjects. These differences were not significant, but with percent coeffients of variation of the order of 100, huge study populations would be required to achieve sufficient power of anlaysis to detect a 30% difference between means. The authors state that one of eight young and one of seven older subjects ". . . exhibited impaired metabolism of metoprolol."

Another hepatically cleared drug in this class includes oxprenolol,[20] which, when given 80 mg orally twice daily to eight young (18–24) and eight older (64–74) women, showed neither age-related differences in concentrations achieved, nor any tendency to accumulate at steady state beyond concentrations predicted by first-order pharmacokinetics. Also, labetalol was studied[8] in 10 patients evenly distributed over 28 to 75 years of age. Clearance was 28.3 ± 5.5 (mean ± S.E.) ml/min per kg in patients younger than 60 years ($n = 5$) and 16.4 ± 2.7 in those older than 60 ($n = 5$). As $p = 0.08$, the authors stated that clearances "tended" to be lower in the older patients.

The disposition of acebutalol, which is cleared both hepatically and renally, and diacetalol, a renally cleared active metabolite, was studied in five young (23.4 ± 0.7 years) men and five older (79.4 ± 3.8 years) women, after i.v. (0.35 mg/kg) and oral (400 mg) doses.[10] Creatinine clearances (ml/min) were 128.4 and 54.4 in the young and older subjects, respectively. Resulting acebutalol half-lives (mean ± S.E., in hr), AUC(0 → ∞) in ng hr/ml, and CLtot (ml/min per kg) were

	Intravenous dose			Oral dose	
	t1/2	AUC (0 → ∞)	CLtot	t1/2	AUC(0 → ∞)
Young	3.1 ± 0.3	640 ± 145	8.8 ± 1.5	7.2 ± 0.7	4350 ± 475
Older	2.9 ± 0.2	818 ± 36	6.2 ± 0.3	11.6 ± 0.8	9420 ± 1070
p	NS	NS	NS	0.01	0.001

For diacetalol, after oral dosing, findings were

	t1/2	AUC (0 → ∞)
Young	12.0 ± 0.6	121 ± 890
Older	14.8 ± 0.7	22970 ± 5480
p	0.01	0.01

The most striking finding in these data is the difference in half-lives between oral and i.v. doses. The most probable explanation is that after a small, rapid i.v. "bolus," serum concentrations are detectable only during the phase in which drug is both cleared from the body and also distributed to the "peripheral" tissues. The apparent serum clearance, during this "distribution phase," is not representative of the total body clearance and should not have been labeled as such. These relatively meaningless clearances were not different in the two populations. The authors wisely refrained from estimating the bioavailability as the dose-normalized ratios of oral to i.v. $AUC(0 \rightarrow \infty)$. They would have been falsely high. A much more important index of clearance in these two populations is the much larger oral $AUC(0 \rightarrow \infty)$, achieved for both compounds, in older subjects. If the i.v. data are held suspect, the study design cannot differentiate between differences in CLtot and/or bioavailability. It is probable that, upon repeated dosing, the concentrations of both compounds would be considerably higher in the older subjects.

Among the primarily renally cleared β-receptor blockers, age-related differences in clearance are not surprising. In the study reviewed above, Barber also gave his subjects atenolol, 50 mg orally daily, for 8 days. Mean ($\pm$ S.E.) creatinine clearances (CCr in ml/min), "total body clearances" ("CLtot" in ml/min), and $AUC(0 \rightarrow 12)$ at steady state (ng hr/ml) reported were

	CCr	"CLtot"	$AUC(0 \rightarrow 12)$
Young	98 ± 17	364 ± 41	1962 ± 260
Older	65 ± 9	183 ± 28	3189 ± 210
p	NS	0.02	0.001

As mentioned for propranalol, although reported as total body clearances, "CLtot" must have been the actual clearance divided by the bioavailability. On the basis of either this measure or the $AUC(0 \rightarrow 12)$, the older subjects achieved significantly higher concentrations.

In contrast, Rubin et al.[9] gave atenolol, 10 mg i.v. and 100 mg orally 1 week apart, to seven young (23–32 years) and seven older (66–78 years) subjects with normal indices of renal function. Creatinine clearances were not measured but were estimated to be 95–144 ml/min in the young and 54–88 ml/min in the older subjects. Median (and range) for CLtot (ml/min per kg), bioavailability (F), and half-lives (in min) after i.v. ($t_{1/2\text{i.v.}}$) and oral ($t_{1/2\text{p.o.}}$) doses were

	CLtot	F	$t1/2$i.v.	$t1/2$p.o.
Young	2.3(0.4–4.4)	0.56(0.1–0.9)	200(135–350)	282(171–417)
Older	2.6(1.5–3.0)	0.55(0.35–0.9)	211(161–275)	350(259–448)

There was no significant correlation ($r = 0.2$, $p < 0.1$) to the regression of CLtot on CCr. None of these differences between groups (young versus older) achieved statistical significance, but the half-lives after the oral dose were significantly longer than after i.v. The authors observe that their $t_{1/2\text{i.v.}}$ was considerably shorter than that previously reported[21,22] after 50-mg i.v. doses but similar to another study that also used 10 mg.[23] They postulate two possible explanations for the disparities in half-lives. One is the phenomenon described previously for acebutalol—that the half-lives after the i.v. dose were reflective of net clearance from the serum into the periphery in addition to elimination from the body. The other is that one or more of the several pathways of atenolol elimination are saturable within the concentration range seen after the oral dose. If the first of these postulates is true, the methods used to calculate CLtot and F are invalid. If the second is true, the method used to determine F was invalid, and the values for Cltot are invalid for larger doses and, perhaps, repeated 10-mg i.v. doses. It is unfortunate that the authors chose not to present the AUCs after the oral dose. That measure may have been the most meaningful reflection of clearances in these two populations that could have been compared. Considering these questions and the differences in concentrations documented by Barber, it would be advisable to interpret these values for CLtot and F with extreme caution.

More predictable results have been documented for sotalol, another primarily renally cleared β-blocking agent. Ishizaki et al.[11] studied 12 male subjects 19–35 years old and nine (four male, five female) hypertensive patients, 60–74 years old. Sotalol, 80 or 160 mg, was administered orally. Blood and urine were collected for 24–48 hr and assayed for drug and creatinine. Terminal half-lives ($t_{1/2}$ in hr), apparent serum clearance ("CLtot"; actual clearance/F, in ml/min per kg), renal sotalol clearance (CLr in ml/min per kg),and creatinine clearance (CCr in ml/min), as mean ± S.E. were

	$t1/2$	"CLtot"	CLr	CCr
Young	7.1 ± 0.9	5.93 ± 1.00	4.10 ± 0.60	118.0 ± 4.0
Older	11.4 ± 1.6	3.32 ± 0.23	1.93 ± 0.32	67.4 ± 5.8
p	0.025	0.05	0.01	0.001

Apparent total body clearances were significantly less in the older subjects owing, primarily, to a decrease in CLr. CLr was approximately twice CCr and 60% of "CLtot" in all patients. The correlation of CLr on CCr was significant ($p < 0.01$, $r = 0.665$, $n = 21$). The (mean ± S.E.) 48-hr urinary excretion of unchanged drug was 87.6 ± 5.8% of the dose in the young and 78.5 ± 6.7% of the dose in the older subjects ($p < 0.05$). The "CLtot" was calculated as DOSE/AUC($0 \rightarrow \infty$) and the renal clearance as the amount excreted unchanged in the urine over 48 hr, divided by AUC($0 \rightarrow 48$). The percent excreted in the urine does not agree with the ratios of CLr/CLtot. Complicating factors may have included a less than total bioavailability and/or systematic assay error. The reported percentage recovery (when drug was added to blank medium and assayed) in serum was 86.2 ± 4.0%, and in urine it was 112.3 + 5.2%.

Finally, in Hitzenberger's study,[15] 27 patients received pindolol, 5 mg orally, three times daily for 3 days. Serum creatinine concentrations did not differ between groups. The mean (± S.E.) AUC($0 \rightarrow 8$) in ng hr/ml was

	Smokers	Nonsmokers
Young	63 ± 9	59 ± 8
Older	122 ± 19	101 ± 23

Analysis of variance suggests there was a significant ($p < 0.0001$) effect of age on steady-state AUC. These results again are consistent with an age-related decrease in renal function, which, because of a decrease in creatinine production, can occur without a change in serum creatinine.

Thus the beta blockers that are primarily excreted by the kidneys, nadolol, sotalol, atenolol, and pindolol, all exhibit lower total body clearances in the elderly, owing to the decrease in renal drug clearance associated with decreasing renal function. Accumulation of these agents in the elderly, associated with normal chronic dosing regimens, could result in higher steady-state serum concentrations. Reductions in the dose or administration frequency of these agents is recommended because of the known alteration in their clearance in this population.

6. SENSITIVITY OF BETA BLOCKERS

The majority of beta blockers, when administered chronically, will accumulate in the serum of the elderly to a greater extent than in younger patients. Such drug accumulation raises the specter of enhanced drug toxicity. There are, however, several reasons why the elderly not only tolerate normal doses of β-blocking agents but have been reported to be less sensitive to the pharmacological effects of beta blockers than have younger patients. Unlike the majority of other cardiovascular agents used to treat the common problems of the elderly, there is no well-established serum-concentration-related response to these agents. Variability in sympathetic tone, reflex adrenergic response, number and responsiveness of β-adrenergic receptors, and degree of disease severity all contribute to the ultimate pharmacological effect induced by a given serum concentration of beta blocker. Many of these physiological determinants of drug response are altered in the elderly.

The relationship between clinical response and β-adrenergic-blocking activity depends in part on the number of β receptors occupied by endogenous agonist (epinephrine or norepinephrine) and the number of β receptors occupied by the beta blocker. Plasma norepinephrine levels have been shown to increase with advancing age.[24] In addition, there are reports that β-adrenergic receptor numbers are either normal or decreased in the elderly. Even in studies that have shown no decrease in the number of β receptors (on lymphocytes) in the elderly, there is a consistent decrease in β-receptor sensitivity to both β agonists and antagonists.[25] Speculation as to whether the decreased number of β receptors is the result of receptor down-regulation, due to higher concentrations of norepinephrine, or whether the decrease in receptor number or sensitivity leads to a compensatory increase in norepinephrine levels has not been resolved. It is most likely that there is a decrease in renal norepinephrine excretion in the elderly contributing to the accumulation of the neurotransmitter. The increased circulating norepinephrine could compete with propranolol for occupancy of β receptors, thereby causing an apparent reduction in the affinity of the receptor for the antagonist and resulting in decreased drug sensitivity.

Vestal et al.[26] demonstrated that the sensitivity of β receptors for propranolol diminishes with age. They reported that the effect of a given free drug concentration to antagonize the pharmacological effect of isoproterenol decreases progressively with age. The hemodynamic effect of a given serum concentration of antagonist therefore depends not only on the free drug concentration but on the affinity of the β receptors for the drug. The observed increase in serum propranolol concentration in the elderly, then, is somewhat offset by the reduced sensitivity of β receptors. Hitzenberger et al.[15] reported a 249% increase in maximum serum propranolol concentration, a 298% increase in AUC, a 106% increase in maximum pindolol concentration, and a 178% increase in AUC in the elderly. This increase in serum concentration does not, however, overcome the 16-fold decrease in β-receptor sensitivity reported by Vestal in a group of patients between 21 and 73 years of age.

In addition to the variability observed in response of elderly patients to beta blockers due to changes in sympathetic tone and receptor response, the response to beta blockers appears to be related to the activity of the renin–angiotensin–aldosterone system.[27] Plasma renin activity has been shown to decrease with age, probably as a result of decreasing function of the juxtaglomerular apparatus of the kidney, which undergoes fibrotic changes or arteriolar hyalinization. This may be especially important in the response of elderly hypertensive patients. Buhler et al.[27] in a study of 315 essential hypertensives, divided the patients into categories based on renin profile and age. They found that the percentage of patients with low renin hypertension increased with age and that the antihypertensive efficacy of beta blockers as monotherapy in 137 patients was less effective in patients with low renin hypertension (4% had pressures decreased to less than 95 mm Hg diastolic) than in patients with high renin (85% success) or normal renin (73% success). He also found antihypertensive efficacy correlated with age with a success rate of 80% in patients under 40 years old but only a 20% success rate in patients over 60 years of age.

7. BETA-BLOCKER EFFICACY

Hypertension in the elderly is a common problem with a variety of pathological causes. The US National Health Survey indicated that in patients over 65 years of age, the incidence of hypertension (blood pressure above 160/95) was about 50%.[28] Whether the hypertension is classified as essential or secondary, the altered physiological processes resulting in blood pressure elevation are multiple, complex, and often interrelated. Geriatric patients may suffer from isolated systolic hypertension, diastolic hypertension, or a combination of both forms. Among the factors associated with diastolic hypertension are abnormalities in vascular responsiveness to sodium, increased peripheral resistance secondary to atherosclerosis, decreased compliance of small arteries and arterioles, and decreased baroreceptor sensitivity. Abnormalities associated with systolic hypertension include decreased compliance and capacitance in the aorta and the large arteries. In addition, numerous neuronal and humoral changes in the elderly are associated with blood pressure elevation. These include alterations in plasma norepinephrine levels, changes in the renin–angiotension–aldosterone system, and changes in body fluid distribution.

There is abundant documentation that whether the blood pressure elevation affects primarily the systolic or diastolic component, chronic hypertension in the elderly results in coronary artery disease, congestive heart failure, and strokes. The Framingham data indicate that for patients between 65 and 74, the incidence of cardiovascular disease is three times that of the normotensive population.[29] In fact, isolated systolic hypertension, a common form seen in the elderly, is prognostically as significant a factor for the occurrence of thromboembolic stroke, ischemic heart disease, and congestive heart failure as is an elevation in diastolic or mean arterial pressure.

The effect of drug treatment on hypertension in the elderly is not clear. Numerous studies, including the Veterans Administration (VA) study and the HDFP study, have demonstrated the beneficial effects of drug treatment in diastolic hypertension in all age groups. The VA study demonstrated a beneficial effect of antihypertensive treatment in patients over 59 years of age with diastolic blood pressures in the 90–104 mm Hg range.[30] Patients with major hypertensive complications before treatment benefited the most. One difficulty with the analysis of these data is that there were a relatively small number of patients in the elderly group. No well-controlled trials have as yet been published to evaluate the effect of antihypertensive treatment on isolated systolic modality. Such a study, the Systolic Hypertension in the Elderly Program, is in progress.

β-Adrenergic blocking agents have been used as one of the main therapeutic modalities in the treatment of hypertension since the introduction of propranolol in 1976. The 1984 Report of the Joint National Committee on Detection Evaluation and Treatment of High Blood Pressure recommends that patients over 60 years of age with diastolic blood pressures over 90 mm Hg be treated, albeit initially with lower doses of antihypertensive agents, and that geriatric patients with isolated systolic hypertension be treated on an individualized basis.[31] They also recommend that the therapeutic objective of a systolic pressure between 140 and 160 should be the initial end point and that low-dose diuretic therapy should be the initial treatment modality. Combinations of beta blockers and diuretics appear to be less effective in treating isolated systolic hypertension than do centrally acting agents such as methyldopa and clonidine. β-Adrenergic blockers when combined with a diurectic are, however, effective in treating diastolic hypertension in the elderly.

Beta blockers remain a mainstay in the treatment of stable and unstable angina. The effect of these agents to decrease myocardial oxygen demand by decreasing heart rate, reducing myocardial contractility, and lowering blood pressure produces a marked reduction in effort-induced angina, and they have been demonstrated to be beneficial in some forms of unstable myocardial ischemia. Varient angina, due to coronary artery vasospasm, is a relative contraindication to beta-blocker therapy as the potential to increase vasospasm from unopposed α-adrenergic stimulation must be considered.

Like the situation related to the data for treatment of hypertension, there are few well-controlled studies that evaluate the efficacy of beta blockers in a homogeneous group of elderly patients with angina. Although the pathophysiology of coronary artery disease is better understood than is the pathogenesis of essential hypertension, the scope of the disease is wide ranging, from single-vessel disease in patients with pure exertional angina to multiple-vessel disease in patients with rest pain and global myocardial ischemia. To date there is no data that evaluate differences in efficacy of the various beta blockers in the treatment of angina in the elderly. Theoretical differences, like the potential disadvantage of ISA in angina or advantages of agents with longer biological half-lives, have often been cited in the literature, but have not been well studied in the geriatric population.

8. BETA-BLOCKER TOXICITY

The elderly are often singled out as a group especially predisposed to drug toxicity. Cardiovascular agents are ranked among the drugs most frequently associated with side effects in this population because of their potency and because of the physiological alterations associated with both the aging process and the disease states so commonly found in this cohort of patients.

For the most part, beta blockers are well tolerated by elderly patients who have satisfactory left ventricular function (ejection fractions greater than 35%). Despite some depression in cardiac output due to age, previous coronary artery disease, diabetic myocardopathy, or hypertensive heart disease, most patients tolerate the effects of β blockade without further myocardial depression. Their ability to tolerate the depressant effects may relate to the decreased sensitivity of β receptors in the elderly, but this has yet to be demonstrated. Beta blockers with intrinsic sympathetic activity such as pindolol or oxprenolol are touted as being advantageous in this regard, but well-controlled trials with large enough numbers of patients to draw such conclusions are lacking. The ethical problems associated with such studies may prevent adequate data from being generated. As a result, continued reliance on case reports may be the only method of comparing the incidence of myocardial depression among the various agents.

Since drug accumulation and increased serum concentrations are partially related to lipophilicity, some investigators prefer beta blockers with simple pharmacokinetics, renal elimination, and no first-pass metabolism, such as nadolol or atenolol. Whether or not there are truly any differences in side-effect incidence between these agents and drugs with potential pharmacokinetic disadvantages has yet to be established.

As with younger patients, beta-blocker therapy predisposes the elderly to bronchospasm, claudication, abnormal glucose utilization, bradycardia, heart block, and a variety of central nervous system (CNS) symptoms, including lethargy, somnolence, depression, and behavioral changes. Patients who are predisposed to these side effects are those with a preexisting history involving one or more of these organ systems. Patients with a history of AV nodal disease are, for example, more prone to develop symptomatic AV block when given β antagonists. Central nervous system toxicities are difficult to evaluate, especially in elderly patients with a previous history of changing mental status. Beta blockers with low lipid solubility are theoretically advantageous in this group, but again, data to substantiate a true decrease in frequency of CNS side effects with less lipophilic compounds are lacking.

9. Suggestions Regarding Pharmacokinetic Study Design

A number of the methodological flaws in study design and performance that were adequately addressed in the earlier sections will not be discussed again here. Instead, a subset of these issues will be more completely discussed and suggestions for improved study design offered.

The most common design flaw in this body of literature is inadequate sample size. In several of these trials, substantial differences between means failed to achieve statistical significance. For any of these differences that are "real," this failure stems from the population variance (which can only somewhat be minimized by perceptive and stringent enrollment criteria) and the sample size (a factor over which investigators have much more control). When evaluating the effects on drug disposition of a factor as nonspecific as aging, a sample size of 5–12 subjects/cell has associated with it an unacceptably high risk of making a type II error (failure to identify a difference between groups that is truly present). The table below illustrates this point. It was derived assuming that equal numbers in two groups would be studied and that means of parameter values differ between populations but that the variance about the means is comparable. "Beta" is the probability of failing to identify a "true" difference between means. "CV%" is 100 times the population standard deviation divided by the (smaller) mean value of a parameter of interest (such as F or CLtot). "Difference" is the percent difference between mean values. Defining M2 as the smaller of two population means, this is calculated as

$$\% \text{ Difference} = \frac{(M1 - M2) \times 100}{M2}$$

The body of the table has the values for the total number (the sum of both groups) of subjects needed to achieve the specified power of analysis (1-beta). An alpha (probability that an observed difference is due to chance alone) of 0.05 was employed throughout.

	CV% = 50			CV% = 75		
Beta	0.05	0.10	0.20	0.05	0.10	0.20
Difference						
25%	208	108	126	470	378	284
50%	52	42	32	118	96	72
75%	24	20	14	58	42	32

As it was rare, in the trials reviewed here, for a sample CV% to be less than 50, it is clear that the probability of making a type II error was quite high in most of these studies.

Use of stable or radioisotope labeling of drugs permits an elegant solution to a number of the pharmacokinetic issues in these references. This technology is not always feasible. To employ it, the means of labeling the drug must be developed. Sufficiently sensitive separation and assay techniques, to quantitate labeled and unlabeled drug and metabolites, must be developed. The labeled compound must be safe, and investigators should demonstrate that disposition in the body of labeled and unlabeled drug is identical. If these prerequisites are achieved, several of the issues regarding bioavailability and clearance in the beta blockers could be better evaluated.

Castleden and George[1] and Barber et al.,[2] studying propranolol, and Quarterman et al.,[3] and Regardh et al.,[4–7] studying metoprolol, reported results following repeated oral doses that may have been affected by one or more factors (age, time, concentration, and/or dose-related changes in absorption, first-pass effect, and/or clearance). If subjects were, instead, given a relatively small i.v. dose of labeled drug simultaneous with an oral dose of unlabeled drug, studied, brought to steady state on repeated doses, and then studied again (around simultaneous i.v. labeled and oral unlabeled doses), several objectives could be acomplished. The systemic clearance, after the first dose and at steady state, could be calculated as

$$CLtot = \frac{DOSE^*}{AUC^*(0 \rightarrow \infty)}$$

where DOSE* is amount of labeled drug given and AUC* is for labeled compound only. The bioavailability of the first dose could be estimated using

$$F(\text{dose 1}) = \frac{AUC(0 \rightarrow \infty)}{AUC^*(0 \rightarrow \infty)} \times \frac{DOSE^*}{DOSEp.o.}$$

where DOSEp.o. and AUC are the dose and AUC of unlabeled drug. The bioavailability after repeated doses would similarly be estimated as

$$F \text{ (steady-state)} = \frac{AUCp.o.(0 \rightarrow Tau)}{AUC^*(0 \rightarrow \infty)} \times \frac{DOSE^*}{DOSEp.o.}$$

where Tau is the dose interval. This design not only permits evaluation of the effects of repeated dosing on AUC, as do the designs employed in these trials, but also permits a determination of two factors that contribute to AUC (F and CLtot). If, as is the case for metoprolol, it is suspected that systemic concentrations of drug may affect F, it is desirable to keep the i.v. dose as small as possible, based on assay sensitivity. If necessary, the i.v. dose could be administered after most of the oral dose had been absorbed (3–4 hr postdose, for example). This usually would necessitate more blood sampling times and longer study periods, however.

If the assay for labeled drug were sufficiently sensitive, the same technique should work well for atenolol and acebutolol, which showed significantly different half-lives after i.v. and oral doses. If needed, because the assay was not sufficient to characterize the terminal half-life, larger i.v. doses of labeled drug could be employed, using the same design. Also, if labeled compound technology were not feasible, more valid comparisons of oral and i.v. disposition could be derived from "traditional" study design (an oral and an i.v. dose on two different occasions), employing larger i.v. doses than those used[9,10] above. If, because of high peak concentrations, larger doses by rapid i.v. infusions were deemed unsafe, a longer infusion could be employed so that more nearly equal AUCi.v. and AUCp.o. could be achieved and, thus, more valid comparisons made.

10. CONCERNS FOR FUTURE INVESTIGATIONS

Much investigation remains to be done in the area of altered therapeutic and pharmacokinetic response for β-adrenergic blockers in the elderly. Goldberg and Roberts[32] indicate two areas in which further investigation is needed. These are (1) studies that relate alterations in drug sensitivity to alterations in drug disposition and (2) studies that relate alterations in drug sensitivity to alterations in the responding target organ. They suggest that the first type of study would best be carried out on intact animals or humans and that the second type is most suited to *in vitro* testing with animal organs or tissues. Relating these needs to the geriatric patient, it is obvious that alterations in physiology have measurable effects on drug disposition in the elderly. Much work needs to be done, however, to characterize the effects of these changes on therapeutic outcomes. How closely correlated, for example, are the decreased sensitivity to the antihypertensive effects of beta blockers to the altered pharmacokinetics of these drugs? How much does receptor regulation depend on the altered disposition of these agents? How much does the change in sympathetic physiology affect the sensitivity of the intact organ to respond to beta blockers? These are only a few of the questions that deserve further investigation.

Additional attention must also be paid to the basic requirements of study design to properly assess the impact of any further investigation of beta blockers in the elderly. The most obvious requirement is that homogeneous groups of patients must be chosen whenever possible. The interaction of multiple etiologies and alterations in physiological response to produce the disease we label hypertension or angina must be recognized and their effects separated to the greatest extent possible. By not doing so, the net result of changes in pharmacokinetic or therapeutic response on clinical outcome may be lost.

Comparative trials to evaluate the efficacy or toxicity of one beta blocker over another in the geriatric population are almost completely lacking. Are the differences in lipophilicity, cardioselectivity, and intrinsic sympathetic activity purely theoretical, or do they in fact contribute to the enhanced safety and decreased risk of toxicity in this age group?

References

1. Castleden CM, George CF: The effect of aging on the hepatic clearance of propranolol. *Br J Clin Pharmacol* 1979; 7:49–54.
2. Barber HE, Hawksworth GM, Petrie JC, Rigby JW, Robb OJ, Scott AK: Pharmacokinetics of atenolol and propranolol in young and elderly subjects. *Br J Clin Pharmacol* 1981; 2:118P–119P.
3. Quarterman CP, Kendall MJ, Jack DB: The effect of age on the pharmacokinetics of metoprolol and its metabolites. *Br J Clin Pharmacol* 1981; 11:287–294.
4. Regardh CG, Landahl S, Larsson M, et al: Pharmacokinetics of metroprolol and its metabolite OH-metroprolol in healthy, non-smoking, elderly individuals. *Eur J Clin Pharmacol* 1983; 24:221–226.
5. Regardh CG, Jordo L, Ervik M, Lundborg P, Olsson R, Ronn O: Pharmacokinetics of metoprolol in patients with hepatic cirrhoses. *Clin Pharmacokin* 1981; 6: 375–377.
6. Hoffmann KJ, Regardh CG, Aurell M, Ervik M, Jordo L: The effect of impaired renal function on the plasma concentration and urinary excretion of metoprolol metabolites. *Clin Pharmacokin* 1980; 6:181–191.
7. Jordo L, Attman PO, Aurell M, Johansson G, Regardh CG: Pharmacokinetic and pharmacodynamic properties of metoprolol in patients with impaired renal function. *Clin Pharmacokin* 1980; 5:169–180.
8. Kelly JG, McGarry K, O'Mally K, O'Brien ET: Bioavailability of labetalol increases with age. *Br J Clin Pharmacol* 1982; 14:304–305.
9. Rubin PC, Scott PJW, McLean K, Pearson A, Ross D, Reid JL: Atenolol disposition in young and elderly subjects. *Br J Clin Pharmacol* 1982; 13:235–236.
10. Roux A, Henry JF, Fovache Y, et al: A pharmacokinetic study of acebutolol in aged subjects as compared to young subjects. *Gerontology* 1983; 29:202–208.
11. Ishizaki T, Hirayama H, Tawara K, Nakaya H, Sato M, Sato K: Pharmacokinetics and pharmacodynamics in young normal and elderly hypertensive subjects: A study using sotolol as a model drug. *J Pharmacol Exp Ther* 1980; 212:173–181.
12. Piafsky KM: Disease-induced changes in the plasma binding of basic drugs. *Clin Pharmacokin* 1980; 5:246–262.
13. Geokas MC, Haverback BJ: The aging gastro-intestinal tract. *Am J Surg* 1969; 117:881.
14. Vestal RE, Wood AJJ: Influence of age and smoking on drug kinetics in man. *Clin Pharmacokin* 1980; 5:309–319.
15. Hitzenberger G, Fitscha P, Beveridge T, Neusch E, Pacha W: Influence of smoking and age on pharmacokinetics of beta-receptor blockers. *Gerontology* 1982; 28(suppl 1):93–100.
16. Castleden CM, Kaye CM, Parsons RL: The effect of age on plasma levels of propranolol and practolol in man. *Br J Clin Pharmacol* 1975; 2:303–306.
17. Schneider RE, Bishop H, Yates RA, Quarterman CP, Kendall MJ: Effect of age on plasma propranolol levels. *Br J Clin Pharmacol* 1980; 10:169–171.
18. Lennard MS, Silas JH, Freestone S, Ramsey LE, Tucker GT, Woods HF: Oxidation phenotype—A major determinant of metroprolol metabolism and response. *N Engl J Med* 1982; 307:1558–1560.
19. Briant RH, Dorrington RE, Ferry DG, Paxton JW: Bioavailability of metoprolol in young adults and the elderly with additional studies on the effects of metaclopramide and probathine. *Eur J Clin Pharmacol* 1983; 25:353–356.
20. Kandall MJ, Quarterman CP: The effect of age on the pharmacokinetics of oxprenolol. *Int J Clin Pharmaco Ther Tox* 1982; 3:101–104.

21. Wan SH, Koda RT, Maronde RF: Pharmacokinetics, pharmacology of atenolol and effect of renal disease. *Br J Clin Pharmacol* 1979; 7:569–579.
22. Reeves PR, McAinsh J, McIntosh DAD, Winrow MJ: Metabolism of atenolol in man. *Xenobiotica* 1978; 8:313–320.
23. Fitzgerald JD, Ruffin R, Smedstad KG, Roberts R, McAinsh J: Studies on the pharmacokinetics and pharmacodynamics of atenolol in man. *Eur J Clin Pharmac* 1978; 13:81–89.
24. Lake CR, Ziegler MG, Coleman MD, Kopin JK: Age-adjusted plasma norepinephrine levels are similar in normotensive and hypertensive subjects. *N Engl J Med* 1977; 296:208–211.
25. Abrass IB, Scarpace PJ: Human lymphocyte beta-adrenergic receptors are unaltered with age. *J Gerontol* 1981; 36:298–301.
26. Vestal RE, Alastair JJ, Wood MC, Shand MB: Reduced beta-adrenoreceptor sensitivity in the elderly. *Clin Pharmacol Ther* 1979; 26:181–186.
27. Buhler RF, Burkart F, Lutold B, Kung M, Marbet G, Pfisterer M: Antihypertensive beta blocking action as related to renin and age: A pharmacologic tool to identify pathogenetic mechanism in essential hypertension. *Am J Cardiol* 1975; 36:653–669.
28. Ostfeld AM: Elderly hypertensive patient: Epidemiologic review. *NY State Med* 1978; 78:1125–1129.
29. Kannel WB, Gordon T: Evaluation of the cardiovascular risk in the elderly: The Framingham Study. *Bull NY Acad Med* 1978; 54:573–591.
30. Veterans Administration Cooperative Study Group in Antihypertensive Agents: Effects of treatment on morbidity in hypertension. III. Influence of age, diastolic pressure, and prior cardiovascular disease; further analysis of side effects. *Circulation* 1972; 45:991–1004.
31. The 1984 Report of the Joint National Committee on Detection, Evaluation, and Treatment of High Blood Pressure. *Arch Intern Med* 1984; 144:1045–1047.
32. Goldberg PB, Roberts J: Age and responsiveness to cardiovascular drugs, in Jarvik LF, Greenblatt DJ, Harmon D (eds): *Clinical Pharmacology and the Aged Patient.* New York, Raven Press, 1981.

CHAPTER 13

ANTIEPILEPTIC DRUGS IN THE ELDERLY

WILLIAM H. THEODORE

1. INTRODUCTION

Epilepsy is a common clinical problem in the elderly. Seizures may occur acutely in the course of systemic disease, due to drug toxicity, or as a consequence of disorders of the central nervous system (CNS) itself. Although seizures appearing in the context of an acute illness such as pneumonia, or even a cerebrovascular accident, may not recur, the incidence and prevalence of chronic seizures ("epilepsy") increase with aging. Hauser and Kurland found an overall epilepsy incidence rate for Rochester, Minnesota of 5.7/1000, but 10.2/1000 for the population over 60.[1]

The underlying causes of epilepsy also change through the lifespan. Seizures due to genetic diseases, intrauterine or neonatal injury, and childhood diseases such as meningitis become less common. In the 50- to 70-year age range, brain tumors are a frequent cause of seizures.[2–4] After age 70, however, the importance of tumors declines. Seizures due to cerebrovascular disease also become more common after 60, perhaps

WILLIAM H. THEODORE • Clinical Epilepsy Section, National Institute of Neurological and Communicative Disorders and Stroke, National Institutes of Health, Bethesda, Maryland 20205.

accounting for as much as 30% of new cases.[2,5–8] In an urban hospital population, the incidence of seizures due to alcohol withdrawal may decrease in the over-60 population, because of earlier-than-expected death of alcoholics: what Courjon described as "la disparation prematurée des grandes ethyliques."[2] Seizures may also occur in the course of CNS degenerations, such as Alzheimer's disease or Creutzfelt-Jakob disease, which are very rare in younger patients. Nevertheless, even in the elderly, the cause of seizures remains unknown in up to 50% of cases.[2,7]

The clinical classification of seizures seen in the elderly also contrasts with those in a younger population. Primary generalized seizures (absence, petit mal) almost invariably begin before age 10, and rarely continue past 30 or 40.[9] Some studies suggest that partial seizures may also become less common and generalized tonic–clonic attacks more frequent.[1,2]

These differences in the clinical manifestations of epilepsy in the aging population naturally have an impact on the design and interpretation of clinical trials. But it is also important to consider difficulties in differential diagnosis. Even patients referred to an epilepsy center for "intractable seizures" may eventually be found to have another diagnosis, such as cardiovascular disease, psychiatric disorder, or autonomic nervous system dysfunction.[10] Any patient entered into a controlled study should have evaluation by simultaneous video and electroencephalographic monitoring to confirm the diagnosis of epilepsy. Clinically significant cardiac arrythmia may be detected in 30–50% of elderly people with syncope suspected of having seizures, suggesting that epilepsy is not the cause of their illness.[11,12]

2. STUDIES OF ANTIEPILEPTIC DRUG EFFICACY AND TOXICITY

Proper choice of subjects is one of the most important prerequisites for clinical trials in epilepsy.[13] Failure in previous clinical trials to reach conclusions about the relative efficacy and toxicity of various antiepileptic drugs (AED) has been due in part to poor patient selection, as well as inadequate definition of seizure subgroups.[13,14] For example, no distinction has been made in some studies between complex partial and absence seizures, which may have superficial clinical similarities but respond to different classes of drugs. Patients with treatable causes of seizures, such as brain tumors, as well as significant nonneurological diseases, and those taking other medications, should be excluded from clinical trials.[15] In these cases the course of the underlying disease itself might influence seizure frequency or lead to symptoms erroneously suggesting AED toxicity or to metabolic derangements complicating interpretation of pharmacokinetic data.

Phase I clinical trials, performed to demonstrate drug safety and obtain pharmacokinetic data, are performed on healthy adult normal volunteers.[15] Older individuals are probably not appropriate for phase I trials. Phase II and III AED trials are performed in patients with uncontrolled seizures. Even studies designed to investigate the pharmacology of AEDs already in use should use patients with uncontrolled seizures as subjects.[15] The need for relatively long inpatient hospital stays, or repeated oupatient visits and laboratory tests not ordinarily needed for clinical monitoring, suggests that investigators should be able to offer patients the chance of therapeutic reevaluation and improved seizure control as a concomitant of participation in a study. Patients should probably be excluded from AED trials in the first year after seizure onset, in order to ensure stability of clinical patterns and exclude a progressive intracerebral lesion (which can remain undetected on initial diagnostic tests.)[15,16] Some studies suggest that seizures appearing after age 60 have a better prognosis and are more likely to be controlled by standard drugs.[5] In aging patients, there is an increased risk of adverse drug reactions not completely explained by their increased exposure to complicated drug regimens involving multiple agents.[17,18] They may have a greater risk of drug toxicity and less possible benefit than younger patients in trials of novel AEDs. In evaluating drug toxicity in particular, abnormal blood counts and chemistries may be more difficult to interpret in the aging, as may be changes in pulse or blood pressure. If neuropsychological test batteries are used to help assess drug toxicity, age-adjusted norms would be needed. Tests used to assess drug effects on motor functions such as reaction times and rotatory pursuit would also have to be standardized on age-matched controls.

Another feature of many AED trials has been an inconsistency in regimen followed by the subjects in addition to the drug being tested. This variability has made interpretation of results more difficult, even in a crossover design, with patients serving as their own controls. Each patient becomes, in a sense, a separate study, vitiating statistical analysis of the data. Moreover, the possibilities for drug interaction are increased. The effect of a putative novel AED may only be to change blood levels of another drug the patient is already taking, leading to the erroneous assumption of a therapeutic or toxic effect. In an ideal design, only the drug being tested and placebo would be used in an initial trial to establish efficacy. However, the more fragile metabolic status and greater risk of cardiovascular illness in older patients increase their vulnerability to the adverse systemic and CNS effects of uncontrolled seizures. For the same reason, randomization into a double-blind placebo-controlled trial may not be appropriate, especially if a two-period crossover design requiring withdrawal of therapy already known to be effective is used.[19] Additional disadvantages of two-period crossover designs include the possibility of a carryover effect from the first to the second treatment period.[19] Since the pathophysiology of epilepsy is not fully understood, the influence of drugs cannot be assumed to end when pharmacological washout is complete.

Epilepsy is a chronic illness with intermittent symptoms, and trials of new AEDs may last several years, even without a crossover design.[13] Participation of patients with a limited life expectancy in such a trial may be inappropriate. Yet a double-blind design is especially important for testing AEDs, owing to the unpredictable timing of seizures, similarity of both drug toxicity and seizures themselves to symptoms of other diseases, and possible effect of emotional factors on seizure frequency. Mattson et al.[20] have proposed a double-blind trial design structured to conform to the goals of clinical practice in obtaining adequate seizure control without unacceptable side effects. Patients are initially randomly assigned to a regimen consisting of an active drug plus a placebo, and therapy is unchanged unless toxicity or inadequate seizure control occurs, leading to crossover to another drug or addition of a second active agent, depending on seizure type.[20] This design avoids, at least initially, the disadvantages of an add-on study in which the subject is taking multiple drugs. However, it is aimed at newly diagnosed, untreated patients, which may make the design inappropriate for the elderly. It is also better for comparing the efficacy of drugs already known to have some antiepileptic activity than for evaluating novel compounds. On the other hand, in an initial test of the study design, patients in the 18- to 45-year age range were more likely to be lost to follow-up than patients 45–65 years old,[20] suggesting that older patients may be good subjects for appropriately designed AED trails. Moreover, after initial screening of potential candidates, the number of patients found to be suitable for a double-blind controlled trial may be small.[15] Arbitrary elimination of older patients may substantially increase the difficulty of obtaining a large enough sample size to ensure statistical significance.

3. AED PHARMACOLOGY IN THE ELDERLY

Changes in drug metabolism with aging may affect both the interpretation of pharmacokinetic data from clinical trials and decisions about seizure treatment. Although drug absorption does not appear to decline in the elderly, decreases in liver weight and function have been noted.[21] The half-life and clearance of antipyrine were significantly increased in an elderly population (mean age 83) compared to young subjects (mean age 28), even when corrected for liver volume.[22] Phenobarbital (PB) induction of cytochrome P-450 and liver microsomal protein was less in aged than young rats even when corrected for liver weight.[21] Alterations in renal structure and function include decreases in kidney weight, decreased number of glomeruli, sclerotic changes in glomeruli and vessel walls, and decreased renal plasma flow and glomerular filtration rate.[23] Changes in the pharmacokinetic parameters of individual drugs are unpredictable, however, owing to the interrelations of variations in volume of distribution, clearance, and protein binding.[24] These difficulties may be exacerbated by disease-related derangements of hepatic, renal, and cardiovascular function. Aging patients probably should not be candidates for studies designed to determine basic pharmacokinetic parameters of novel AEDs or to refine our understanding of the general pharmacology of established drugs. On the other hand, it is clear that studies of the special pharmacology of each AED in aging patients are necessary for proper treatment of their seizures.

Phenytoin (PHT) has been the most frequently studied AED. Several reports have failed to show any change in the dose/blood level relationship with aging.[25–27] Houghton and Richens[28] found a weak positive correlation ($r = 0.31$) for increasing PHT levels with age when patients were given 300 mg/day. Most of the patients were also on PB, or primidine, or several additional drugs. The difference was attributed to an increased V_D, and was not felt to be clinically significant. Since the study was performed on outpatients, compliance factors may have played a role.[20] In one study of PHT, Michaelis–Menton parameters were unchanged compared to reported values in the literature on 17 patients aged 67–96.[26] Bauer and Blouin,[29] however, found a decrease in V_{max} with age from 7.5 ± 2.2 to 6.0 ± 1.9 mg/kg per day in 92 patients aged 21–78. K_m did not change. The authors suggested that elderly patients might need a lower PHT dose. Bach et al.[22] found no significant differences in serum half-life ($t_{1/2}$) and clearance of total PHT, but did show decreased free PHT clearance in 14 elderly patients. Since several studies reported an increase in free PHT levels in the elderly, a change in free PHT clearance may be important.[30–32] Decreased PHT protein binding and increased total have been attributed to decreased serum albumin.[31] Some of these studies were performed at subtherapeutic plasma levels and non-steady-state conditions, however. Accurate measurement of PHT clearance rates depends in part on relative levels of tissue and protein binding, and should be performed at steady state. Other investigators have failed to show a relation between PHT binding and patient age, even when serum albumin was decreased.[33,34] Even in studies showing increased free PHT levels in the elderly, the increase has been small, from 10 to 13% free PHT.[30–32] Moreover, distribution of PHT may be less affected by plasma protein binding than by ability to enter the CNS, avid tissue binding, and metabolism.[35] Phenytoin is highly bound to brain subcellular fractions: brain concentration in humans and laboratory animals range up to 3 times total and 10 times free plasma concentrations.[36] Evidence that PHT kinetics may fit a multicompartment model also implies binding changes may not have a clinically significant effect on drug distribution.[37,38]

Although no evidence for increased CNS PHT sensitivity has been reported in humans, patients over 60 may be more sensitive to the hypotensive effects of intravenous PHT used to treat status epilepticus.[39] Kitani et al. reported that minimal effective plasma and brain PHT concentrations for abolition of the toxic hindlimb extensor component of the maximal electroshock seizure were significantly lower in older (24–30 month) than 6-month-old (young adult) mice.[40] Pharmacodynamic studies in aging humans would be extremely difficult to perform, owing to the "random" fluctuation of seizure frequency observed clinically, disparate etiologies, and variation in plasma PHT levels even on frequent dosing regimens.

Phenobarbital, in contrast to PHT, is only 40–50% protein bound, and as much as 50% may be excreted unchanged in the urine.[41] The relatively few studies of the effect of aging on human barbiturate metabolism that have been performed showed no changes in dose/blood level relationship or plasma protein binding.[25,33]

Higher plasma and brain PB concentrations were found in aged rats after bolus injection or during continuous administration, related to decreased plasma clearance; the ratio of plasma/brain levels was increased in older animals.[42] Hexobarbital-induced sleeping times were also prolonged in aged rats, a phenomenon attributed to decreased elimination.[43] A number of reports, however, suggest that elderly patients may be more sensitive to the toxic effects of barbiturates. Paradoxical excitement, insomnia, and suppression of rapid-eye-movement sleep, followed by rebound nightmares, are often due to barbiturate hypnotics in older patients.[44] The incidence of nocturnal femoral fractures appears to be strongly correlated with barbiturate sedation.[45] Although descriptions of "barbiturate pseudodementia" in the elderly have been "anecdotal,"[46–48] the well-documented adverse neuropsychological effects of these drugs would, in any case, discourage their use.[49]

Age-induced changes in the metabolism of carbamazepine (CBZ) and valproic acid (VPA) have not been as fully studied. Mesdjian et al. reported no significant effect of age on the dose/blood level relationship or protein binding of valproic acid, but the mean patient age in their study was only 22 ± 17 years.[50] Bryson et al.[51] reported an increase in single-dose intravenous VPA half-life from 7.2 to 14.9 hr in elderly (75–87 years) patients on a long-stay geriatrics ward compared to young, healthy controls. Clearance did not differ between the two groups, and the increased half-life in the older subjects was attributed to increased V_D. Perucca et al.,[52] on the other hand, found no difference in single-dose oral half-life, V_D, or clearance between healthy elderly and young controls, although the former did show increased free VPA fraction (9.5 versus 6.6%) and decreased free drug clearance. These results were attributed to a decrease in both plasma protein binding and liver metabolizing capacity for VPA.

Since VPA is used mainly to treat absence or atonic attacks in children, and only as a "second-line" agent for generalized tonic–clonic and partial seizures, the lack of data on CBZ is more serious. CBZ induces its own metabolism, a process that may take several weeks to complete.[53] Studies of this drug should be performed in patients on chronic therapy. Further studies of PHT are also needed to define pharmacokinetic parameters and refine dosage schedules; the available evidence suggests that, in the elderly, some reduction, at least in initial doses, may be warranted.

REFERENCES

1. Hauser WA, Kurland LT: The epidemiology epilepsy in Rochester, Minnesota 1935 through 1967. *Epilepsia* 1975; 16:1–66.
2. Courjon J, Artru F, Zeskov P: A propos des crises d'epilepsie apparaissant apres 60 ans observees en clientele de neurologie dans une service de neuro-chirurgie. *Sem Hop Paris* 1970; 48:3129–3132.
3. Wookcock S, Cosgrove J: Epilepsy after the age of 50. *Neurology* 1964; 14:34–40.
4. Raynor R, Payne RS, Carmichael EA: Epilepsy of late onset. *Neurology* 1959; 9:111–117.
5. Feuerstein J, Weber M, Kurtz D, et al: Etude statistique des crises epileptique apparaissant apres l'age de 60 ans. *Sem Hop Paris* 1970; 48:3125–3128.
6. Hyllestead K, Pakkenberg H: Prognosis in epilepsy of late onset. *Neurology* 1963; 13:641–644.
7. Schold C, Yarnell PR, Earnest MP: Origin of seizures in elderly people. *JAMA 1977; 238:1177–1178.*

8. Hildrick-Smith M: Epilepsy, in Caird FL (ed): *Neurological Disorders in the Elderly.* Bristol, Wright PSG, 1982, pp 146–162.

9. Sato S, Dreifuss FE, Penry JK: Prognostic factors in absence seizures. *Neurology* 1976; 26:788.

10. Mattson RH: Value of intensive monitoring, in Wada J, Penry JK, (eds): *Advances in Epileptology.* Xth International Symposium. New York, Raven Press, 1980, pp 43–51.

11. Johansson BW: Long term ECG in ambulatory clinical practice. *Eur J Cardiol* 1977; 5:39.

12. Silverstein, MD, Singer DE, Mulley AG et al: Patients with syncope admitted to medical intensive care units. *JAMA* 1982; 248:1185–1189.

13. Delgado Escueta AV, Mattson RH, Smith DB, et al: Principles in designing clinical trials for antiepileptic drugs. *Neurology* 1983; 33 (suppl 1):8–13.

14. Smith DB, Delgado Escueta AV, Cramer JA, et al: Historical perspective on the choice of antiepileptic drugs for the treatment of seizures in adults. *Neurology* 1983; 33 (suppl 1):2–7.

15. Cereghino JJ: Clinical considerations of drug testing, in Woodbury DM, Penry JK, Pippenger CE (eds): *Antiepileptic Drugs.* New York, Raven Press, 1982, pp 141–157.

16. Theodore WH, Schulman EA, Porter RJ: Intractable seizures: Long term follow-up after prolonged inpatient treatment in an epilepsy unit. *Epilepsia* 1983; 24: 336–343.

17. Vestal RE: Drug use in the elderly: A review of problems and special considerations. *Drugs* 1978; 16:358–382.

18. Massoud N: Pharmacokinetic considerations in geriatric patients, in Benet LZ, Massoud N, Gambertoglio JF (eds): *Pharmacokinetic Basic for Drug Treatment.* New York, Raven Press, 1984, pp 283–310.

19. White BG: Drug testing: Statistical considerations, in Woodbury DM, Penry JK, Pippenger CE (eds): *Antiepileptic Drugs.* New York, Raven Press, 1982, pp 159–166.

20. Mattson RH, Cramer JA, Delgado Escueta AV, et al: A design for the prospective evaluation of efficacy and toxicity of antiepileptic drugs in adults. *Neurology* 1983; 33(suppl 1):14–25.

21. Stevenson IH, Salem SAM, Shepherd AMM: Studies in drug absorption and metabolism in the elderly, in Crooks J, Stevenson IH (eds): *Drugs and the Elderly.* Baltimore, University Park Press, 1979, pp 51–63.

22. Bach B, Hansen JM, Kampman JP, et al.: Disposition of antipyrine and phenytoin correlated with age and liver volume in man. *Clin Pharmacokinet* 1981; 6:389–396.

23. Rowe J: Aging, renal function, and drugs, in Jarvik LF, Greenblatt DJ, Harman B (eds): *Clinical Pharmacology and the Aged Patient.* New York, Raven Press, 1981, pp 115–130.

24. Mitchard M: Drug distribution in the elderly, in Crooks J, Stevenson IH (eds): *Drugs and the Elderly.* Baltimore University Park Press, 1979, pp 65–76.

25. Furlanut M, Benetello P, Testa G, et al: The effects of dose, age, and sex on the serum levels of phenobarbital and diphenylhydantoin in epileptic patients. *Pharmacol Res Comm* 1978; 10:85–89.

26. Lambie DC, Caird FL: Phenytoin dosage in the elderly. *Age Aging* 1977; 6:133–137.

27. Sherwin AL, Loynd JS, Bock GW, et al: Effects of age, sex, obesity and pregnancy on plasma diphenyhydantoin levels. *Epilepsia* 1974; 15:507–521.

28. Houghton, G, Richens A: Effect of age, height, weight, and sex on serum phenytoin concentration in epileptic patients. *Br J Clin Pharmacol* 1975; 2:251–256.

29. Bauer LA, Blouin RA: Age and phenytoin kinetics in adult epileptics. *Clin Pharmacol Ther* 1982; 31:301–304.
30. Hooper WD, Bochner F, Eadie MJ, et al: Plasma protein binding of diphenylhydantoin: effects of sex hormones, renal, and hepatic disease. *Clin Pharmacol Therap* 1974; 15:276–282.
31. Hayes MJ, Langman MJS, Short AH: Changes in drug metabolism with age: Phenytoin clearance and protein binding. *Br J Clin Pharmacol* 1975; 2:73–79.
32. Patterson M, Heazelwood R, Smithurst B, et al: Plasma protein binding of phenytoin in the aged: in vivo studies. *Br J Clin Pharmacol* 1982; 13:423–425.
33. Bender AD, Post A, Meier JP, et al: Plasma protein binding of drugs as a function of age in human subjects. *J Pharmaceu Sci* 1975; 64:1711–1713.
34. Theodore WH, Yu L, Price B, et al: The clinical value of free phenytoin levels. *Ann Neurol* 1985; 18:90–92.
35. Martin BK: Potential effect of plasma protein on drug distribution. *Nature* 1965; 207:274–276.
36. Woodbury DM: Phenytoin: Absorption, distribution, excretion, in Woodbury DM, Penry JK, Pippenger CE (eds): *Antiepileptic Drugs*, New York, Raven Press, 1982, pp 191–208.
37. Coffey JJ, Bullock FJ, Schoenemann PT: Numerical solution of nonlinear pharmacokinetic equations: Effect of plasma protein binding on drug distribution and elimination. *J Pharmaceu Sci* 1971; 60:1623–1628.
38. Theodore WH, Qu Z-P, Tsay J-Y, et al: Phenytoin: the pseudo-steady state phenomenon. *Clin Pharmacol Ther* 1984; 35:822–825.
39. Cranford RE, Leppik IE, Patrick B, et al: Intravenous phenytoin: clinical and pharmacokinetic aspects. *Neurology* 1978; 28:874–880.
40. Kitani K, Masuda Y, Sato Y, et al: Increased anticonvulsant effect of phenytoin in aging BDF mice. *J Pharmacol Exp Ther* 1984; 229:231–236.
41. Whyte MP, Dekaban AS: Metabolic fate of phenobarbital. *Drug Metab Dispos* 1977; 5:63–70.
42. Kapetanovic IM, Sweeney DJ, Rapaport SI: Phenobarbital pharmacokinetics in rat as a function of age. *Drug Metab Dispos* 1982; 10:586–589.
43. Hewick DS: Barbiturate sensitivity in aging animals, in Crooks J, Stevenson IH (eds): *Drugs and the Elderly*. Baltimore, University Park Press, 1979, pp 211–227.
44. Kales A, Kales JD: Sleep disorders: Recent findings in the diagnosis and treatment of disturbed sleep. *N Engl J Med* 1974; 280:487–489.
45. MacDonald JB, MacDonald ET: Nocturnal femoral fracture and continuing widespread use of barbiturate hypnotics. *Br Med J* 1977; 2:483–485.
46. Rudd TN: Prescribing patterns and iatrogenic situations in old age. *Gerontol Clin* 1972; 14:123–128.
47. Gibson IJM: Barbiturate delirium. *Practitioner* 1966; 197:345–347.
48. Exton-Smith AN: The use and abuse of hypnotics. *Gerontol Clin* 1967; 9:264–269.
49. Trimble ME: Anticonvulsant drugs and mental symptoms: A review. *Psychol Med* 1976; 6:169–178.
50. Mesdjian E, Dranet C, Roger J: Sodium valproate plasma levels in epileptic patients; influence of dose, age, and associated therapy, in Levy RH, Pitlick WH, Eichelbaum M, et al (eds): *Metabolism of Antiepileptic Drugs*. New York, Raven Press, 1984, pp 115–128.
51. Bryson SM, Verma N, Scott PJW, et al: Pharmacokinetics of valproic acid in young and elderly subjects. *Br J Clin Pharmacol* 1983; 16:104–105.

52. Perucca E, Grimaldi R, Gatti G, et al: Pharmacokinetics of valproic acid in the elderly. *Br J Clin Pharmacol* 1984; 17:665–669.
53. Faigle JW, Feldmann KF: Carbamazepine: Biotransformation, in Woodbury DM, Penry IK, Pippenger CE (eds): *Antiepileptic Drugs*. New York, Raven Press, 1982, pp 483–495.

CHAPTER 14

Pharmacokinetics and Bioavailability of Corticosteroids in the Treatment of Neurological Diseases of the Elderly

Marinos C. Dalakas

1. Introduction

Geriatric patients frequently present with perplexing multiple disorders and complaints, and it is often difficult to differentiate those symptoms related to disease from those due to aging.[1] This is particularly important in deciding whether, for a suspected clinical syndrome, the clinician should start therapy with corticosteroids, a group of drugs that if not given cautiously, even in younger patients, may result in significant side effects. From our experience with administration of steroids in the treatment of neuromuscular diseases in the elderly, it has been clear that corticosteroids can be lifesaving and prevent permanent disability in those patients who have one of the corticosteroid-responsive neuromuscular disorders. There is, however, quite significant variability in the clinical response of these patients to administered corticosteroids with often a wide range in pharmacological doses needed for therapeutic

MARINOS C. DALAKAS • Infectious Diseases Branch, National Institute of Neurological and Communicative Disorders and Stroke, National Institutes of Health, Bethesda, Maryland 20892.

benefit even for the same neuromuscular illness. Some patients can be managed with small doses of corticosteroids whereas others require much higher doses; in addition, others exhibit multiple severe complications whereas others have relatively little adverse effects despite similar doses (M. C. Dalakas, unpublished observation).

In the elderly, these variations to beneficial and adverse effects of corticosteroids are more apparent and appear to be related to the bioavailability of the drugs influenced by the impaired absorption, distribution, rate of metabolic degradation, and impairment of liver or renal function and to other drugs that the elderly are concurrently receiving. Although little work has been done in characterizing the pharmacokinetics of therapeutic doses of prednisone in different populations, information regarding the specific handling of these drugs by the elderly is lacking perhaps because high doses of corticosteroids have not often been used in this group of patients for long-term therapy. In this chapter we shall discuss our experience in administering high doses of prednisone for the treatment of certain neuromuscular disorders that affect the elderly, and we shall detail the factors that can influence the bioavailability and pharmacokinetics of prednisone in the older population.

2. FACTORS THAT INFLUENCE CLINICAL RESPONSES TO ADMINISTERED STEROIDS

2.1. NORMAL BIOAVAILABILITY AND PHARMACOKINETICS

The free non-protein-bound fraction of the circulating steroid is the biologically active form of the drug. Most of the hydrocortisone in the circulation is in an inactive form bound to plasma protein (transcortin and albumin) with only 5–8% free to interact with cells.[2] Prednisone, the most widely used oral preparation, is an 11-ketosteroid that requires reduction to the 11-β-hydroxyl form, prednisolone, to be biologically active.[3] Prednisolone in plasma binds to albumin and to a specific α and β globulin (transcortin).[4] Although transcortin (the human corticosteroid-binding globulin) shows a high affinity for synthetic steroids, its relatively low concentration in plasma (approximately 10^{-7} M), results in saturation of its binding sites as the steroid concentration increases above physiological levels.[4] Albumin, on the other hand, has a low affinity constant for prednisolone but a greater steroid-binding capacity than does transcortin because of its greater plasma concentration (approximately 10^{-4} M).[3,4] These factors cause the distribution of steroids to differ in comparisons of physiological versus pharmacological doses. Because the biologically active form of the orally administered prednisone is in the unbound form of prednisolone (considered to be 30–40%

of the administered prednisone),[2] changes in the prednisolone plasma protein binding or total plasma protein influence the bioavailability of the administered steroid including its half-life, metabolic clearance, and variance in volume of distribution.[5] This indicates that patients with low serum albumin level could be more susceptible to steroid side effects when treated with high doses of prednisone (presumably because of increased levels of free prednisolone). The percentage of free prednisolone has also been found to be significantly higher in patients with liver disease who are also hypoalbuminemic and have a lower total body clearance of prednisolone[6] being thereby predisposed to higher incidence of steroid side effects.[4,7] Another factor influencing the bioavailability of steroids is their rate of metabolic degradation, which is impaired in conditions influencing metabolic activity such as thyroid diseases.[8]

The rate of steroid clearance can be also affected by the concomitant administration of other drugs that are capable of stimulating liver microsomal enzymes such as phenobarbital, diphenylhydantoin, and rifampin by inducing a more rapid hepatic clearance of prednisolone.[9–11] This can result in failure to achieve an appropriate clinical response with the doses used.

Prednisolone can exhibit dose-dependent pharmacokinetics, so that with increasing doses (from 5 to 200 mg), values for volume distribution, plasma clearance, and half-life may increase. Although the exact reasons for these changes have not been established, they are believed to be related to changes in the plasma protein binding of prednisolone.[12] Because prednisone is rapidly metabolized to its active form prednisolone, the bioavailability of prednisone is generally measured in terms of plasma levels of prednisolone. In normal subjects all tablets of prednisone appear to have equivalent bioavailability, and its absorption does not appear to be dissolution-rate controlled although small differences may exist between different commercial brands of prednisone.[12,13] The average peak plasma prednisolone level after oral administration of prednisone ranges from 200 to 280 ng/ml in several studies with a mean time of peak level ranging from 1 to 3 hr.[4,13,14] Prednisone tablets with low dissolution rates and clinical ineffectiveness have, however, been reported to occur rarely for different generic brands of prednisone, with proper clinical benefit when proprietary brands of the same prednisone dose were given to the same patients.[4,12]

Serum prednisolone levels after oral administration of prednisone appear to be unaffected by the concomitant administration of antacids. Meal ingestion appears to cause a delay of up to 7–10 hr before achieving peak levels after oral administration of prednisone, and when given to fasting subjects, the drug achieves much higher peak concentrations compared to those after meals.[15]

Steroids are transported to the target tissues via the blood stream and tissue fluids and enter the cells by simple diffusion or facilitated transport. Each molecule of steroid is then bound to a tissue-specific cytoplasmic steroid receptor.[16] The hormone–receptor complex is then translocated to the nucleus in an "activated" form where it is bound to the target cell genome. The target cell responds by increased RNA synthesis with the transcription of specific mRNAs. Nuclear binding to the receptor is an absolute requirement for hormone action and occurs only when the receptor is complexed to an active steroid molecule.[2] Cells contain a finite number of receptors that are 40% saturated at physiological glucocorticoid concentrations.[17] The absolute number of receptors determines the magnitude of response to a steroid.[2] Cells in which receptors are absent or abnormal do not respond to steroids.[2] Glucorticoid cell receptors bind a number of different steroid molecules including the synthetic analogs, and their affinity of binding is highly correlated with the potency of a given steroid. These aspects of steroid receptors are very relevant to the function and efficacy of the administered steroids because most of the alteration in glucorticoid sensitivity is absolutely correlated with the number of the glucocorticoid receptors in the cell.[2] This is especially important in the elderly because aging cells have decreased amounts of glucocorticoid receptor sites.[18,19]

2.2. PHARMACOKINETIC CONSIDERATIONS OF STEROIDS ADMINISTERED IN GERIATRIC PATIENTS

Studies regarding the altered pharmacokinetic parameters, i.e., absorption, distribution, metabolism, and/or excretion, involving the handling of therapeutic high doses of corticosteroids in the aging are lacking. The concerns regarding the handling of therapeutic doses of corticosteroids in the elderly can be, however, addressed if one combines the factors affecting the drug disposition and responses in aging[20] along with the previously described factors influencing the bioavailability and kinetics of administered steroids.

Drug disposition and responses in geriatric patients depend on possible changes in the absorption, body composition, protein binding, metabolism, excretion, altered receptor sensitivity, and multiple disease states or drugs.[20] A decrease in lean body mass, total body water, serum albumin concentration, cardiac output, hepatic and cerebral blood flow, and decline in glomerular and tubular functions that occur in the elderly[21] can substantially influence the disposition and kinetics of corticosteroids.

In the elderly there is reduction in the number of absorbent cells, decreased gastrointestinal motility, elevated gastric pH, and a reduced blood flow in the gut,[22] which can theoretically interfere with the absorption of the oral corticosteroids. Although specific data for cortico-steroid absorption are lacking, recent evidence suggests that any changes in passive or active absorption of a drug in the elderly can be balanced by a concomitant alteration in its binding, metabolism, excretion, or distribution.[23] If this is further substantiated, the absorption per se may not be a crucial factor affecting the bioavailability of steroids in the elderly.

With regard to distribution of an absorbed drug, it should be re-membered that in aging much of the metabolically active tissue is slowly replaced by fat with a reduction of lean body mass in proportion to total body weight and a concomitant substantial reduction in total body water.[21,24] These factors suggest that the maximum safe therapeutic steroid dosage in the elderly may be lower than the one administered in the younger age groups. For example, for the treatment of myasthenia gravis, polymyositis, or immune polyneuropathies,[25-27] we usually rec-ommend an empirical high starting dose of 100 mg of prednisone or 1.4 mg/kg of body weight every day for a period of 3–4 weeks with subsequent reduction to every other day.[25,26] For the successful treat-ment of these diseases in patients above the age of 70, we prefer to start on lower doses of 1 mg/kg of body weight prednisone, providing that other concomitant medical conditions do not interfere with the relative safe administration of corticosteroids, as will be discussed.

Two of the most important factors that influence the distribution of corticosteroids in the elderly are the carrier proteins and the amount of available steroid receptors. Albumin concentrations fall in the elderly by approximately 0.4–0.6 g/dl whereas a rise in gamma globulin con-centration occurs.[24,28] This will result in an increased amount of un-bound form of prednisolone (the biologically active form of the admin-istered oral prednisone[2]), which could conceivably result in higher levels of active steroid predisposing the elderly to higher incidence of side effects if administered at the same doses as those used in younger in-dividuals. Although the rise in α-globulin concentration (the antibody-bearing proteins) that occurs with aging[20,24,28] may theoretically require more active amounts of prednisone to suppress antibody production, there are no experimental or clinical data pointing to the appropriate amount of prednisone needed for immunosuppressive use in the elderly. The decreased amounts of glucocorticoid receptor sites that occur in aging cells,[18,19] along with the fact that the potency of a steroid synthetic analog is correlated with its binding affinity for the available receptor cites, as described previously, are additional factors for considering lower-than-normal safe therapeutic doses of cortocosteroids in the elderly.

The conversion of oral prednisone to its active component pred-nisolone as well as the further metabolic degradation of corticosteroids usually takes place in the liver[7] and may be altered in the elderly. This is because the aging liver undergoes cellular changes with diminished liver blood flow, loss in weight, and reduction in the degree of binding for certain drugs.[20] The concomitant administration of other drugs, so common in the elderly, may also affect the clearance of steroids especially if these drugs are among the enzyme "inducers," i.e., phenobarbital, dephenylhydantoin, or rifampin, as discussed earlier.

Excretion by the aging kidneys may also have an effect on the bioavailability of corticosteroids. Although data for the prednisolone excretion are not available, any drug that is unbound within the plasma will be filtered by the kidneys.[20] In the case of prednisone administered orally, the amount of converted free unbound active prednisolone to be filtered is between 30 and 40%, as described earlier. With advancing age, the kidneys' glomerular filtering process is reduced, and their capacity to concentrate water, reabsorb phosphate, and reserve sodium is diminished, enhancing the potential for electrolyte disturbance and toxicity induced by high levels of active unfiltered prednisolone.

In summary, some of the factors described earlier can influence the pharmacokinetics of therapeutic doses of corticosteroids in the elderly, often necessitating lower doses to avoid side effects. No work has been done, however, to clearly delineate those parameters and provide figures of safe therapeutic doses, perhaps because the usefulness of immunosuppressive doses of corticosteroids in treating certain diseases of the elderly has been largely overlooked. Today the number of individuals above the age of 70 is constantly increasing, and the chances that some of them may require therapy with cortocosteroids for an immunological disorder are higher. This is particularly true for certain neuromuscular diseases responsive to high doses of steroids, as will be outlined.

3. ADMINISTRATION OF CORTICOSTEROIDS IN NEUROMUSCULAR DISEASES OF THE ELDERLY

3.1. THE AGING NEUROMUSCULAR SYSTEM

The aging neuromuscular system is so intimately linked to its supporting musculoskeletal elements (skeletal, vascular, and connective tissue) that it is difficult to focus on age-related factors affecting the nerves and muscles alone without considering the impact of the aging musculoskeletal system. Although advancing years often increase the chance of disability resulting from diminished mobility, the focus has been on disorders affecting primarily the central nervous sytem (i.e., strokes, tremors, Parkinson's disease). Other, more insidious but ultimately equally disabling disorders affecting primarily the muscles and nerves, including their diagnosis and management, have been often overlooked or accepted by the aging individual without complaint or without consulting a physician. Perhaps one of the reasons is that a person 65 years or over suffers, on the average, from five different disease entities[29] and takes at least one medication masking the concept of "a single-disease entity" that is so clearly addressed in younger adults. Before the neuromuscular diseases that occur in aging are discussed as well as the administration and pharmacology of steroids that are required for their treatment, some known neuromuscular changes that occur in senescence will be briefly reviewed.

Man reaches his full size about the middle twenties with muscle mass constituting over 40% of body weight. Above the age of 60 the total muscle mass decreases to about 25% of total body weight[30] with an age-related shift toward anaerobic metabolism, diminished resting oxygen consumption, and decrease in number of type 2A muscle fibers.[31,32]

The number of motor units decreases after mid-life.[33,34] This, noted by others,[35,36] has also been confirmed in our detailed histochemical study of muscle biopsies in 10 healthy control individuals where a few angulated (denervated) fibers and fiber type grouping were noted (N. Cutler and M. Dalakas, unpublished observation). A 20–30% reduction of muscle strength has been observed in normal individuals between the ages of 60 and 90[37,38] potentially related to the reduction of the number and size of the muscle fibers. In older individuals, loss of muscle bulk is more pronounced in the thigh, calf, and intrinsic hand muscles unrelated to disease and unaccompanied by fibrillations or fasciculations.[39] Loss of bulk in the first dorsal interosseus is an especially common finding in individuals above the age of 70, as we have very often observed (M. C. Dalakas, unpublished observation). Tendon reflexes diminish with aging,[39] and mild structural changes in the sensory ganglia and myelin along with mild dropout of the large myelinated fibers are reported.[40] Lumbosacral roots are estimated to lose 350 fibers per decade as a consequence of anterior horn cell loss[41] and some deterioration of motor fiber function has been suggested on the basis of electrophysiological criteria.[42] The latter has been estimated as slowing of motor nerve conduction at the rate of 1 meter/sec per decade, after the second decade.[39] Similarly, sensory nerve action potentials seem to decrease in amplitude and propagate at slower velocities between the third and eighth decades. The mean sensory velocity in digital nerves steadily declines from 57 to 48 meters/sec and the amplitude drops by 50%.[43] The needle electromyography is essentially normal with aging other than a slightly increased number of polyphasic units on voluntary activation. The structure of the muscle fibers in 10 healthy individuals above the age of 70 studied with muscle enzyme histochemistry showed no architectural or enzymatic abnormalities other than slight fiber type grouping and increased frequency of tubular aggregates (M. C. Dalakas and N. Cutler, unpublished observation).

3.2. COMMON NEUROMUSCULAR PROBLEMS IN THE AGED THAT NEED STEROID THERAPY

Many elderly people complain of fatigue associated with physical or mental stress but not true muscle weakness. This situation should be distinguished from coexisting depression or other medical illness. Fatigability may, however, be the beginning of weakness due to an upper or lower motor neuron lesion, and careful neurological examination is needed to distinguish the two. Falling in the elderly is not necessarily due to muscle weakness. In fact, only 3% of 500 cases of patient falls were related to predisposing weakness[44,45] whereas the most common cause was tripping or accidental falls followed by drop attack and dizziness. Patients with subacute or chronic neuromuscular problems usually tend to be more aware of their disability, exercise reasonable caution in their activities,[45] and probably fall less often. The most common neuromuscular problems in the aged requiring steroids are the following:

3.2.1. POLYMYOSITIS/DERMATOMYOSITIS (PM/DM). This is a spectrum of inflammatory muscle diseases that does not spare the aged.[46] In fact, in one review[47] patients between 65 and 80 accounted for 16% of 118 patients. In the aged there is a high incidence of associated malignancy especially in patients with dermatomyositis, which varies from 18 to 29%.[45,46] Clinical evidence of dermatomyositis can often precede that of a neoplasm.

The diagnosis of PM/DM is made on the basis of a combined clinical, electrophysiological, and histological study, as has been described previously.[25,46] Patients develop subacute onset of muscle weakness (greater in proximal muscles), weakness of the neck flexors, and elevated creatinine kinase (CK) along with the other sarcoplasmic muscle enzymes. Cranial nerves are spared. A typical skin rash of redness of the face, eyelids, upper chest, and knees may occur in patients with DM. The diagnosis is confirmed with muscle biopsy, which shows active muscle involvement with necrosis and phagocytosis along with interstitial and perivascular inflammatory infiltrates.

Treatment with steroids or immunosuppressants is indicated based on the underlying immune abnormalities implicated in the disease such as the presence of killer T lymphocytes sensitized against muscle cells, a possible rhabdomyocytoxic immunoglobulin, and its frequent association with collagen vascular diseases.[25,46] PM/DM responds to corticosteroids especially if treatment is begun early.[25,46] The suggested steroid doses in the elderly should be lower than the dose we previously reported[25] for the reasons outlined earlier. The method of tapering the dosage, duration of therapy, and the needed collateral therapeutic regime have been described.[25,46]

3.2.2. MYASTHENIA GRAVIS (MG). This is a dysimmune disease that causes weakness resulting from a defective neuromuscular transmission. It presents with fatigability of the cranial nerve musculature and the muscles of the trunk and extremities. It responds to anticholineresterase medication, worsens with d-tubocurarine, and shows a decremental response on repetitive electric stimulation at low frequencies (2–5 Hz). It is due to a circulating IgG antibody against the nicotinic acetylcholine receptor (nAchR) at the end plate. Antibodies to nAchR are detected in the circulation in 90% of MG patients. In MG the thymus plays a central role. It shows hyperplasia with germinal center formation in 70% of the patients, thymoma in 10%, and contains active-for-thymosin-α1 epithelial cells. We have speculated that in MG, the thymic epithelial cells may also be hypersecretory for thymosin α1.[48,56]

Myasthenia in the elderly is not unusual,[49,50] and we have treated and followed-up patients until the age of 85 (M.C. Dalakas, unpublished observation). Elderly myasthenics represented 18% in one series[51] and 7% in another.[49] In another series 45% of patients with MG ranged in age from 60 to 82 years.[52] Men predominate in this age group by at least 2 : 1,[53] and the incidence of thymoma in patients above the age of 50 years may be up to 19%.[54] Clinical response to thymectomy occurs in patients above the age of 40.[55] This is also supported by our observations that in this age group the thymus contains few but active epithelial cells rich in thymosin α1.[48,56,57] The clinician should be careful when performing edrophonium tests in the elderly to confirm the diagnosis of MG. He should have atropine ready to treat severe bradycardia or, as some suggested,[45] should pretreat the patient with atropine (before administering endrophonium) to avoid bradycardia.

MG responds well to prednisone, especially in the elderly, following the treatment plan we have previously described.[25] The kinetics of steroids discussed earlier and the potential side effects in Section 2.2.3 should be considered before deciding on the appropriate therapeutic dose. Other therapeutic modalities including immunosuppressants, or plasmapheresis, can be considered in difficult cases, bearing always in mind the pharmacodynamic and pharmacokinetic concerns in the elderly, as discussed in several chapters.

3.2.3. POLYMYALGIA RHEUMATICA (PR). This is a syndrome of the elderly, and more than 75% of the patients are older than 60 years. It presents with myalgia, muscle tenderness, and arthralgia. Less frequent are malaise, fever, and weight loss. Sedimentation rate is often more than 70 mm/hr, which is important to differentiate PR from the nonspecific aches and muscle pains of the elderly. In several series there is a relationship or association with temporal arteritis. PR responds to low doses of steroids (20–30 mg every day).

3.2.4. CHRONIC RELAPSING POLYNEUROPATHY (CRP). This is a chronic demyelinating polyneuropathy of slow onset, progressive or relapsing–remitting course, elevated cerebrospinal fluid protein, marked slowing of nerve conduction velocity, segmented demyelination demonstrable in sural nerve biopsies, and absence of systemic illness or abnormal serum immunoglobulins.[26] The disease is of immune nature because of immunoglobulin deposits in the patient's myelin sheath and abnormal protein patterns in the cerebrospinal fluid.[58,59] CRP is steroid responsive, and often the patients become steroid dependent requiring a small, maintenance, low-dose prednisone to prevent recurrences.[26] Other therapeutic modalities, i.e., immunosuppressive therapy or plasmapheresis, can be considered in difficult cases.

Several elderly patients may be affected with CRP, and we have seen and treated successfully at least five such patients above the age of 70. All required high doses of prednisone to regain their muscle strength and subsequent low doses to maintain it (M. C. Dalakas, unpublished observation).

3.2.5. POLYNEUROPATHY WITH PARAPROTEINS IN THE SERUM. In these patients the slowly progressive, axonal or demyelinating polyneuropathy is associated with an abnormal serum paraprotein (IgG or IgM monoclonal gammopathy).[27] The bone marrow is normal and there is no evidence of myeloma or lymphoma.[27] Thus, the gammopathy is benign. The IgM paraprotein is an antibody to myelin and peripheral nerve ganglioside.[60] Many patients with IgM monoclonal gammopathy and neuropathy are males above the age of 70 (M. C. Dalakas, unpublished observations). Amyloid should be excluded because in our experience, amyloid neuropathies do not respond to immunosuppressive therapy.[61] Although patients with IgM paraproteinemic polyneuropathy do not respond to corticosteroids as well as the previous group with CRP, steroid therapy has been beneficial in a few patients.[27] Such a treatment in the elderly should, however, be exercised with considerable caution.

3.2.6. ACUTE GUILLAIN-BARRÉ SYNDROME (GBS). This is an acute polyradiculoneuropathy with a self-limited course which does not spare the elderly. The effectiveness of steroids has not been established, and we personally do not use them routinely for GBS patients above the age of 70. Intense supportive care and patience are the most important factors in the successful management of these patients. Plasmapheresis, however, can be of help if started early.

3.2.7. BELL'S PALSY.   This is an acute mononeuropathy not uncommon in the elderly. The clinician should be cautious to rule out other coexistant neurological illness such as metastatic tumor in the basal meninges, systemic polyneuropathy, and sarcoid or infectious illnesses. Although the efficacy of steroids has not been established, a short, 2–3 weeks' course of moderate doses carefully administered has been well tolerated by the elderly.

3.2.8. OTHER DISORDERS.   Some of the other neurological diseases of the elderly that may necessitate treatment with steroids include temporal arteritis, acute transverse myelitis, and cerebral edema or spinal cord compression resulting from injuries, abscesses, or tumors requiring treatment with dexamethasone for a shorter period.

3.3. INCIDENCE OF SIDE EFFECTS FROM THERAPEUTIC DOSES OF CORTICOSTEROIDS IN THE ELDERLY

Although the exact pharmacokinetics of therapeutic doses of corticosteroids in the elderly are unknown, the generally impaired capacity for absorption, distribution, and rate of metabolic degradation associated with aging, together with the increased frequency of multiple illness such as cardiac disease, diabetes, cerebrovascular illness, emphysema, osteoporosis, malignancy, kidney or prostate difficulties and the multiplicity of drugs they receive, are factors that can enhance the incidence of corticosteroid-induced side effects. The long list of complications and side effects of corticosteroids has been previously discussed in detail.[46] The interference of corticosteroids with the function of certain normally aging organs or abnormal tissues may enhance some of the known complications of corticosteroids. The clinician should, therefore, be alert to avoid or recognize them promptly and make therapeutic recommendations aimed at lowering the incidence of toxicity while optimizing therapy and compliance. Some of the steroid side effects seen more frequently in elderly patients treated with high steroid doses are the following:

3.3.1. ELECTROLYTE DISTURBANCES.   Because of less potent homeostatic mechanisms, and frequent use of diuretics along with often poor dietary intake of potassium, the elderly may be more prone to disturbances in electrolyte balance and potassium loss when they are receiving steroids. This may also be facilitated by other drugs or renal disease and can aggravate the preexisting muscle weakness if administered to patients with a neuromuscular illness. For these reasons administration of potassium supplements is essential together with steroid therapy. As total-body potassium depletion is only poorly reflected in serum potassium levels, less dramatic alterations resulting in mild weakness and fatigue may go undiagnosed if not suspected and prevented.

3.3.2. GASTROINTESTINAL DISTURBANCES. Although the possible higher incidence of gastrointestinal bleeding with corticosteroids is debated,[62,63] it is possible that in geriatric populations high doses of prednisone (therapeutic–immunosuppressive doses) may increase the chances of a bleeding tendency because of the concomitant atherosclerosis and higher incidence of duodenal diverticula or diverticulitis.

3.3.3. OSTEOPOROSIS. Corticosteroids induce osteoporosis by decreasing bone formation and enhancing bone resorption.[64,65] The corticosteroid-induced osteoporosis will worsen the osteoporosis due to aging, especially in a female patient with neuromuscular weakness who is physically inactive, stays in bed for long periods, and has an increased urinary calcium output.

3.3.4. DIABETES. The incidence of diabetes is higher in the elderly, and the chances that therapeutic doses of corticosteroids could trigger a preexisting borderline diabetes in the elderly should be considered.

3.3.5. GLAUCOMA AND POSTERIOR SUBCAPSULAR CATARACTS. Because their incidence increases with age, corticosteroids, which can induce these changes even in the younger patient,[46] may facilitate their manifestation.

3.3.6. CARDIOVASCULAR CHANGES. In the aging heart there is a decrease in the speed of contraction, the valves are stiffer, and the cardiac output, which decreases by 0.9% per year after the age of 20, is reduced.[20] By the eighth decade cardiac output has decreased 40%.[66] Corticosteroids with their mineralocorticoid properties and water retention may increase the load for an already weak heart. This can be clinically significant in an elderly patient who has already suffered previous myocardial infarction, is hypertensive, or has early congestive heart failure. The clinician should be able to make the right therapeutic adjustments of the steroid dosage to lower toxicity and maximize therapy in a cardiac patient who has one of the corticosteroid-responsive debilitating illnesses such as myasthenia, polymyositis, temporal arteritis, or relapsing neuropathy, as discussed earlier.

3.3.7. BEHAVIOR ABNORMALITIES OR PSEUDODEMENTIA. Pseudodementia due to depression and other mood disorders are frequent in the elderly especially in those patients who live alone or are developing a disabling physical condition. In addition, in the aged brain there is a gradual loss of functional neuronal tissue and changes in the sensitivity of brain receptors that can be responsible for an increased sensitivity of the brain to the pharmacological action of drugs. These factors can potentially enhance the known behavioral and psychiatric complications of corticosteroids,[67] which occur in approximately 5–10% of younger patients.[67,68] The clinician should be alert in recognizing psychotic symptoms in the elderly and adjust the corticosteroid doses or accelerate the pace of tapering from every day to alternate days.[26,45]

3.3.8. IMMUNOLOGICAL DISTURBANCES. Glucocorticosteroids generally suppress the immunological functions of T lymphocytes with a rather differential effect on helper versus suppressor cells.[69] These changes may alter the already impaired immune function in aging individuals and cause reduction in the competence of the body's defensive apparatus to cope with exposure to new microbial antigens.[70] Elderly patients who receive high doses of corticosteroids could therefore have a much higher incidence of infection, both bacterial and viral. In our experience, *herpes zoster* has been seen more frequently in the elderly who receive therapeutic doses of corticosteroids (M. C. Dalakas, unpublished observation).

4. CONCLUSIONS

Corticosteroids in high doses (up to 1 mg/kg) may be required in the management of certain diseases in the elderly. In particular, patients with certain neuromuscular diseases respond favorably to corticosteroids, which, if administered properly, can be lifesaving. Knowledge of the pharmacokinetics of corticosteroids used in therapeutic doses for the management of several diseases in the elderly is lacking and should be a topic of future studies. Bioavailability and pharmacokinetic parameters of steroids in the elderly may depend on the patient's mobility, associated illnesses, the level of the total serum protein and albumin, the function of liver and kidney, and interactions with the other drugs. Although the number and severity of steroid side effects can be higher in the elderly, cautious administration aimed at lowering the incidence of toxicity while optimizing therapeutic benefit can result in the successful management of certain disabling neurological disorders considered corticosteroid responsive.

REFERENCES

1. Kenny AD: Designing therapy for the elderly. *Drug Ther* July 1979; 49–64.
2. Morris HG: Factors that influence clinical responses to administered corticosteroids. *J Allergy Clin Immunol* 1980; 66:343–346.

3. Wagner JG, Wexler D, Agabeyoglu T, et al: Plasma protein-binding parameters of prednisone in immune disease patients receiving long term prednisone therapy. *J Lab Clin Med* 1981; 97:487.

4. Cambertoglio JG, Amend WJC, Benet LZ: Pharmacokinetics and bioavailability of prednisone and prednisolone in healthy volunteers and patients: A review. *J Pharmacokin Biopharmaceutics* 1980; 8:1–52.

5. Meikle AW, Weed JA, Tyler FH: Kinetics and interconversion of prednisolone and prednisone studied with new radioimmunoassays. *J Clin Endocrinol Metab* 1975; 41:717.

6. Uribe M, Summerskill WHJ, Go VLW: Why hyperbilirubinemia and hypoalbuminemia predispose to steroid side effects during treatment of chronic active liver disease. *Gastroenterology* 1977; 72:1143.

7. Powell LW, Axelsen E: Corticosteroids in liver diseases: Studies on the biological conversion of prednisone to prednisolone and plasma protein binding. *Gut* 1972; 13:690.

8. Lipsett MB: Factors influencing the rate of metabolism of steroid hormones in man. *Ann NY Acad Sci* 1971; 179:442.

9. Jubiz W, Meikle AW, Levinson RA, et al: Effect of diphenylhydantoin on the metabolism of dexamethasone. *N Engl J Med* 1970; 283:11–14.

10. Brooks PM, Buchanan WW, Grove M, et al: Effects of enzyme induction on metabolism of prednisolone. *Ann Rheum Dis* 1976; 35:339–343.

11. Van Marle W, Wooks KL, Beeley L: Concurrent steroid and rifampicin therapy. *Lancet* 1979; 1:1020.

12. Tanner A, Bochner F, Caffin J, et al: Dose-dependent prednisone kinetics. *Clin Pharmacol Ther* 1979; 25:571–578.

13. Sullivan TJ, Sakman E, Albert KS, et al: *In vitro* and *in vivo* availability of commercial prednisone tablets. *J Pharmacol Sci* 1975; 64:1723–1725.

14. Tembo AV, Hallmark MR, Sakmar E, et al: Bioavailability of prednisolone tablets. *J Pharmacokin Biopharm* 1972; 5:257–270.

15. Henderson RG, Wheatley T, English J, et al: Variation in plasma prednisolone concentration in renal transplant recipients given enteric-coated prednisolone. *Br J Med* 1979; 1:1534–1536.

16. Chan L, O'Malley BW: Steroid hormone action: Recent advances. *Ann Intern Med* 1978; 89(1):694–701.

17. Ballard PL: Delivery and transport of glucocorticoids to target cells. *Monogr Endocrinol* 1978; 12:25.

18. Roth GS, Adelman RC: Age-related changes in hormone binding by target cells and tissues: Possible role in altered adaptive responsiveness. *Exp Gerontol* 1975; 10:1.

19. Roth GS, Livingston JN: Reduction in glucocorticoid inhibition of glucose oxidase and presumptive glucocorticoid receptor content in rat adipocytes during aging. *Endocrinology* 1976; 99:831.

20. Massoud N: Pharmacokinetic considerations in geriatric patients, in Benet ZB, Nassoud N, Cambertogero (eds): *Pharmacokinetic Basis of Drug Treatment*. New York, Raven Press, 1984, p 283.

21. Vestal RE: Drug use in the elderly. *Drugs* 1978; 16:358–382.

22. Bender AD: Effect of age on intestinal absorption: Implications for drug absorption in the elderly. *J Am Geriatr Soc* 1968; 16:1331–1339.

23. Ritschel WA: Pharmacokinetic approach to drug dosing in the aged. *J Am Geriatr Soc* 1976; 24:344–354.

24. Crooks J, O'Malley K, Stevenson IH: Pharmacokinetics in the elderly. *Clin Pharmacokinetics* 1976; 1:280–296.

25. Engel WK, Dalakas MC: Treatment of neuromuscular diseases, in, Wiederholt WC (ed): *Therapy for Neurologic Disease.* New York, John Wiley & Sons Inc, 1982, pp 51–101.

26. Dalakas MC, Engel WK: Chronic relapsing (dysimmune) polyneuropathy: pathogenesis and treatment. *Ann Neurol* 1981; 9(suppl):134–145.

27. Dalakas MC, Engel WK: Polyneuropathy with monoclonal gammopathy. Studies of 11 patients. *Ann Neurol* 1981; 10:45–52.

28. Wallace S, Whiting B: Factors affecting drug binding in plasma of elderly patients. *Br J Clin Pharmacol* 1976; 3:327–330.

29. Cape R: *Aging: Its Complex Management.* Hagerstown, MD, Harper & Rowe, 1979.

30. Korenchevski V: *Physiological and Pathological Aging.* Basel, Karger, 1961.

31. Ermini M: Aging changes in mammalian skeletal muscle: Biochemical studies. *Gerontology* 1976; 22:301–316.

32. Tanchi H, Yoshioke T, Kabayashi H: Age change of skeletal muscles of rats. *Gerontology* 1971; 17:219–227.

33. Brown WF: A method for estimating the number of motor units in thenar muscles and the changes in motor unit count with aging. *J Neurol Neurosurg Psychiatry* 1972; 35:845.

34. Campbell MJ, McComas AJ, Petito F: Physiological changes in aging muscle. *J Neurol Neurosurg Psychiatry* 1973; 36:174.

35. Jeunekens FGI, Tomlinson BE, Walton JN: Histochemical aspects of five limb muscles in old age: An autopsy study. *J Neurol Sci* 1971; 14:259.

36. Tomonaga M: Histochemical and ultrastructural changes in senile human skeletal muscle. *J Am Geriatr Soc* 1977; 3:125.

37. Damon A, Seltzer CC, Stoud T, et al: Age and physique in healthy white veterans at Boston. *J Gerontol* 1972; 27:202.

38. Larson L, Karlsson J: Isometric and dynamic endurance as a function of age and skeletal muscle characteristics. *Acta Physiol Scand* 1978; 104:129.

39. Schaumberg HH, Spencer PS, Ochoa J: The aging human peripheral nervous system, in Katzman R, Terry RD (eds): *The Neurology of Aging.* Philadelphia, FA Davis Co, 1983, p 111.

40. Dyck PJ: Pathologic alterations in the peripheral nervous system of man, Dyck PJ, Thomas MC, Lambert ED (eds): *Peripheral Neuropathy,* Philadelphia, WB Saunders Co, 1975, p 296.

41. Kawamura Y, Okazaki H, O'Brien PC, et al: Lumbar motoneurons of man. I. Number and diameter histogram of alpha and gamma axons of ventral roof. *J Neuropath Exp Neurol* 1977; 36:853.

42. Stevens JC, Lofgren EP, Dyck PJ: Histometric evaluation of branches of peronal nerve: A technique for combined biopsy of muscle nerve and cutaneous nerve. *Brain Res* 1973; 52:37.

43. Buchtal F, Rosenfalck A: Evoked action potentials and conduction velocity in human sensory nerves. *Brain Res* 1956; 3:1.

44. Sheldon JH: On the natural history of falls in old age. *Br Med J* 1960; 2:1685–1690.

45. Kula RW: Neuromuscular disorders in geriatric practice, in Slade WR (ed): *Geriatric Neurology.* Mt. Kisco, NY, Futura Publishing Co Inc, 1981, p 253.

46. Dalakas MC: *Polymyositis/Dermatomyositis.* Boston, Butterworths Publishers, 1987 (in press).

47. De Vere R, Bradley WG: Polymyositis: Its presentation, morbidity and mortality. *Brain* 1975; 98:637–666.

48. Dalakas MC, Engel WK, McLure JE, et al: Thymosin α1 in myasthenia gravis. *N Engl J Med* 1980; 302:1092–1093.

49. Hokkanen F: Epidemiology of myasthenia gravis in Finland. *J Neurol Sci* 1969; 9:463–478.
50. Herishano Y, Abramsky O, Feldman S: Myasthenia gravis in the elderly. *J Am Geriatr Soc* 1976; 24:228–231.
51. Osserman KE, Kornfeld P, Cohen F et al: Studies in myasthenia gravis: Review of 282 cases at the Mount Sinai Hospital, New York City. *Arch Intern Med* 1958; 102:72–81.
52. Gluz L, Jerusalem F, Mumenthaler M: Myasthenia gravis in presenium and serium: eine retrospektive klinische studie von 57 patenten. *Schweiz Med Wochenschr* 1976; 106:1001–1005.
53. Stern FH: Myasthenia gravis; an often overlooked disease in the geriatric group. *J Am Geriatr Soc* 1966; 14:1052–1057.
54. Drachman DB: Myasthenia gravis. *N Engl J Med* 1978; 298:136–142, 186–193.
55. Perlo VP, Arnason BWG, Castleman B: The thymus gland in elderly patients with myasthenia gravis. *Neurology* 1975; 25:294–295.
56. Dalakas MC, Engel WK, McLure JE et al: Thymosin α1 in thymic epithelial cells of normal and myasthenia gravis patients and in thymic cultures. *J Neurol Sci* 1981; 50:239–247.
57. Dalakas MC, Engel WK, McLure JF et al: Identification of human thymic epithelial cells with antibodies to thymosin α1 in myasthenia gravis. *Ann NY Acad Sci* 1981; 477–485.
58. Dalakas MC, Engel WK: Immunoglobulin deposits in chronic relapsing polyneuropathies. *Arch Neurol* 1980; 37:637–640.
59. Dalakas MC, Houff SA, Engel WK, et al: CSF monoclonal bands in chronic relapsing polyneuropathy. *Neurology* 1980; 30:864–867.
60. Ilyas A, Quarles RH, MacIntosh TD et al: IgM paraproteins associated with peripheral neuropathy bind to a ganglioside and to oligosaccharide moieties of the myelin-associated glycoprotein. *Proc Natl Acad Sci USA* 1984; 81:1225–1229.
61. Dalakas MC, Fujihara S, Askanas V et al: Nature of amyloid deposits in hypernephroma. Immunocytochemical studies in 2 cases associated with amyloid polyneuropathy. *Am J Pathol* 1984; 116:447–454.
62. Conn HO, Blitzer BL: Nonassociation of adrenocosteroid therapy and peptic ulcer. *N Engl J Med* 1976; 294:473–479.
63. Messer J, Reitman D, Sacks HS et al: Association of adrenocorticosteroid therapy and peptic ulcer disease. *N Engl J Med* 1983; 309:21–24.
64. Whedon GD: Osteoporosis. *N Engl J Med* 1983; 305:397–399.
65. Issekutz B Jr, Blizzard JJ, Birkhead NC, et al: Effect of prolonged bed rest on urinary calcium output. *J Appl Physiol* 1966; 21:1013–20.
66. Rossman I: *Clinical Geriatrics.* Philadelphia, JB Lippincott, 1979, pp 23–52, 132–137, 224–229.
67. Glazer GJ: Psychotic reactions induced by corticotropin (ACTH) and cortisone. *Psychosom Med* 1953; 15:280–291.
68. Falk WE, Mahnke MW, Poskanzer DC: Lithium prophylaxis of corticotropin-induced psychosis. *JAMA* 1979; 241:1011–1012.
69. Bradley LM, Mishell RI: Differential effects of glucocorticoids on the functions of helper and suppressor T lymphocytes. *Proc Natl Acad Sci USA* 1981; 78:3155–3159.
70. Craddock CG: Corticosteroid-induced lymphopenia, immunosuppression, and body defense. *Ann Intern Med* 1978; 88:564–566.

CHAPTER 15

PHARMACOLOGICAL TREATMENT OF PARKINSON'S DISEASE

DONALD B. CALNE AND ARTO LAIHINEN

1. INTRODUCTION

Drugs used to treat Parkinson's disease fall into six main categories: (1) levodopa; (2) extracerebral decarboxylase inhibitors; (3) artificial dopamine agonists; (4) monoamine oxidase inhibitors; (5) anticholinergic agents; and (6) amantadine. Levodopa is the cornerstone of Parkinsonism disease treatment. The other five drug classes are useful supplements or occasional alternatives to levodopa.

Because of the increase in the incidence and prevalence of Parkinson's disease in the elderly (Fig. 1), many of the pharmacokinetic problems characteristic of later life are encountered in the routine treatment of Parkinson's disease. Of special concern are the many patients who experience abrupt and often profound fluctuations in response to therapy, which are often related to the time of administration of medication. Because of the obvious pharmacokinetic origin of these deteriorations in mobility, they are often termed "wearing-off" phenomena or "end-of-dose" reactions. These fluctuations are similar to those of certain

DONALD B. CALNE AND ARTO LAIHINEN • Division of Neurology, Department of Medicine, Health Sciences Centre Hospital, University of British Columbia, Vancouver, British Columbia, Canada V6T 1W5.

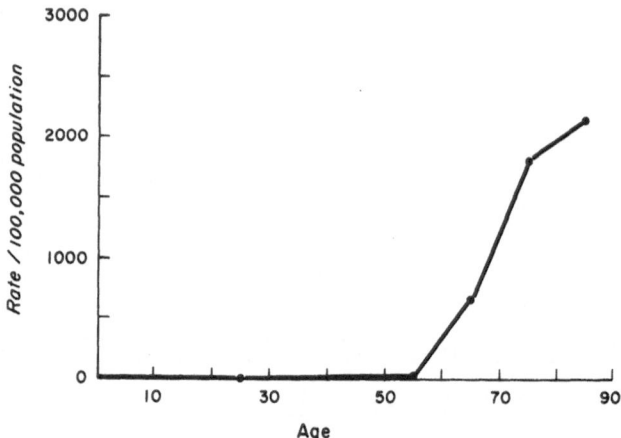

FIGURE 1   Age-specific prevalence rates per 100,000 population for Parkinsonism.[25]

diabetic and myasthenic patients, to whom the term "brittle" is applied to emphasize their therapeutic instability.

Because of wearing off reactions, antiparkinson drug regimens are often complex and demanding. Some drugs may be taken six or even eight times a day. Such schedules are difficult for anyone to follow; in the case of elderly patients, who may have impairment of recent recall, compliance is a major difficulty.

Other general problems of geriatic medicine affect many patients with Parkinson's disease. One such problem is difficulty in managing slowly advancing disease in the elderly. As the disease progresses, higher doses of medications are required. However, as the patients become older, they have increasing difficulty in tolerating medications because of declining hepatic and renal function. In Parkinson's disease there is the additional risk of adverse reactions developing because of the inexorable increase of the underlying neuropathology. This last factor is probably responsible for the increasing dyskinesia, dementia, and "on–off" reactions (unpredictable, sudden episodes of decreased mobility that are not related to the time of administering levodopa). There is no answer to this dilemma, except for the obvious practical approach of seeing patients frequently to determine the best regimen to provide an optimal balance between wanted and unwanted effects.

Another problem of elderly parkinsonian patients is that because of their age, they are subject to several chronic diseases, often requiring long-term pharmacotherapy. The risk of drug interactions is therefore enhanced; particular care is required with such frequently administered agents as hypnotics.

The special needs of elderly patients will be addressed as we discuss each category of antiparkinsonian drug treatment, beginning with the most important, levodopa.

2. LEVODOPA

Levodopa's major site of entry is the jejunum. Absorption is delayed by slowing of gastric emptying and by the presence of large amino acids, which compete for the same active transport system. Levodopa's absorption is enhanced by antacids, which hasten transit through the stomach. Evans et al.[1,2] have reported increased absorption of levodopa in elderly patients (Fig. 2). This augmentation may be a consequence of age-related decreases in the activity of dopa decarboxylase in the enteric mucosa.

After absorption, levodopa is widely distributed throughout the body. It undergoes decarboxylation, 3-O-methylation, transamination, and partial oxidation to form ultimately homovanillic acid and dihydroxyphenylacetic acid; these metabolites are excreted in the urine.

Only a small fraction of the administered levodopa, probably less than 0.1%, actually enters the brain.[3] One important factor limiting the passage of levodopa into the brain is the blood–brain barrier. The same active transport system that operates in the gut plays a similar role in the brain; so again, large amino acids can delay and attenuate the response to levodopa.[4] The half-life of levodopa after oral administration is from 0.77 to 1.08 hr, and approximately 70–80% of an administered dose appears in the urine in the form of metabolites within 24 hr.[5]

The practical question emerges—should levodopa be taken on an empty stomach, or after food? If taken without food, levodopa is absorbed, distributed, metabolized, and excreted rapidly. It therefore achieves a therapeutic response quickly, but this is of short duration and evanescent. If levodopa is taken after a large meal rich in protein, the presence of an overwhelming quantity of large neutral amino acids may so delay absorption from the jejunum and crossing of the blood–brain barrier that metabolism may proceed more rapidly than penetration of levodopa to the striatum of the brain; so the drug will have decreased efficacy, or even none at all. A compromise is necessary in which levodopa should be taken after small meals or snacks, sufficient to prolong but not significantly attentuate transport across the gut and brain capillaries.

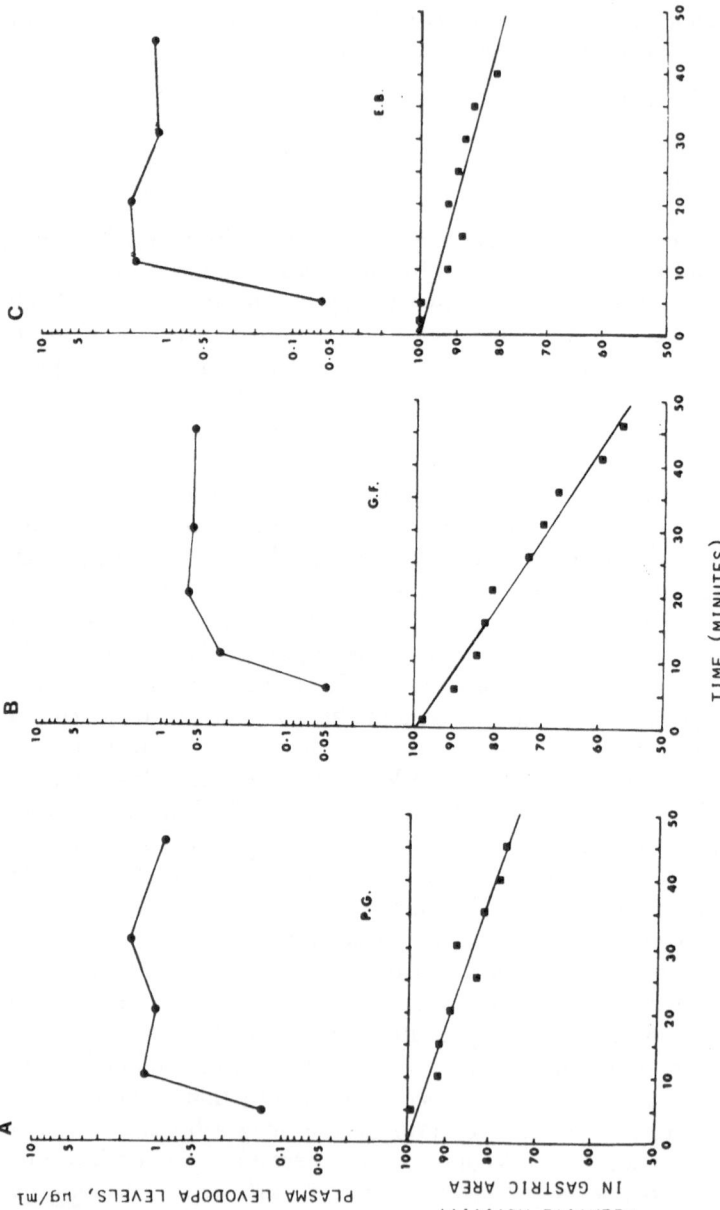

FIGURE 2   Relative disappearance of the dose of levodopa from the stomach in relation to the initial appearance of unchanged levodopa in the systemic circulation in young and elderly subjects. (A) Elderly parkinsonian patient; (B) young healthy volunteer; (C) elderly nonparkinsonian subject. Oral dose of levodopa 500 mg.[1,2]

In this way a reasonable balance between therapeutic response and adverse reactions can be achieved.

3. Extracerebral Decarboxylase Inhibitors

Carbidopa and benserazide are inhibitors of L-aromatic amino acid decarboxylase that do not readily cross the blood–brain barrier. They are employed to suppress certain unwanted effects of decarboxylation products of levodopa, such as dopamine. Because these drugs do not penetrate to the striatum, they do not abolish the therapeutic action of levodopa. However, they do gain access to certain regions of the brain, such as the chemoreceptor trigger zone, where there is no significant permeability barrier. By blocking the conversion of levodopa to dopamine in this region, they suppress anorexia, nausea, and vomiting.

Approximately 100–200 mg of carbidopa or benserazide is required daily for optimal inhibition of dopa decarboxylase. This can be obtained from preparations containing either 4:1 or 10:1 ratios of levodopa to inhibitor, depending on the total daily intake of Sinemet (levodopa/carbidopa) or Madopar (levodopa/benserazide).

Oral carbidopa achieves extensive inhibition of extracerebral decarboxylase in 1 hr, and that effect persists for 4–6 hr.[6]

4. Artificial Dopamine Agonists

The earliest artificial dopamine agonists to be employed for the treatment of Parkinson's disease were apomorphine[7] and N-propyl-noraporphine.[8] Although these were found to have a definite therapeutic action, they proved too toxic for routine treatment.

The only group of dopamine agonists that have, up to now, proved to have a satisfactory therapeutic index for treating parkinsonian patients are tetracyclic ergot derivates (Fig. 3). The most widely used has been the ergopeptine bromocriptine. The ergoline lisuride has also been used extensively. The relative doses and profiles of activity are shown in Table 1. These drugs are all metabolized, with only trace amounts being excreted. Most of the degradation occurs on first pass through the liver, the extraction ratios being 0.94 for bromocriptine[9] and 0.85–0.90 for lisuride.[10,11] The most important determinant of hepatic metabolism is hepatic blood flow, which varies widely between individuals (Fig. 4).

The dopaminomimetic ergot derivatives are extensively bound to plasma proteins (90% bromocriptine; 70% lisuride). They have a more

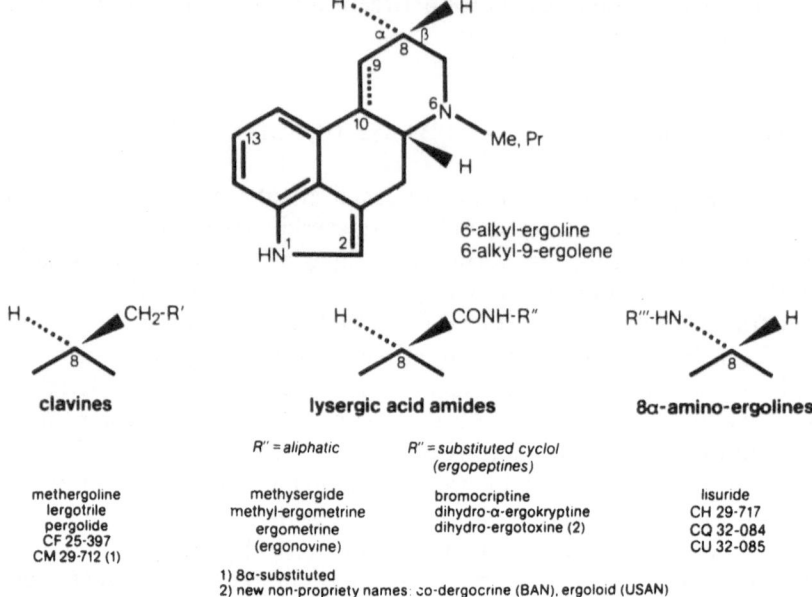

6-alkyl-ergoline
6-alkyl-9-ergolene

clavines	lysergic acid amides	8α-amino-ergolines
	R″ = aliphatic R″ = substituted cyclol (ergopeptines)	
methergoline lergotrile pergolide CF 25-397 CM 29-712 (1)	methysergide methyl-ergometrine ergometrine (ergonovine)	bromocriptine dihydro-α-ergokryptine dihydro-ergotoxine (2)	lisuride CH 29-717 CQ 32-084 CU 32-085

1) 8α-substituted
2) new non-propriety names: co-dergocrine (BAN), ergoloid (USAN)

FIGURE 3   Structural characteristics of ergot derivates.[26]

prolonged plasma half-life than levodopa (bromocriptine 3.0 hr; lisuride 1.0 hr) and a correspondingly extended therapeutic action. They are therefore particularly useful in patients with wearing off reactions.

Dopaminomimetic ergot derivates also produce less dyskinesia than levodopa but they produce more frequent and more severe psychiatric reactions, especially in the elderly.

Extracerebral decarboxylase inhibitors cannot be employed to de-

TABLE I   PROFILES OF ACTIVITY AND EQUIVALENT DOSES
OF LEVODOPA AND DOPAMINE AGONISTS

	Dose (mg/day)	D1 receptors	D2 receptors
L-Dopa	4.0 g	+	+
L-Dopa with carbidopa/benserazide	1.0 g	+	+
Bromocriptine	40 mg	−	+
Lisuride	3.5 mg	−	+
Pergolide	3.0 mg	+	+

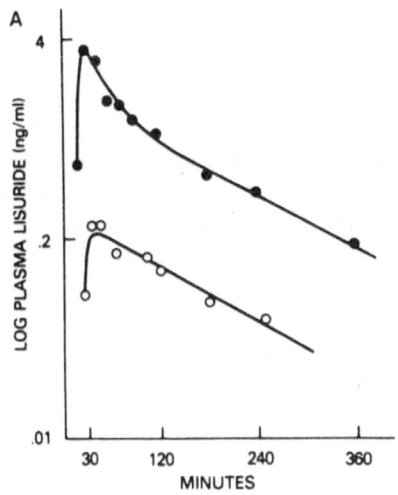

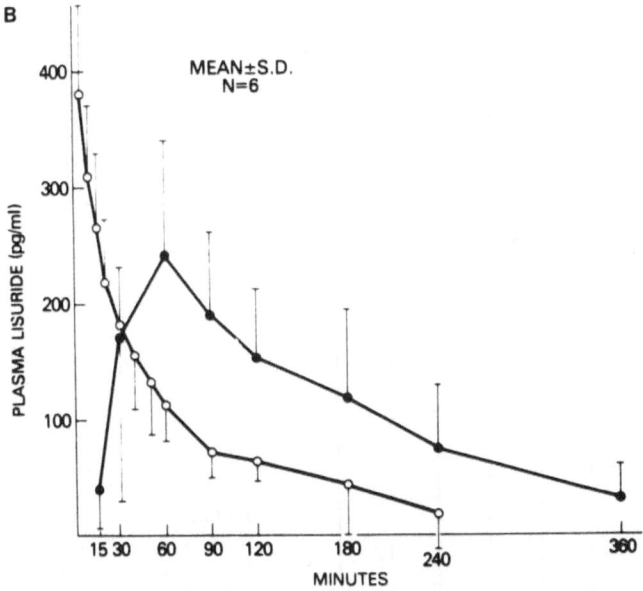

FIGURE 4 Interindividual variation in plasma levels of lisuride following oral and intravenous administration.[10] (A) ●, Patient F.L.; ○ patient L.S. (B) ●, 300 μg P.O.; ○, 25μg I.V.

crease emesis when artifical dopamine agonists are administered, because these drugs do not depend on decarboxylation to an active metabolite. Domperidone is a dopamine receptor antagonist that does not easily cross the blood–brain barrier; so it can be used instead of a decarboxylase inhibitor when nausea and emesis are troublesome.

Because of the age of the typical patient, the same problems arise with artificial dopamine agonists that have already been mentioned with levodopa. As time passes, the disease progresses and more medication is required, but conversely, older patients do not tolerate the agonists as well as younger patients. These points are illustrated in Figs. 5–7.

5. MONOAMINE OXIDASE INHIBITORS

Monoamine oxidases (MAOs) play an important role in the intra-neuronal catabolism of neurotransmitters. Type A MAO (MAO-A) deaminates by oxidation serotonin (5-hydroxytryptamine) and nor-adrenaline predominantly. Clorgyline is a selective inhibitor of MAO-A.

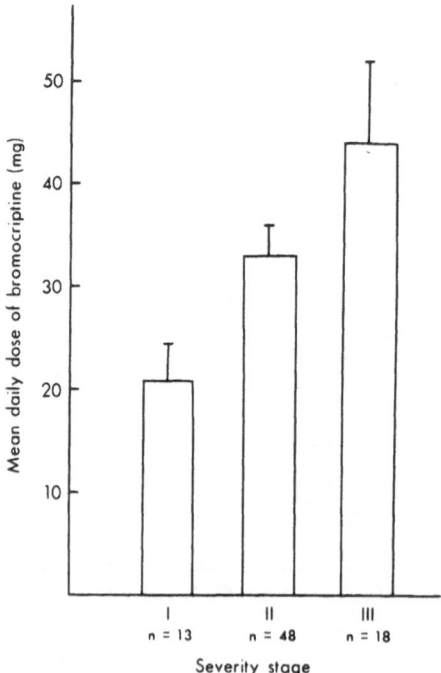

FIGURE 5 The relationship between the severity of Parkinson's disease and the mean daily intake of bromocriptine. The bars denote 1 SEM. Stage I is equivalent to Hoehn and Yahr stage 1; Stage II is equivalent to Hoehn and Yahr stages 2 and 3. Stage III is equivalent to Hoehn and Yahr stages 4 and 5.[27]

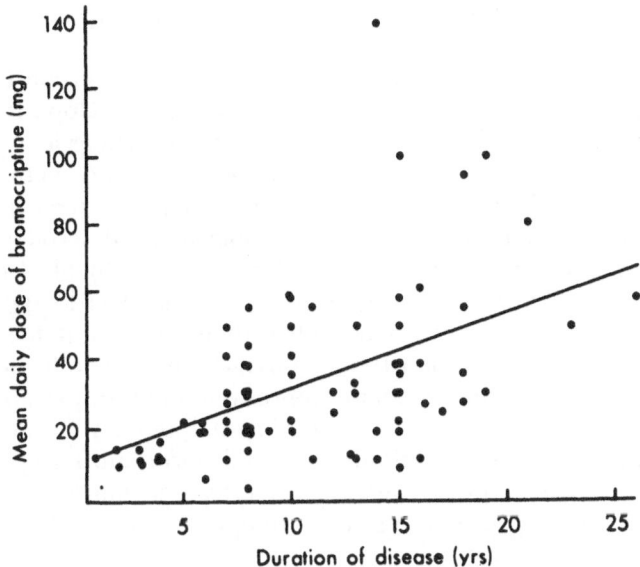

FIGURE 6   Correlation of the duration of Parkinson's disease to the mean daily dose of bromocriptine. The correlation is significant ($p < 0.001$). The regression line is drawn by a computer[27] ($r = 0.50$).

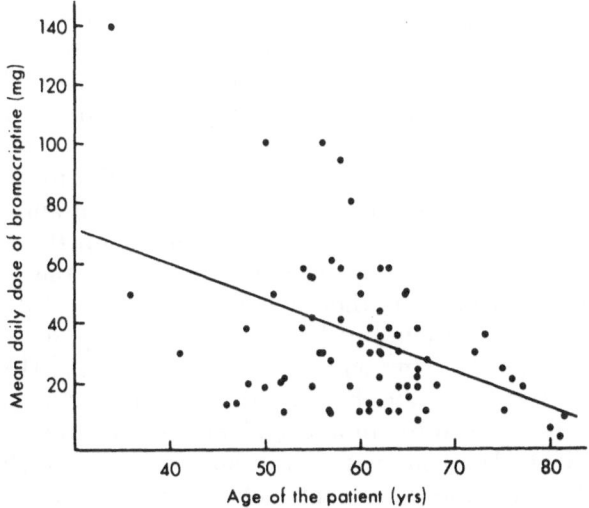

FIGURE 7   Correlation of the age of the patients to the mean daily intake of bromocriptine. The negative correlation is significant ($p < 0.001$). The regression line is drawn by a computer.[27] ($r = 0.46$).

Type B MAO (MAO-B) deaminates phenylethylamine, benzyl-amine, and dopamine.[12]

L-Deprenyl, an MAO-B inhibitor, is marketed in certain European countries, where it is considered an adjuvant to levodopa treatment; there is very little reliable evidence available concerning its pharmacokinetics. It is partly metabolized to amphetamine and methylamphetamine.[13] No unchanged L-Deprenyl is recovered from the urine.[13] Adverse reactions to L-Deprenyl, given in combination with levodopa, derive from the possible formation of excessive dopamine by blockade of its degradation. Clinically, toxicity therefore resembles overdosage with levodopa. L-Deprenyl may be given safely with levodopa; it potentiates the tissue concentrations of the latter, so dosage may have to be decreased. Recent observations indicate that L-Deprenyl inhibits the extrapyramidal syndrome induced by MPTP in experimental animals.[14] The possible clinical implications of this finding are still uncertain. L-Deprenyl has been reported to improve wearing-off effects in more than 50% of patients.[15,16] It may also have some antidepressant action.[17]

6. ANTICHOLINERGICS

Belladonna alkaloids had been the traditional drug treatment for Parkinson's disease throughout the nineteenth and twentieth centuries until the levodopa era of the last 25 years. The natural alkaloids, e.g., atropine and hyoscine, were replaced 40 years ago by the synthetic anticholinergics, such as benzhexol (trihexyphenidyl), benztropine, biperiden, methizene, cycrimine, ethopropazine, orphenadrine, and chlorphenoxamine.

Their antiparkinsonian effect relates to their ability to block central muscarinic acetylcholine receptors.[18] However, certain anticholinergic drugs, especially benztropine, antagonize dopamine uptake into central dopaminergic neurons.[19] It remains unsettled whether this presynaptic mechanism plays a significant therapeutic role.

Before levodopa, certain anticholinergics were supposed to be better than others for specific symptoms of Parkinson's disease. However, no double-blind crossover studies have been undertaken to investigate the comparative effects of different anticholinergics in Parkinson's disease. Although individual patients may tolerate one anticholinergic better than others, the efficacy of the various anticholinergics is essentially the same. Tremor and rigidity are improved more than hypokinesia.

Most anticholinergic drugs act for about 1–6 hr with a peak clinical response from 2 to 4 hours.[20] Burke and Fahn[21] found that the plasma half-life of trihexyphenidyl was 1.7 ± 0.2 hr in normal volunteers who were administered a small dose while in dystonic patients on higher doses the half-life was 3.7 ± 0.3 hr. In normal subjects, cerebral side effects, including confusion, correlated more closely with age than with peak serum levels.

In elderly people, the psychiatric side effects of anticholinergics are prominent. Loss of the cholinergic system originating in the nucleus basalis of Meynert[22] has been claimed to be linked to the memory loss in Alzheimer's disease. It has been argued that the greater the preexisting impairment of memory, the easier it is for the anticholinergic drugs to cause confusional states; in the elderly there may be increased chronic deterioration of intellectual capacity. However, it is not known whether anticholinergics contribute to the damage of cholinergic neurons in demented patients. Besides psychiatric side effects, there may be urinary retention, especially in elderly male patients who have prostatic hypertrophy. Other problems include impairment of ocular accommodation and dryness of the mouth.

Anticholinergics may still have a limited role in the treatment of Parkinson's disease especially during the initial stages of the disorder, though caution is necessary when treating elderly patients; these drugs should be avoided in confused patients.

7. AMANTADINE

Amantadine, which was originally employed as an antiviral agent, was shown to have antiparkinsonian properties some 20 years ago. Although the clinical efficacy of amantadine is significantly less than that of levodopa, it can still be used as adjuvant therapy in Parkinson's disease.[20]

The side effects of amantadine are most often edema, livedo reticularis, and psychiatric reactions. Doses of 200 mg daily should not be exceeded in the elderly.

Amantadine is readily absorbed, and the maximum concentration in blood is reached in 1–4 hr. The plasma half-life is 10 ± 28.5 hr.[23] Most amantadine is excreted unchanged in the urine (92%). According to Montanari et al.,[24] the half-life of urinary excretion in elderly people is approximately 34 hr.

8. CONCLUSIONS

The unique pharmacokinetics of the elderly must be understood by those who treat elderly parkinsonian patients. Though there is evidence that the intestinal absorption of levodopa is increased in the elderly,[12] the pharmacokinetics of levodopa do not change radically with advancing age. During long-term levodopa treatment there are often fluctuations in response and disability. Reducing the dose of levodopa and commencing a dopamine agonist can be helpful. In many cases of wearing-off reactions, L-Deprenyl and amantadine have also been used as adjuvant forms of therapy. In the elderly, the most frequent problems with antiparkinsonian therapy are psychiatric reactions, hypotension, and retention of urine.

References

1. Evans MA, Triggs EJ, Broe GA, et al: Systemic availability of orally administered L-Dopa in the elderly parkinsonian patient. *Eur J Clin Pharmacol* 1980; 17:215–221.
2. Evans MA, Broe GA, Triggs EJ, et al: Gastric emptying rate and the systemic availability of levodopa in the elderly parkinsonian patient. *Neurology* 1981; 31:1288–1294.
3. Wurtman RJ, Hirsch MJ, Growdon JH: Lecithin consumption raises serum free choline levels. *Lancet* 1977; 2:68.
4. Nutt JG, Woodward WR, Hammerstad JP, et al: The "on-off" phenomenon in Parkinson's disease. *N Engl J Med* 1984; 310:483–488.
5. Nutt JG, Fellman JH: Pharmacokinetics of levodopa. *Clin Neuropharmacol* 1984; 7:35–49.
6. Ward CD, Trombley IK, Calne DB, et al: L-dopa decarboxylation in chronically treated patients. *Neurology* 1984; 34:198–201.
7. Schwab RS, England AC, Poskanzer DC et al: Amantadine in the treatment of Parkinson's disease. *JAMA* 1969; 208:1168–1170.
8. Cotzias GC, Papavasiliou MD, Tolosa ES, et al: Treatment of Parkinson's disease with Aporphines. *N Engl J Med* 1976; 294:567–572.
9. Schran HF, Bhuta SI, Schwarz HJ, et al: The pharmacokinetics of bromocriptine in man, in Goldstein M, Calne DB, Lieberman A, Thorner ML (eds): *Ergot Compounds and Brain Function: Neuroendocrine and Neuropsychiatric Aspects.* New York, Raven Press, 1980, pp 125–139.
10. Burns RS, Calne DB: Disposition of dopaminergic ergot compounds following oral administration, in Calne DB, Horowski R, McDonald RJ, et al (eds): *Lisuride and Other Dopamine Agonists.* New York, Raven Press, 1983, pp 153–160.
11. Humpel M, Nieuweboer B, Wendt H: Radioimmunoassay of plasma lisuride in man following intravenous and oral administration of lisuride hydrogen maleate: Effects on plasma prolactin level. *Eur J Clin Pharmacol* 1981; 20:47–51.
12. Glover V, Sandler M, Owen F, et al: Dopamine is a monoamine oxidase B substrate in man. *Nature* 1977; 265:80–81.
13. Schachter M, Marsden CD, Parkes D, et al: Deprenyl in the management of response fluctuations in patients with Parkinson's disease on levodopa. *J Neurol Neurosurg Psychiatry* 1980; 43:1016–1021.
14. Chiba K, Trevor A, Castagnoli N: Metabolism of the neurotoxic tertiary amine, MPTP, by brain monoamine oxidase. *Biochem Biophys Res Commun* 1984; 120:574–578.
15. Lees AJ, Kohout LJ, Shaw KM, et al: Deprenyl in Parkinson's disease. *Lancet* 1977; 2:791–795.
16. Rinne UK, Siirtola T, Sonninen V: L-Deprenyl treatment of on-off phenomena in Parkinson's disease. *J Neural Transmission* 1978; 43:253–262.

17. Eisler T, Teravainen H, Nelson R, et al: Deprenyl in Parkinson's disease. *Neurology* 1981; 31:19–23.
18. Duvoisin RC: Cholinergic–anticholinergic antagonism in Parkinsonism. *Arch Neurol* 1967; 17:124–136.
19. Farnebo L, Fuxe K, Hamberger B, et al: Effect of some antiparkinsonian drugs on catecholamine neurons. *J Pharm Pharmacol* 1970; 22:733–737.
20. Quinn NP: Anti-parkinsonian drugs today. *Drugs* 1984; 28:236–262.
21. Burke RE, Fahn S: Pharmacokinetics of trihexyphenidyl after acute and chronic administration. *Ann Neurol* 1982; 12:94.
22. Whitehouse PJ, Price DL, Clarke AW, et al: Alzheimer's disease: evidence for selective loss of cholinergic neurons in the nucleus basalis. *Ann Neurol* 1981; 10:122–126.
23. Pacifici GM, Nadini M, Ferari P: Effect of amantadine on drug-induced parkinsonism: relationship between plasma levels and effect. *Br J Clin Pharmacol* 1976; 3:883–889.
24. Montanari C, Ferrari P, Bavazzano A: Urinary excretion of amantadine by the elderly. *Eur J Clin Pharmacol* 1975; 8:349–351.
25. Kurland LT, Kurtzke JF, Goldberg ID: *Epidemiology of Neurologic and Sense Organ Disorders*. Cambridge, MA, Harvard University Press, 1973, pp 436.
26. Fluckiger E, Briner U, Enz A, et al: Dopaminergic ergot compounds: an overview, in DB Calne et al (eds): *Lisuride and Other Dopamine Agonists*. New York, Raven Press, 1983, pp 1–9.
27. Larsen TA, Newman R, LeWitt P, et al: Severity of Parkinson's disease and the dosage of bromocriptine. *Neurology* 1984; 34:795–7.

CHAPTER 16

COGNITIVE ENHANCERS IN ALZHEIMER'S DISEASE

NEAL R. CUTLER AND PREM K. NARANG

1. INTRODUCTION

Cognitive enhancers are compounds that have been purported to improve memory. Several drugs from various classes have been investigated clinically in the elderly population with different types of dementia. It is important to clarify that the dose–response relationships with these agents in the elderly population are almost nonexistent. This difficulty associated with the clinical trials in dementia patients is primarily due to problems related to quantification of the drug effect, as neuropsychological assessments have a large degree of associated measurement error. Pharmacokinetic and dynamic studies with these drugs have been few. Therefore, rather than choosing to discuss kinetics, as expected from the contributions in this part, we have decided to address methodological issues that pertain to all agents being evaluated in the elderly with some form of dementia.

Dementia is the most common cause of disability in the older population. Approximately 4 million individuals over the age of 65 years in the

NEAL R. CUTLER • Department of Geriatrics, Cedars–Sinai Medical Center, University of California–Los Angeles School of Medicine, Los Angeles, California 90048.   PREM K. NARANG • Clinical Pharmacokinetics Research Laboratory, Pharmacy Department, Clinical Center, National Institutes of Health, Bethesda, Maryland 20892.

TABLE I   NEUROPATHOLOGICAL AND NEUROCHEMICAL OBSERVATIONS
IN ALZHEIMER'S DISEASE.

Findings	Ref.
Neurofibrillary tangles in cerebral cortex and hippocampus	Constantinidis[2]
Preferential loss of large cortical neurons in midfrontal and temporal regions	Terry et al.[12]
Neuritic plaques in cerebral cortex and amygdala	Constantinidis[2]
Reductions in choline acetyltransferase and acetylcholinesterase activity	Coyle et al.[11]
Decreased brain concentrations of dopamine and norepinephrine	Carlsson et al.[14] Gottfries et al.[17]
Reduction in muscarinic, cholinergic, GABAergic and dopaminergic receptors	Gottfries et al.[17]
Reduction in levels and receptors of somatostatin in cerebral cortex	Davies et al.[31] Rossor et al.[32] Beal et al.[33]

United States (or 15% of the elderly population) have some degree of dementia, and some individuals may suffer from Alzheimer's disease (AD).[1]

Age-associated decline in cognition that is consistent with the diagnosis of AD is both a clinical and a pathophysiological syndrome. Its diagnosis usually rests on the exclusion of several possible etiological factors. Among the foremost are (1) infectious causes (e.g., neurosyphilis and chronic meningoencephalitis; (2) nutritional (e.g., vitamin B_{12} deficiency, chronic malabsorption syndrome); (3) neoplastic (e.g., primary or metastatic cerebral lesions); (4) endocrinological causes (e.g., hypotholamic dysfunction, disorders of pituitary); (5) vascular etiologies, e.g., multiple cerebral infarctions and/or hemorrhage or chronic subdural hematoma; (6) toxic causes (i.e., alcohol- and drug-related events, heavy metal poisoning); and (7) normal-pressure hydrocephalus. If early detection and diagnosis are made, several of these dementias may be reversed. After the secondary causes of acute organic brain syndromes are excluded, cognitive declines with AD etiology reveal insidious onset and progressive deterioration in mental functioning. Therefore, AD is presently regarded as a progressive disorder involving deterioration of both intellect and personality. The neuropathological features of AD are shown in Table I of which the neurofibrillary tangles and neurocortical senile plaques are the most predominant. The actual diagnosis can be made only at autopsy; clinical diagnosis has been shown to be accurate 70–80% of the time.[2] Attempts at pharmacological intervention have been palliative, i.e., those involving the treatment of signs and symptoms of AD, as the disease progressed through its various stages. In recent years, as some of the known neuropathological and neurochemical deficits associated with AD have become known, drug therapies and studies designed to compromise these deficits have evolved. However, these promising treatment rationales have not yet produced the anticipated treatment success.

In this overview, we shall initially review the history of AD treatment including a brief review of some nonneurochemically based treatments in practice. Second, we shall provide the reader of the current treatment strategies that are based on neurochemical findings with a brief comment on other avenues presently being explored in halting or modifying the progression of the disease. Finally, we shall discuss some of the methodological concerns in conducting clinical trials with these agents.

Therefore, understandably, these cognitive enhancers and their trials have to be treated as unique as compared to the other class of compounds discussed in Part II. However, one can still gain information regarding the design of clinical trials in AD and the related dementias.

2. EMPIRICAL THERAPY

Although there can be considerable debate, for practical purposes AD can be classified in four different phases: mild (phase I); moderate (phase II); severe (phase III); and final (phase IV) (Table II).[3]

TABLE II   CLASSIFICATION OF VARIOUS PHASES OF ALZHEIMER'S DISEASE
AND THEIR ASSOCIATED SYMPTOMS

Phase I	Phase III
1. Emotional/mood	1. Language/motor
Depression	Aphasia
Anxiety	Apraxia
Fatigue	Agnosia
Decreased activity	Echolalia
2. Memory deficits	2. Psychiatric
Recent	Hallucinations/delusions
Delayed	3. Disrupted sleep
Phase II	Reversed circadian rhythms
Language aphasias	Phase IV
Tangential	1. Uncontrollable behavioral disturbances
	2. Decreased arousal
	3. Urinary/fecal incontinence

In the mild phase, central nervous system (CNS) stimulants, such as amphetamines, methylphenidate, pemoline, and caffeine derivatives, have been used. These compounds produce CNS stimulation and thereby increase motor activity, concentration, and alertness. Increased central adrenergic activity and improved memory retention seen in animal studies have led to such trials. Clinicians and investigators have assumed that the lethargy and decreased arousal associated with AD and the resulting decreased capacity to concentrate and learn can be overcome by CNS stimulants. Unfortunately, this has not been the case with the elderly AD patient. The potential side effects of increased cardiac activity, irritability, decreased appetite, exacerbated psychosis, sleep disturbances, and the eventual development of tolerance to these agents do not make them ideal therapeutic agents for an elderly AD patient.[4,5]

Antidepressants, such as amitriptyline, imipramine, maprotiline, and trazodone, have also been used to treat AD patients. They have shown effectiveness in treating pseudodementia, a depressive episode masking as dementia, or simply dementia accompanied by or exacerbated by depression. Controlled trials of antidepressant drugs for use as cognitive enhancers of AD patients have been few and primarily unsuccessful[6] because of their anticholinergic properties that can induce a confusional episode or worsen the already impaired cognition. The dosage, duration of action, and side effects of these agents, which include orthostatic hypotension, tachycardia, urinary retention, and conduction defects, have limited their use in the geriatric AD patient.[7]

Vasodilators, e.g., papaverine, cyclandelate, nifedipine, and isoxysuprine, have been tried in the initial or early stages of AD in an attempt to improve cerebral blood flow (CBF) and with the hope of improving mental function or cognition. Numerous behavioral variables, including alertness, depression, anxiety, confusion, and even the severity of dementia, have shown improvement. These investigations have been based on the concept of decreased CBF as a possible factor in the etiology of AD. This supposition does not appear to be well founded, as healthy elderly volunteers with normal mental function have exhibited decreased CBF. The approach of cerebrovasodilation has been specifically useful in the treatment of multiinfarct dementia rather than AD.[4,5] In fact, vasodilators can potentially direct blood away from the ischemic brain areas. Systematic investigations in controlled homogeneous AD patients with this class of compounds have yet to be undertaken.

Only one vasodilator, cyclandelate, has been found to improve AD mental dysfunction. This finding however, still requires further exploration and support. Ergotamines and other ergot derivatives, compounds with vasodilatory properties, have also been administered to AD patients with mixed results.[8] The studies that have indicated positive results have focused on improving symptoms of motivation–initiation, hostility, dizziness, and emotional lability.[9]

Anxiolytic agents, e.g., diazepam, lorazepam, and alprozolam, have been evaluated with regard to their effect on the agitated AD patient. These agents have shown some promise but, like the others, seem to be useful only in the early stages of AD. As the underlying disease progresses, these compounds become ineffective, and major tranquilizers are often needed for control. In addition, the side effects, such as drowsiness, paradoxical reactions, and stupor, associated with this class of compounds[7,10] usually make them less desirable.

The antipsychotic agents, e.g., chlorpromazine and fluphenazine, from the class of the phenothiazines, and others such as haloperidol or from the class of thioxanthenes, have been used principally for their antipsychotic effects.[10] These agents predominantly interact with both noradrenergic and dopaminergic neurons and are targeted to ameliorate such symptoms as paranoid delusions and auditory and visual hallucinations, which are usually associated with the late-moderate-to-severe-stage AD patient. These agents have shown some effectiveness in controlling anxiety, agitation, and psychotic symptoms. The major concern regarding the use of these agents in the elderly AD patient has been the potential for extrapyramidal side effects, e.g., acute dystonia, parkinsonism, and possibly tardive dyskinesia. It is believed that reduction in dopamine activity, expressed postsynaptically, is responsible for extrapyramidal effects.

3. PHARMACOLOGICAL BASIS FOR NEUROCHEMICAL INVESTIGATIONS

Postmortem studies of cell counts and neuropathological changes in brains of patients with AD have demonstrated a number of morphological and neurochemical alterations. Morphological alterations, as indicated earlier, include neurofibrillary tangles in the cerebral cortex and preferential loss of large cortical neurons in midfrontal and temporal regions along with neurotic plaques in cerebral cortex and amygdala. Alterations in choline acetyltransferase activity; in brain concentrations of dopamine, norepinephrine, and serotonin; and changes in the muscarinic, cholinergic, and dopaminergic receptors have all been based on research done in the past decade on brain neurotransmitters.[11–18] These investigations have attempted to identify functions that may be related to the clinical presentation of AD. Research has revealed possible implications of several neurotransmitter systems in AD. Glycolytic pathway enzymes such as phosphofructokinase aldolase and phosphoglucose isomerase have also been implicated in this disease along with a reduction in brain consumption oxygen and glucose.[19,20]

Recently, several major hypotheses of AD etiology have been proposed based on neurochemical deficiencies found in the cholinergic, noradrenergic, serotoninergic, dopaminergic/γ-aminobutyric acid (GABAergic), and neuropeptidergic transmitter systems. These alterations along with those associated with the enzymes of the glycolytic pathway form the major foundation for the pharmacological approaches that have been investigated to develop new treatment strategies in AD.

3.1. CHOLINERGIC SYSTEM

Although 11 major neurotransmitter systems have been implicated in AD, the most dramatic and consistent findings are related to deficits in the cholinergic system. As can be seen from Fig. 1, the cholinergic neuron converts choline to acetylcholine, in the presence of enzyme choline acetyltransferase (CAT), which is then released into the synaptic cleft for interaction with postsynaptic receptor.[21] The cholinergic hypothesis is based on the consistent finding of a deficit in CAT activity, primarily in the hippocampus and frontal cortex, regions of the brain associated with memory and cognition. Negative correlation of deficiency in CAT with increasing plaque count in the brains of AD patients, observed by Perry and Perry,[21] suggests the involvement of the cholinergic system. Other morphometric and neurochemical findings of loss of cell bodies in the substantia inominata[11] and deficiency in acetylcholinesterase (ACHE), an enzyme that degrades acetycholine, as well as in acetycholine content and/or its synthesis, provide compelling evidence in support of the cholinergic system's implication in AD. Based on this argument, pharmacological agents that augment cholinergic neurotransmission have been investigated in trials to improve memory and cognition. Based on this deficiency model, as used in Parkinson's disease with L-dopa, choline and lecithin have been administered as precursors of acetylcholine in an attempt to increase brain acetylcholine brain concentrations.[22] This "transmitter substitution" technique has not achieved overall success. One reason for the failure of these precursors to improve cognition in AD patients may be a preferential loss of presynaptic terminals with an intact postsynaptic muscarinic receptor.

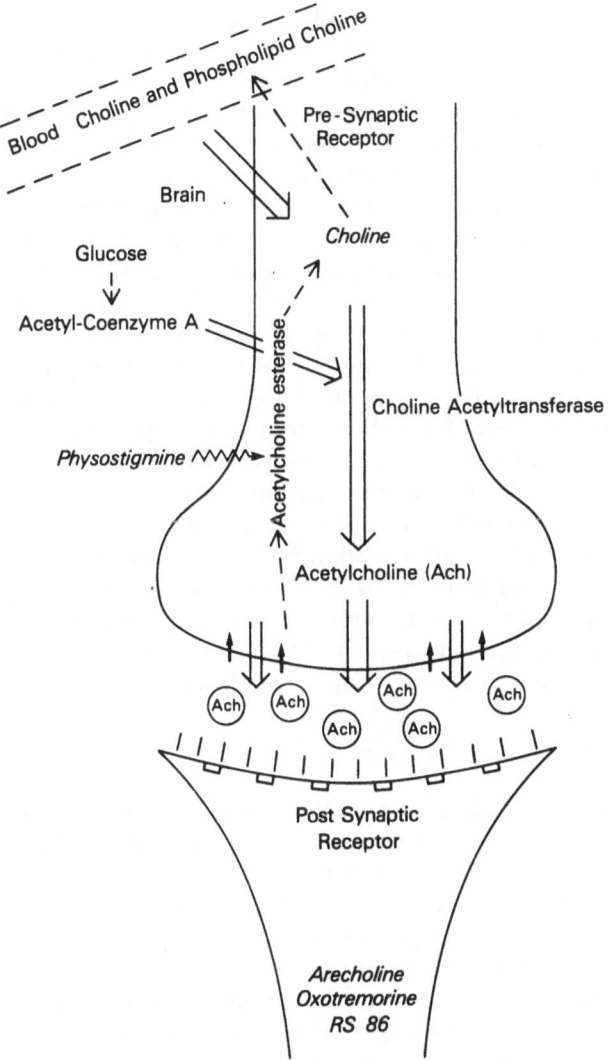

FIGURE 1   Biosynthesis of acetylcholine and drugs that modulate the cholinergic system.

Another alternative that has been explored for correcting cholinergic deficits has been to reduce the degradation of acetylcholine by administering ACHE inhibitors such as physostigmine. This strategy has been demonstrated to be useful in reversing dementia induced in normals by scopolamine, and its efficacy in AD therapy seems promising.

In a number of studies in which physostigmine was administered, either alone or in combination with lecithin, small improvement in memory was demonstrated by evaluating cognitive measures.[23] Although the reports of improvements have been few, this minority has found significant increases in memory function. However, there are limitations to the administration of physostigmine, such as its narrow therapeutic index, peripheral cholinergic side effects, and poor oral absorption. More controlled clinical studies, especially those involving the use of physostigmine in combination with lecithin, appear to be warranted in AD.

The cholinergic system deficiency in AD has also been attacked by the administration of cholinergic agent with direct muscarinic properties, i.e., arecholine, a drug that has shown promise in improving memory in both normals and AD patients.[24] However, investigations in AD patients with arecholine have shown minimal improvement, much less in magnitude than found in normal subjects. In addition, the narrow dose range, undesirable administration route (intravenous) for chronic dosing, and frequent side effects have made this treatment strategy less desirable for further investigation.[23] A new muscarinic agent, RS 86, still under investigation, appears to be well absorbed orally and has shown some ability in improving cognition and behavior in some AD patients in a preliminary report.[25] However, a recent chronic double-blind study failed to show a similar beneficial effect.[26]

3.2. NORADRENERGIC SYSTEM

Postmortem studies have provided evidence that the noradrenergic neurotransmitter system may also play some role in AD. Cortical dopamine hydroxylase activity has been found to be reduced along with the number of noradrenergic cell bodies in the locus ceruleus.[18] Modulation of the noradrenergic system in AD patients with pharmacotherapeutic agents has not been attempted. Various modifications to enhance noradrenergic function such as the administration of clonidine (α^2 agonist) or agents with similar properties need to be explored in order to carefully evaluate the role of this system in AD.[27]

3.3. DOPAMINERGIC/GABAERGIC SYSTEMS

Measures of homovanillic acid, a metabolite of dopamine, and dopamine concentration in postmortem in AD brains reveal deficiencies of the dopaminergic neurotransmitter system.[17] Dopamine and its metabolite levels in the caudate nucleus of AD patients were found to be 54 and 70%, respectively, of those of normal controls.[18] Results from initial studies of replacement therapy with L-dopa in AD patients have been mixed. Only a few studies have been well controlled. In one double-blind crossover study[28] of 6 months' duration, both clinical and neuropsychological measures demonstrated improvement in the early stages of AD. The improvement was found following the administration of a total of 200 mg/day of L-dopa and a peripheral decarboxylase inhibitor. This study does, however, suggest a need for further investigation with other dosages of L-dopa and other dopaminomimetic agents to determine whether the underlying pharmacomodulation reveals a therapeutic role in AD treatment.

The GABAergic system has also shown decrements in the temporal lobes of postmortem AD brains.[18] There appears to be an interrelationship between GABAergic and cholinergic neurotransmitter system, but to date, trials of GABA-mimetic drugs, such as muscimol and the benzodiazepines, as cognitive enhancers in AD patients have not been undertaken. Piracctam, a GABA derivative and "metabolic enhancer," in combination with choline may have some utility in the treatment of AD.[29]

3.4. SEROTONINERGIC SYSTEM

Neurotransmitter, serotonin (5-HT), remains historically the most intimately involved with neuropsychopharmacology. Several theories exist regarding its role in various forms of mental illness which are based on biochemical abnormalities in its synthesis. Only 1–2% of the serotonin in the whole body is found in the brain. Brain cells are capable of synthesizing 5-HT, as it does not cross the blood–brain barrier. As can be seen from Fig. 2, the primary substrate for its synthesis, amino acid tryptophan, is taken up actively from the systemic pool and is then converted to serotonin following a hydroxylation and decarboxylation step.

Serotonin has also been implicated in AD, based on the biochemical changes observed in autopsied brain tissue.[16,17] Earlier studies have shown decrements in serotonin level in the caudate nucleus. Very few, if any, drug trials have attempted to evaluate the effects of modulation of the serotoninergic system in AD. The effects of Alaproclate, a specific serotonin inhibitor, were recently evaluated in a double-blind study in AD. Overall, there was no significant effect on memory or cognition; however, 20% of the patients in the study did show positive improvement. Because of the lack of clinical trials exploring this system, we designed a double-blind, placebo-controlled, crossover trial to assess neuropsy-

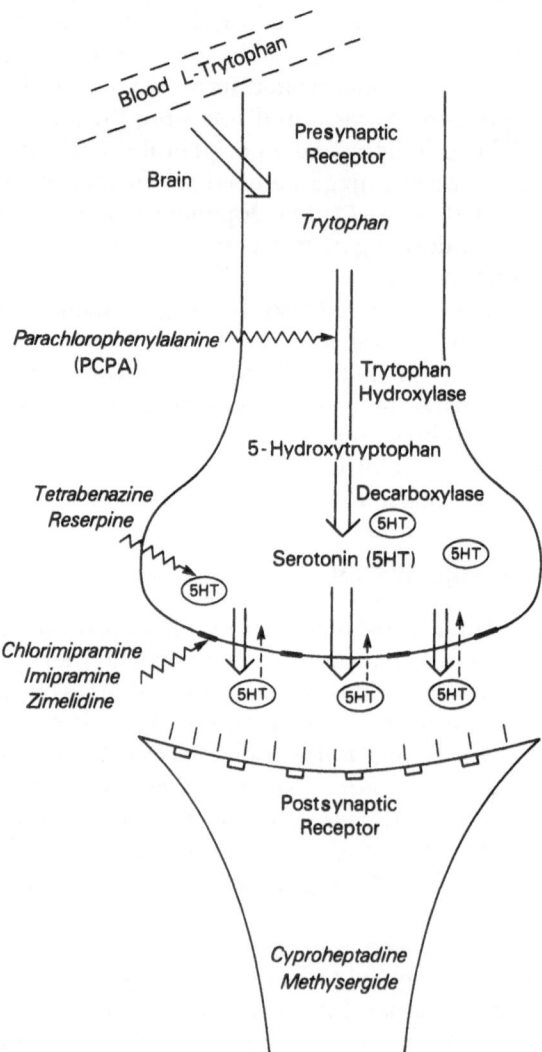

FIGURE 2   Neuronal biosynthesis of serotonin and drugs that modulate the serotoninergic system.

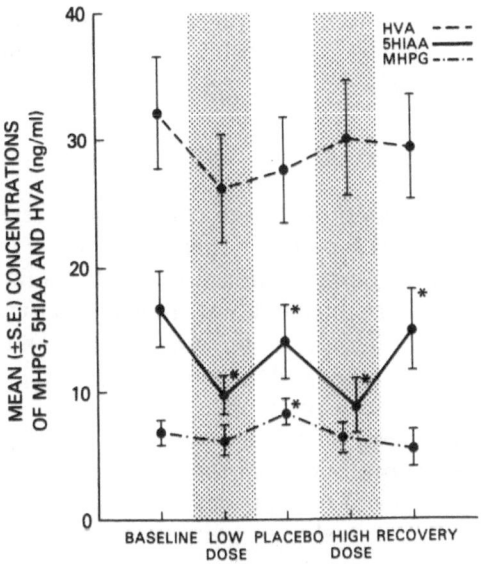

FIGURE 3 Effect of Zimelidine on neurotransmitter metabolites during the placebo-controlled trial.[30] ▨, Active drug; *, $p < 0.05$.

chological and neurochemical effects of Zimelidine* in four patients with clinically diagnosed AD. Individualized doses of Zimelidine, a relatively specific serotonin reuptake blocker, were administered to achieve a target plasma Zimelidine concentration of approximately 50 (low) or 100 (high) ng/ml. Overall, there was no significant effect of Zimelidine on memory or reaction time measures as compared to placebo ($p > 0.1$). The drug, however, significantly reduced (by up to 38%) 5-hydroxy-indolacetic acid concentration in the cerebrospinal fluid (Fig. 3) and almost abolished (90% reduction) platelet serotonin uptake ($p < 0.05$). Cerebrospinal fluid concentrations for both 3-methoxy-4-hydroxyphenylglycol, a major metabolite of norepinephrine, and homovanillic acid, the major metabolite of dopamine, were not altered. Our findings indicate that alterations in central and peripheral serotoninergic function by a specific serotonin reuptake blocker (Zimelidine) do not appear to be accompanied by measurable changes in memory and/or reaction time in AD patients.[30]

The serotonin neurotransmitter system still needs further exami-

* The drug was withdrawn from the world market in 1983 by the manufacturer.

nation because other serotoninergic agents may modulate the system differently than Zimelidine and possibly produce beneficial effects on memory function in AD.

3.5. NEUROPEPTIDERGIC SYSTEMS

Neuropeptides are another class of compounds presently being investigated clinically for their potential therapeutic value in treating AD. These neuropeptides are either synthesized in the brain or are transported there from the hypothalamus and are believed to function either as neurotransmitters or as their modulators. Neuropeptides related to vasopressin have shown effects of enhancing attention, memory, and cognitive deficits associated with aging.

To date, only one neuropeptide somatostatin, a peptide that inhibits the release of growth hormone from the pituitary gland, has been reported to be reduced in AD patients. Its levels have been only 30% of controls in all areas of cerebral cortex of AD patients.[31,32] Recently, the number of somatostatin receptors has also been found reduced in the frontal and temporal cortex of AD autopsied brain tissue.[33] In addition, corticotropin-releasing factor (CRF) hormone concentrations have been found to be reduced in CSF and postmortem brain tissue from AD patients. However, concentrations of six other neuropeptides, including vasoactive intestinal peptide, arginine vasopressin, cholecystokinin, thyrotropin-releasing hormone, luteinizing-releasing hormone, and substance P, have been reported to be unchanged in AD.[18] In addition, a recent report has suggested the possible coexistence of the cholinergic and somatostatinergic systems in individual neurons from cultured rat cortex.[34]

Clinical trials with vasopressin and analogs in AD patients have been based on the positive observations of memory improvement in animals and normal volunteers. A trial conducted by Ferris[35] in 20 mildly to moderately impaired AD patients, with lysine vasopressin, demonstrated a small but statistically significant ($p < 0.05$) improvement in certain measures of memory. However, some studies, which generally speaking employed more severely impaired patients, did not observe such improvements.

We have recently evaluated an analog of somatostatin (L 363,586) in a double-blind, placebo-controlled, crossover manner,[36] in an attempt to modulate the peptidergic and/or cholinergic neurotransmission in 10 mild AD patients with a mean age of 68 (range 58–78) years and of mild severity (Mean, Mini-Mental State Exam = 22, range 11–26), out of a possible 30^{37} (the greater the score, the less demented). Our results indicated no significant improvement on either serial or paired associate learning ($p > 0.1$) but showed a small, significant increase in arithmetic scores ($p < 0.05$). Although our early findings do not appear to suggest memory enhancement, we feel that because of extensive distribution of somatostatin (and, we assume, its analog), a higher dosage sufficient to achieve CNS and/or cerebrospinal fluid (CSF) levels above a certain threshold level may be necessary to modulate the system and therefore may require further exploration of its neuromodulatory properties in AD.[38] In addition, we examined CSF, MHPG, 5HIAA, HVA, and corticotropin levels, as response measures, between the placebo and L 363,586 conditions and found no differences ($p > 0.05$). Some investigators have postulated the involvement of the opioid system. Naloxone has been shown to improve memory in animals, and it has been postulated that it may benefit AD patients by blocking brain opioids.[39] Reisberg and others,[40] in an open study, demonstrated moderate improvement in AD patients following the administration of 1 mg of naloxone. This study was followed by a double-blind, placebo-controlled study that demonstrated a positive effect over placebo. Further open studies conducted at various centers around the country have failed to substantiate similar effect of naloxone.[41] Several investigations are presently underway at various centers to resolve this discrepancy.

3.6. GLYCOLYTIC ENZYMES

Brain metabolism and deficiencies in the glycolytic pathway enzymes and their role in AD are presently being explored extensively. One etiological hypothesis currently under examination is that an increased brain metabolic state may be able to prevent a hypoxic condition and thereby prevent decrements in metabolic-dependent neurotransmitter systems. Ergoloids have been tried but with minimal improvement, as these compounds primarily affect behavior and mood. High doses ranging from 6 to 12 mg/day and longer trials up to 1 year are currently under evaluation to test this hypothesis, and preliminary results appear to be encouraging.[8,9] Other agents, such as Piracetam, which enhance metabolic function and prevent cell hypoxia still need further exploration. Animal studies of Bartus et al.[22] have shown improvement of memory in monkeys following the administration of agents such as Piracetam, ergoloid mesylate, and vincamine. This information, gained from an intact, functioning, normal subhuman brain, needs to be validated in the degenerating AD human brain. Detailed information on some investigations using positron emission tomography (PET) in studying cerebral metabolism in aging and AD can be found in Chapter 8, "The Aging Brain."

4. Brain Transplantation

Other directions are also being explored. The work of Bjorklund and colleagues,[42] who successfully transplanted embryo septal grafts into denervated hippocampal formation in rats, attests to these approaches. Activity of the graft transplant was monitored by computerized tomography activity. At the end of 6 months, overall ACh synthesis was found to be restored to normal levels in these animals. By this method "physiological" functioning following complete denervation appears to have been maintained. Thus, transplantation may provide yet another avenue to explore and correct for the deficits associated with AD.

5. Methodological Concerns

Methodologically, a number of issues must be addressed so that the outcome response from clinical trials may be appropriately and correctly determined. The variables that need to be considered are description of subjects, the role of the pilot study, the appropriate duration of a clinical trial, and end-organ measures of response.

5.1. Subject Description

When clinical trials in AD patients are designed, a number of variables must be described and considered in detail. These descriptive variables include the age and the sex of the subjects. The chronological age, age of onset of the disease process, and its severity or the degree are important for subject description and matching. For discussion of problems associated with the description of the variable age, see Chapter 2, "Age: A Complex Variable." The distinction between an early- and a late-onset AD patient is extremely important in classification of the patient and the disease process. It has been demonstrated that AD patients with the early and late onset of disease differ both neurochemically and neuropathologically.[43] The early-onset AD patient's brain has been characterized as having a more severe and malignant course with more changes both neurochemically and neuropathologically as compared to the late-onset AD brain. In addition, patients with early onset of AD show a clinical course of short duration which deteriorates rapidly. Late-onset AD patients appear to be much closer clinically, neurochemically, and neuropathologically to normal aging individuals. Because of these known differences it is important to characterize these patients separately into their respective groups, particularly when designing clinical trials evaluating new drugs or treatment approaches so as to be able to make comparisons among several clinical trials.

Sex differences have not been related to the AD process; however, it is important to classify patients by sex in order to minimize sex-related variability associated with the end-organ measures.

Another reason for considering appropriate classification by age, disease, and sex of these subjects could be to assist in defining healthy control groups. The examination of drug response in healthy age-matched controls is important to assess in order to rule out any age-related alterations in brain function as compared to a patient with primary AD.

5.2. DESCRIPTION OF SEVERITY

Another classification necessary for the description of AD patients is the degree of severity of the disease. This is usually determined by several severity scales which are commonly administered, including the Blessed Memory Information and Concentration Test[44] (a total score of 37; the higher the score, the less demented); the Mini-Mental State Exam[34] (a total score of 30; the higher the score, the less demented), and the Mattis dementia scale score[45] (a total score of 144; the higher the score, the less demented). These scales are used to determine the degree of dementia for an individual patient. Classification addressing this issue is important in conducting clinical trials with AD or other dementia patients. It seems inappropriate to assume that all AD patients with various degrees of dementia progression, although age and sex-matched, respond similarly.

5.3. SCREENING

All subjects should be carefully screened in order to exclude any other medical, psychiatric, or neurological disorders besides their primary AD. The careful assessment of subjects assists in reducing the variance and establishing a homogeneous subject population in order to assess drug effects.

5.4. Duration of a Therapeutic Trial

The duration of trial is another important variable to consider. How long must a trial be conducted before one can appropriately distinguish a response to a drug? A study by Thal *et al.*[46] that evaluated the response of AD patients to coadministration of lecithin and physostigmine found that some patients showed improved memory after 2–3 days,[46] whereas another investigator who examined the same agents over a 6-week trial in AD patients did not find a comparably favorable response.[47] Although the doses of drugs administered were different in the two studies, it is evident that comparison between acute and chronic trial design may reveal potentially different clinical responses. The duration of a trial must be considered in any determination of the effectiveness of types of these agents. The administration of these agents for periods of time up to several weeks to months may be more appropriate to determine the therapeutic efficacy of agents used in AD dementia and related disorders.

5.5. Role of Pilot Studies

A small pilot study, with a small number of subjects, may be wise prior to embarking on a long, chronic, double-blind trial in order to determine dose–response relationships and/or potential side effects of new therapeutic agents. In addition, the pilot study may also provide some information on the response time for the new cognitive enhancers. The dose and route of administration may also play a role in determining the response to a drug, especially if metabolic products possess activity. One must consider both acute and chronic side effects and the route of administration in order to appropriately assess the therapeutic potential of an agent. The choice of an agent that must be intravenously administered chronically in order to elicit an effect is impractical particularly in this special population, e.g., AD. Administering these agents orally is fine, but being reassured of the patient's compliance is another major stumbling block for chronic dosing trials.[30]

5.6. Neuropsychological Measures

Neuropsychological measures to determine cognitive changes as response measures are also extremely important. Memory function needs to be adequately measured by simple tests for immediate, delayed, verbal, and visual memory function. These subtypes of memory may include either various forms of standardized memory tests such as the Wechsler

Memory Scale[48] or Benton Visual Retention Test.[49] Other measures of cortical function such as calculations, visual spatial constructive ability, and frontal lobe function[50] should also be measured in addition to memory function. A drug may elicit one type of cognitive response and not another. It is important to measure as many types of cortical-mediated functions as possible, in order not to overlook a potential drug effect.

5.7. END-ORGAN MEASURES

In addition to cognitive function measures, other CNS end-organ measures should also be assessed. This can help investigators in elucidating or postulating a potential underlying mechanism of action of the agent. Such measures could include the assessment of CSF neurotransmitters and their metabolites, e.g., norepinephrine and its metabolite, serotonin and its metabolite, or dopamine and its metabolite. These CSF measures are particularly useful when conducting clinical trials with cognitive enhancers as most of them alter or modulate the neurotransmitter systems mentioned earlier.

Measurements of these substances may provide clues regarding the alterations within various neurotransmitter systems. We have previously attempted such end-organ measures in two drug studies with AD patients discussed earlier in this chapter (see Sections 3.4 and 3.5). Other end-organ measures could also include PET scanning with ^{18}FDG (see Chapter 8, "The Aging Brain") in order to examine brain metabolic changes following administration of these agents.

6. CONCLUSION

A number of neurochemical and neuropathological changes occur in the brains of patients with AD. Neurofibrillary tangles occur in the cerebral cortex and hippocampus, as well as loss of large cortical neurons in midfrontal and temporal regions. Neurochemical changes include deficits in brain cholinergic, noradrenergic, dopaminergic, GABAergic, serotoninergic, and neuropeptide neurotransmitter systems. With a better understanding of the altered underlying biochemistry and formulation of rational hypotheses based on neurochemical findings, researchers are exploring newer pharmacological strategies for treating AD. Although trials with the cholinergic agents, to date, have revealed minimal success and other drugs affecting other neurotransmitter systems have not been extensively explored, much has been learned. Avenues exploring the neuropeptidergic systems and metabolic enhancers require further study. Brain transplantation of embryonic hippocampal neurons into adult brain has been shown to compensate for degenerated brain and improve cholinergic function. It is clear that the research efforts of the past decade have led to a better understanding of the bio- and neurochemistry, which can now allow clinicians and researchers alike to attack the problem of AD with greater force by designing appropriate clinical trials.

In order to examine drug therapeutics in AD or dementia patients the following must be considered: (1) careful selection and description of the subjects, excluding other types of medical, neurological, or psychiatric disorders; (2) careful grading of severity of disease; (3) appropriate duration of the trials that assess immediate versus long-term effects; (4) appropriate neuropsychometric testing; and finally, (5) a sound understanding of AD and aging processes.

REFERENCES

1. Alzheimer A: Uber eine eigenartige Erkrankung der Hirnrinde. *Allg Z Psychiatrie.* 1907; 64:1460–1480.
2. Constantinidis J: Is Alzheimer's disease a major form of senile dementia? Clinical, anatomical, and genetic data, in Katzman R, Terry RD, Bick KL (eds): *Alzheimer's Disease: Senile Dementia and Related Disorders.* New York, Raven Press, 1978, vol 7, pp 15–25.
3. Roth M: Diagnosis of senile and related forms of dementia, in Katzman R, Terry RD, Bick KL (eds): *Alzheimer's Disease: Senile Dementia and Related Disorders.* New York, Raven Press, 1978, vol 7, pp 71–85.
4. Petrie WM and Ban TA: Drugs in geropsychiatry. *Psychopharmacol Bull* 1978; 14:7–19.
5. Salzman C: Stimulants in the elderly, in Raskin A, Robinson DS, Levine J (eds): *Age and the Pharmacology of Psychoactive Drugs.* New York, Elsevier North Holland Inc, 1981, pp 171–180.
6. Reding MJ, Young R, DiPonte P: Amitriptyline in Alzheimer's disease. *Neurology* 1983; 33:522–523.
7. Cutler NR, Narang PK: Implications of dosing tricyclic antidepressants and benzodiazepines in geriatrics. *Psychiatr Clin North Am* 1984; 7:1–17.
8. Hollister LE, Yesavage J: Ergoloid mesylates for senile dementias: Unanswered questions. *Ann Intern Med* 1984; 100:894–898.
9. Van Loveren-Huyben CMS, Engelaan HFW, Hermans MBM, Van der Bom JA, Leering C, Munnichs JMA: Double-blind clinical and psychologic study of ergoloid mesylates (Hydergine) in subjects with senile mental deterioration. *J Am Geriatr Soc* 1984; 32:584–588.
10. Epstein LJ: Anxiolytics, antidepressants, and neuroleptic patients, in Lipton MA, DiMascio A, Killam KF (eds): *Psychopharmacology: A Generation of Progress.* New York, Raven Press, 1978, pp 1517–1523.
11. Coyle JT, Price DL, DeLong MR: Alzheimer's disease: A disorder of cortical cholinergic innervation. *Science* 1983; 219:1184–1190.

12. Terry RD, Peck A, Deteresa R, Schechter R, Horoupian DS: Some morphometric aspects of the brain in senile dementia of the Alzheimer type. *Ann Neurol* 1981; 10:184–192.

13. Terry RD, Davies P: Some morphologic and biochemical aspects of Alzheimer's disease, in Samuel D, Algeri S, Gershon S, Grimm VE, Toffano G (eds): *Aging of the Brain*. New York, Raven Press, 1983; vol 22, pp 47–59.

14. Carlsson A, Adolfsson R, Aquilonius SM, et al: Biogenic amines in human brain in normal aging, senile dementia, and chronic alcoholism, in Goldstein M, Calne DB, Lieberman A, Thorner MO (eds): *Ergot Compounds and Brain Function: Neuroendocrine and Neuropsychiatric Aspects*. New York, Raven Press, 1980, pp 295–314.

15. Tomlinson BE, Blessed G, Roth M: Observations on the brains of demented old people. *J Neurol Sci* 1970; 11:205–242.

16. Yamamoto T, Hirano A: Nucleus raphe dorsalis in Alzhemer's disease: Neurofibrillary tangles and loss of large cortical neurons. *Ann Neurol* 1985; 17:573–77.

17. Gottfries OG, Adolfsson R, Aquilonius SM: Biochemical changes in dementia disorders of Alzheimer's type (AD/SDAT). *Neurobiol Aging* 1983; 4:261–271.

18. Rossor MN: Neurotransmitters in CNS disease: Dementia. *Lancet* 1982; 2:200–204.

19. Iwangoff P, Armbruster R, Enz A, Meier-Ruge W, Sandoz P: Glycolytic enzymes from human autoptic brain cortex: Normally aged and demented cases, in Roberts PJ (ed): *Biochemistry of Dementia*. London and New York, John Wiley & Sons, 1980, pp 258–262.

20. Iwangoff P, Reichlmeier K, Enz A, Meier-Ruge W: Neurochemical findings in physiological aging of the brain. *Interdiscipl Topics Gerontol* 1979; 15:13–33.

21. Perry EK, Perry RH: The cholinergic system in Alzheimer's disease, in Roberts PJ (ed): *Biochemistry of Dementia*. New York, John Wiley & Sons, 1980, pp 135–183.

22. Bartus RT, Dean RL, Beer B, Lippa AS: The cholinergic hypothesis of geriatric memory dysfunction. *Science* 1982; 217:408–417.

23. Brinkman SD, Gershon S: Measurement of cholinergic drug effects on memory in Alzheimer's disease. *Neurobiol Aging* 1983; 4:261–271.

24. Christie JE, Shering A, Ferguson J, Glen AIM: Physostigmine and arecholine: effects of intravenous infusion in Alzheimer presenile dementia. *Br J Psychiatry* 1981; 138:46–50.

25. Wettstein A, Spiegel R, Koppel-Hefti A: Therapeutic trial with the muscarinic agonist RS-86 in patients with senile dementia of Alzheimer type. Presented at the Collegium Internationale Neuro-Psychopharmacologicum Congress, Florence, Italy, June 18–23, 1983, pp 983.

26. Bruno G, Mohr E, Gillespie MM, Fedio P, Chase TN: RS-86 therapy of Alzheimer's disease. *Neurology* 1985; 35(suppl 1): 265.

27. Zornetzer SF: Neurotransmitter modulation and memory: A new neuropharmacological phrenology? in Lipton MA, DiMascio A, Killam KF (eds): *Psychopharmacology: A Generation of Progress*. New York, Raven Press, 1978, pp 637–649.

28. Jellinger K, Flament H, Riederer P, Schmid H, Ambrogi L: Levodopa in the treatment of (pre) senile dementia. *Mech Aging Dev* 1980; 14:253–264.

29. Ferris SH, Reisberg B, Friedman E: Combination choline-piracetam treatment of senile dementia. *Psychopharmacol Bull* 1982; 18:94.

30. Cutler NR, Haxby J, Kay AD, et al: Evaluation of zimelidine in alzheimer's disease: cognitive and biochemical measures. *Arch Neurol* 1985; 42:744–748.

31. Davies P, Katzman R, Terry RD: Reduced somatostatin-like immunoreactivity in cerebral cortex from cases of Alzheimer's disease and Alzheimer senile dementia. *Nature* 1980; 288:279–280.

32. Rossor MN, Emson PC, Mountjoy CQ, Roth M, Iversen LL: Reduced amounts of immunoreactive somatostatin in the temporal cortex in senile dementia of Alzheimer's type. *Neurosci Lett* 1980; 20:373–377.

33. Beal MF, Mazurek MF, Tran VT, Chattha G, Bird ED, Martia JB: Reduced numbers of somatostatin receptors in cerebral cortex in Alzheimer's disease. *Science* 1985; 229:289–291.
34. Delfs JR, Zhy CH, Dichter MA: Co-existence of acetylcholinesterase and somatostatin-immunoreactivity in neurons cultured from rat cerebrum. *Science* 1984; 223:61–63.
35. Ferris SH: Neuropeptides in the treatment of Alzheimer's disease, in Reisberg B (ed): *Alzheimer's Disease.* New York, The Free Press, 1983, pp 369–373.
36. Veber DF, Saperstein R, Nutt RF, et al: A super active cyclic hexapeptide analog of somatostatin. *Life Sci* 1984; 314:1371–1378.
37. Folstein MF, Folstein SE, McHugh PR: Mini-Mental State, a practical method for grading the cognitive state of patients for the clinical. *J Psychiatr Res* 1975; 12:189–198.
38. Cutler NR, Haxby JV, Narang PK, May C, Burg C, Reines SA: Evaluation of an analogue of somatostatin (L-363,586) in Alzheimer's disease. *N Engl J Med* 1985; 312:725.
39. Kastin AJ, Olson GA, Sandman CA: Possible role of peptides in senile dementia, in: Crook T, Gerson S (eds): *Strategies for the Development of an Effective Treatment for Senile Dementia.* New Canaan, CT, Mark Powley Associates, 1981, pp 139–152.
40. Reisberg B, Ferris SH, Anand R: Effects of naloxone in senile dementia: a double blind trial. *N Engl J Med* 1983; 308:721–722.
41. Blass JP, Reding MJ, Drachman E: Cholinesterase inhibitors and opiate antagonists in patients with Alzheimer's disease. *N Engl J Med* 1983; 309:556.
42. Bjorklund A, Gage FH, Schmidt RH, Stenevi U, Dunnet SB: Intracerebral grafting of neuronal cell suspensions VII. Recovery of choline acetyltransferase activity and acetylcholine synthesis in the denervated hippocampus reinnervated by septal suspension implants. *Acta Physiol Scand* 1983; (suppl 522):59–66.
43. Rossor MN, Iverson LL, Reynolds GP, Mountjoy CQ, Roth M: Neurochemical characteristics of early and late onset types of Alzheimer's disease. *Br Med J* 1984; 288:961–964.
44. Blessed G, Tomlinson BE, Roth M: The association between quantitative measures of dementia and of senile change in the cerebral grey matter of elderly subjects. *Br J Psychiatry* 1968; 114:797–811.
45. Mattis S: Mental status examination for organic mental syndrome in the elderly patient, in Bellak L, Katasu T (eds): *Geriatric Psychiatry: A Handbook for Psychiatrists and Primary Care Physicians.* New York, Grune & Stratton, 1976, vol 7, pp 77–121.
46. Thal LJ, Fuld PA, Masur DM, Sharpless NS: Oral physostigmine and lecithin improve memory in Alzheimer's disease. *Ann Neurol* 1983; 13:491–495.
47. Wettstein A: No effect from double-blind trial of physostigmine and lecithin in Alzheimer's disease. *Ann Neurol* 1983; 13:210–212.
48. Wechsler D: A standardized memory scale for clinical use. *J Psychol* 1945; 19:87–95.
49. Benton AC: *The Revised Visual Retention Test,* New York, Psychological Corporation, 1974.
50. Benson DF: Aphasia, in Heilman EM, Valenstein E (eds): *Clinical Neuropsychology.* New York, Oxford University Press, 1979, pp 22–58.

PART III

General Perspectives

CHAPTER 17

DOSAGE FORM CONSIDERATIONS IN CLINICAL TRIALS INVOLVING ELDERLY PATIENTS

R. GARY HOLLENBECK AND PETER P. LAMY

1. INTRODUCTION

We are in the era of drug delivery systems. Indeed, the potential impact—favorable or unfavorable—of the dosage form on drug delivery is beginning to be appreciated by most members of the medical community. Even the lay public has been educated in the rudiments of pharmaceutics. Among other things, they are asked to choose an antacid based on its acid-neutralizing capacity, to comprehend the benefits of enteric-coated aspirin tablets, and to realize that they can depend on a patch placed behind the ear or on the chest to provide their medicine. Major pharmaceutical companies have recently formed novel drug delivery research groups charged with the task of creating unique methods of drug administration for new and old drugs, ostensibly to achieve more effective therapy with fewer and less severe side effects, but not inconsequently to provide for patent extension or a competitive marketing advantage.

R. GARY HOLLENBECK AND PETER P. LAMY • School of Pharmacy, University of Maryland, The Center for the Study of Pharmacy and Therapeutics for the Elderly, Baltimore, Maryland 21201.

Specific attention to drug delivery in elderly patients is imperative, considering that 31% of all prescription drugs now go to elderly patients, and that it is estimated that within 15 years half of all prescription drugs will be for the elderly.[1] This need is not going to be satisfied by simply testing new and existing products in older patients, although this is an essential beginning. Product development pharmacists must begin to recognize those attributes of elderly patients which differentiate them from the so-called normal group and must design dosage forms with these considerations in mind.

The extent to which a contemporary appreciation for pharmaceutics has influenced the early stages of drug testing, in both animals and in humans, is unclear. Certainly in the industry there is early input from a research or product development pharmacist on a team responsible for a new chemical entity, and every effort is made to go to clinical trials with the final dosage form. This degree of collaboration, and specifically pharmaceutics input, may not be as evident in studies at research institutions conducted to evaluate the therapeutic moiety itself.

The development of small and simple-to-use, rate-controlled drug delivery devices has resulted not only in a whole new approach to the study of biologically active agents in animals, but also the recognition and validation of regimen-dependent actions. It is increasingly common to find that early pharmaceutical research on a drug includes tests to actually determine whether a constant rate delivery and achievement of a steady-state plasma level is optimal. For some drugs, intermittent administration with consequent surges in the drug level is more efficacious than a constant plasma concentration.[2]

If there is a general philosophy for the design of dosage forms for clinical studies and subsequent use, it is: Don't let the dosage form compromise the efficacy of the drug. This philosophy is analogous to type I statistical error and reflects the overriding concern that a potentially useful therapeutic agent might be incorrectly deemed ineffective because of poor bioavailability. Administration of a drug in a solution in which it is unstable, in a tablet that does not disintegrate, as an inhalation with particles too large to penetrate the lung, or in any other dosage form that fails to present the drug for absorption is just as devastating to the integrity of a study as any unacceptable aspect of absorption, distribution, metabolism, or excretion. Before consideration can be given to obtaining the benefits of a dosage form, such as controlled release or even targeted drug delivery, therapeutic value and baseline pharmacodynamic and pharmacokinetic data must be documented.

The examples of pharmaceutical deficiency just cited are obvious, but equally important considerations may be more subtle. Apparent variability of response to a drug may actually be patient-related variability in release of drug from a dosage form due to anything from circadian rhythm to altered physiology due to disease or aging. Table I contains some examples of age-related physical, physiological, pathophysiological, and behavioral factors that have the potential to affect dosage form performance and bioavailability in elderly patients.

The concept of designing a dosage form for a particular population depends on identification of a common denominator in that population. At this time, the design of dosage forms for the elderly resolves itself into an exercise in avoiding problems apparent in a large number of patients, rather than a positive proposition based on a physiological or disease-related attribute present in all elderly patients. Perhaps it would be appropriate to term this exercise "no-fault formulation." The elderly are a group whose only common denominator is advanced age. It is hard to find a basis for classification that results in a more diversified group. Considering the unique combination of disease state and specifically

TABLE I CONSIDERATIONS FOR THE DESIGN OF DOSAGE FORMS
FOR ELDERLY PATIENTS

Area	Age-related factor	Potential patient impact
Physical	Reduction in sensorium: visual and hearing decrements	Cannot read and/or understand instructions
	Tremor	Cannot measure liquids
	Weakness (muscular)	Cannot open container
Physiological	Reduction in renal drug clearance	
	Increase in body fat/ decrease in lean body mass	Abnormal drug distribution leading to inordinately high or low steady-state plasma levels
	Decrease in plasma albumin concentration	
	Variable and unpredictable gastric pH and emptying time	Possible variability in rate and extent of drug release from the dosage form
Pathophysiological	Multiple diseases	Altered drug disposition Increased incidence of
	Multiple drugs	adverse drug reactions Gastrointestinal erosion
Behavioral		Unpredictable compliance Intentional noncompliance

compromised physiology that each patient presents, there is actually no group for which individualized dosing and dosage forms are more appropriate.

In essence, formulators are left with three primary focuses that can serve as a basis for designing dosage forms for the elderly: (1) dosage flexibility, (2) patient compliance, and (3) age-, disease-, and therapy-related factors that can adversely affect performance of a drug delivery system. These are not necessarily presented in order of importance, but rather in a sequence that permits logical determination of the best dosage form for a given situation.

In proceeding, we make the least limiting assumption that there is little difference in designing dosage forms for clinical trials and for the general marketplace. Although the clinical scientist involved in early studies may not be directly concerned with issues of patient compliance, for instance, this is a consideration that must be addressed at some point for any drug that is to be self-administered. For those clinicians who have little understanding or appreciation of the significance of an appropriate dosage form, we make the case for recognition of this issue; for those actually involved in testing the attributes of one dosage form relative to another, we recognize that much of the following will seem very basic.

2. DOSAGE CONTROL AND FLEXIBILITY

In cases where the drug has a narrow therapeutic index or where there is considerable patient-to-patient dose–response variability, a dosage form that permits administration in a continuous dosage range, rather than in discrete quantities, is necessary. Dosing based on blood levels and applied pharmacokinetics requires that the drug can be administered in the frequency and quantity deemed necessary. Several therapeutic agents require control and dosage flexibility to the extent that an infusion pump is a requisite for safe administration. Considering the potency of many new drugs and the increased incidence of adverse drug reactions in the elderly, dosage control and flexibility are highly desirable dosage form attributes.

The most flexibility and greatest control of drug administration are associated with a solution dosage form given intravenously as a bolus injection or infusion. A solution is homogeneous, assuring content uniformity, dosage can be varied continuously, and is limited only by solubility of the drug and the accuracy of volumetric measurement, and the i.v. route ensures immediate action and complete bioavailability.

Indeed, the modern i.v. infusion pump represents the ultimate in control of drug administration in the clinic. Reasonably constant blood levels of a drug may be achieved because of the constant rate administration, avoiding peaks and troughs and associated super- and subtherapeutic levels. Variation of the flow rate and drug concentration of the infusion permit titration of the patient by monitoring therapeutic response or taking blood levels. Dobutamine, dopamine, and nitroprusside are examples of drugs given parenterally which require use of an infusion pump; heparin, theophylline, and i.v. nitroglycerin are also routinely administered in this fashion. Except for drugs that are more effective through periodic administration, the attributes and performance of the i.v. infusion pump represent the standard that all other systems attempt to emulate.

Historically, the need for administration by a health professional and the absolute requirement for sterility have been cited as severe constraints on this route, in general limiting it to the hospitalized patient. However, there are many epidemiological and technological changes that require a reassessment of this limitation. The increasing number of patients, age, and severity of illness encountered in nursing home residents is forcing a major change: use of i.v.'s, chemotherapy, and total parenteral nutrition (TPN) in nursing homes. Home care patients are now being taught to self-administer i.v.'s, and new dosage forms allow admixture in the home. More dramatically, the introduction of externally portable infusion pumps and implantable drug delivery systems portends an era of parenteral drug use in ambulatory patients.

The implantable vascular access system (e.g., Vascular-Access Port™ Port-A-Cath™) represents the simplest parenteral drug delivery approach available for repetitive i.v. administration or blood sampling in outpatients. These devices, which consist of a port, reservoir, and catheter, make injections relatively easy and practically noninvasive. However, many elderly patients may not be able to use such a device, and compliance still remains a potentially serious problem.

Approximately 30 different portable external pumps are available for insulin delivery alone, and many have been or are currently being tested for indications other than diabetes.[3] The advantages of these devices in the clinical testing of drugs for determination of pharmacodynamic parameters is obvious: all aspects of drug delivery are controlled, and compliance is relegated to a nonissue. The willingness of a patient to tolerate a portable pump as the ultimate delivery system for a drug depends primarily on the nature and term of its indication. Short-term applications (days to months) are reasonably well tolerated, but the spector of long-term dependence on the device diminishes its acceptability, even when there is obvious therapeutic benefit. Use of the device is usually gladly accepted in lieu of extended hospitalization.

Despite the obvious conceptual appeal and real advantages of portable infusion pumps, they will never be a delivery system of first choice. Generally the cost/benefit ratio is high, and drugs that must be given parenterally to be effective or that require extremely close control over administration are the only likely candidates. Indeed, use of these pumps will undoubtedly diminish as progress is made in developing small, re-fillable, implantable devices to accomplish the same purpose.

Implantable pumps are currently available. For example, a fixed-rate pump (The Infusaid Company) is on the market. It depends on pressure exerted by a low-boiling-point liquid/vapor system to collapse the bellows and expel liquid at a rate controlled by a flow restrictor. The device can be refilled through a septum with a needle; expansion of the bellows to accommodate the drug solution causes condensation of the propellant. Reports in the literature indicate that considerable variation in flow rate may occur as a consequence of environmental temperature or pressure changes, yet even a temporary 50% increase or decrease in flow rate may be acceptable in comparison to b.i.d., t.i.d., or q.i.d. dosing with a discrete oral dosage form. Implantable pumps with externally programmable rate control have been investigated in humans for insulin therapy and for morphine and cytostatic agents.[3] Though these pumps will undoubtedly improve, their use now must be viewed as experimental and not without problems.

The value of the infusion devices in the context of this chapter lies in their ability to provide a desired steady-state plasma level without concern about patient compliance. The collection of pharmacodynamic data, particularly in the less compliant elderly patient, would benefit from this method of drug delivery. The feasibility of developing conventional dosage forms would logically follow. As discussed later, it is likely that performance of a conventional dosage form is more variable in elderly patients as a group. When a conventional dosage form is used in a clinical study, clinicians may erroneously ascribe response variability to the drug rather than the dosage form.

Although apparently simple, the formulation of a solution dosage form, particularly an injectable product, is often a real challenge. Achieving acceptable solubility and stability may require solubilizing agents, cosolvents, buffers, and chemical and biological preservatives. The more complicated the system, the less comfortable one feels about injecting it directly into the systemic circulation; the agents necessary to achieve acceptable solubility and stability are not always pharmacologically inert.

Formulation of solutions for administration by one of the aforementioned devices presents new challenges to the pharmaceutical scientist. Since the drug is usually for long-term administration, the solution must necessarily be concentrated, and the solution is exposed to higher temperature (37°C) for the entire time period. Both these factors tend to accelerate drug decomposition and place considerable demand on the design of a stable system. The solutions must also be compatible with the materials of construction of the device itself.

A satisfactory approach for drugs that must be injected and are unstable in solution may be preparation of a product for reconstitution through lyophilization, yet this is an expensive process on a large scale.

Oral liquids—solutions, suspensions, and emulsions—all offer dosage flexibility, but in addition to the concerns mentioned previously for solutions, care must be taken to ensure that the dispersions are physically stable so that a consistent concentration of drug is contained in each volume administered, throughout the lifetime of the product. Formulation of a stable suspension or emulsion may be a greater challenge than formulation of a solution, and these dosage forms should be considered as a last resort, when administration of a large amount of solid or oily drug is necessary.

Complete assurance of accurate oral dosing is achieved only when the product is measured by a member of the clinical research team, and when the product is administered through a nasogastric tube. Even then, care must be taken to ensure that drug does not stick to the tube. When the patient is responsible for self-administration, an oral syringe, or at least a reasonably accurate volumetric container, should always be used for administration of oral liquids. The use of a teaspoon is unacceptable; it is foolish to enter into a carefully controlled study with a procedure for drug administration that has a possible error of nearly 100%.[4] In addition, elderly patients often have tremor and extreme difficulty in accurately measuring liquid products.

Because many elderly patients do not tolerate well a large total fluid intake, some companies have tried to introduce concentrated liquid dosage forms. It should be emphasized that this is not a wise approach for drug products targeted to the elderly; volumetric measurement errors of a concentrated liquid represent large mistakes in dosing.

Although seldom employed, the advantages of a solid-dosage form combined with dosage flexibility can be achieved by volumetric measurement of a divided solid. Admittedly, powder papers are anachronistic, and even hand-packed, hard gelatin capsules are only feasible when a limited number of patients are involved. However, modern technology now permits the reproducible manufacture of granular solids and coated beads, with relative ease. A patient capable of measuring a liquid is certainly capable of measuring a quantity of granular solid. This solid may be dispersed in a glass of water or juice or in a foodstuff such as applesauce, offering a distinct advantage for patients with difficulty swallowing integral dosage forms.

Elderly patients are certainly familiar with drug products of this form. Natural laxatives such as Metamucil® have always been administered in this fashion; Perdiem® is a recently developed granular form of psyllium that is an excellent example. Granular forms of drug-containing products such as Questran® (cholestyramine), Mandelamine Granules® (methenamine mandelate), and TheoDur Sprinkle® (sustained-release anhydrous theophylline) exist, but the attribute of dosage flexibility is not promoted or advocated. In the case of TheoDur Sprinkle®, quite the opposite is true. Several strengths (different-sized capsules containing the product) are available, and subdividing the contents of a capsule is not recommended.

Nevertheless, with a series of "scoops" of different volume available, a bulk granular solid can represent an elegant dosing system: the composition and availability of the bulk product is consistent for all patients, dosage flexibility is maintained, the chemical stability of the drug is generally better than in a liquid system, and problems in swallowing tablets or capsules are avoided. This approach represents one of the few feasible methods for dosage flexibility in sustained-release or enteric-coated products. Once again, absolute control of the dose requires administration by supervisory personnel involved in the study; however, doses can be prepared ahead of time by technical personnel by filling an appropriate amount of the solid into a suitable unit dose package.

Individualized dosing may also be conveniently accomplished by using liquid-filled hard gelatin capsules. Liquid-filled soft gelatin capsules are familiar to most consumers, but specialized manufacturing equipment not generally available to the clinical scientist or small contractor is required for their preparation. On the other hand, no specialized equipment other than a variable volume pipette is required to fill hard gelatin capsules. The historical problems of leakage of the non-aqueous liquid fill and separation of the capsule body have been eliminated through use of thixotropic or thermal-setting excipients and self-locking capsules. Thus, the product is a liquid during the filling operation that eventually sets into a viscous semisolid, providing a physically stable dosage form.

Relatively new excipient systems are available for use as vehicles for preparation of liquid-filled hard gelatin capsules. Gelucires®, for example, represent a family of excipients derived from natural hydrogenated fats and oils with specific melting temperatures and hydrophilic lipophilic balance (HLB) values. Selection of the amphiphilic character of the Gelucire® permits solubilization or emulsification (water in oil) of the drug or drug solution, thereby accommodating therapeutic agents with any degree of polarity. Apparently, through a rational combination of melting temperature and HLB value, release patterns ranging from rapid to sustained may be achieved. Some companies even provide technical assistance in the development of a suitable formulation. The content uniformity of the liquid, the accurate fill and dose flexibility that can be achieved through liquid volumetric measurement, and the ability to easily modulate the release profile make this dosage form worthy of consideration.

An interesting approach to solving the problem of dosage flexibility in the tablet dosage form is represented by the Dividose® delivery system, developed by Mead-Johnson and utilized for anhydrous theophylline. The "system" is a rectangular tablet scored on one side so that it may be broken in half, and on the other side so that it may be broken in thirds. The result is that the 300-mg tablet may be used as such, or divided into 100-, 150-, or 200-mg doses. Although this is an innovative and inexpensive option, an elderly patient who has trouble simply complying with a regimen may have trouble remembering how to use the system or may actually have trouble breaking the tablets and handling the pieces.

Although it is not possible to discuss all possible routes of administration here, it is appropriate to address one more: transdermal drug delivery. This route is currently enjoying a period of renewed interest. An ointment, cream, or gel offers inherent dosage flexibility since the quantity applied may easily be varied. However, the amount applied may bear little relationship to the actual dose of drug received. The area over which the substance is applied is the single most important factor in determining actual dose. A requisite for clinical studies involving topically applied systems intended to deliver a drug to the systemic circulation is control over the amount applied and the area over which it is spread. The rate of drug release from these semisolid systems is a complicated function of the physical chemistry of the drug and the dispersion medium (ointment base), and the nature of the skin over which it is applied.

Patches, such as the systems used for nitroglycerin, scopolamine, clonidine, and estradiol, certainly facilitate control of dose and drug release. There is, at least initially, a direct correlation between dose and area of the patch. Systems that contain a reservoir of drug and depend on a permeable membrane for control of the rate of drug release must be regarded as discrete dosage systems. Systems such as Nitrodur® and

Nitrodisk℠ depend on diffusion from a uniform polymer matrix, and these could ostensibly be cut to any size to vary dosage. Design of a polymer matrix system represents a method seemingly ideal for clinical studies of topical drug delivery by presenting the ability to individualize dosing while retaining control over the rate of drug release.

Although the proposition of dosage flexibility is appealing, a valid argument can be made that there are only a few drugs where it is a necessity. There are not that many drugs with a narrow therapeutic index, and within certain limits, steady-state plasma levels may be adjusted by varying the frequency of administration. Although there may not be many drugs with narrow therapeutic indices, the elderly use a large proportion of those that exist: theophylline, phenytoin, lithium, anticoagulants, cardiovascular agents, and antidepressants, for example. In addition, dosage adjustment is not infrequent, even in chronic care, owing to ineffectiveness, interactions, or adverse effects.

Perhaps the first decision to be made in choosing a dosage form for administration to elderly patients should be based on the need for dosage flexibility. Although the economic advantages of fixed-strength products (tablets, capsules, suppositories) are substantial, it is essential that this decision be made on the basis of the patient's ability to comply with the prospective regimen.

3. COMPLIANCE

Within the context of this discussion, the issue of compliance may seem inappropriate. After all, a requisite for a well-designed study is complete assurance that the drug is administered according to protocol. Absolute compliance can only be achieved when the drug is administered with an implantable device or by a member of the clinical research team; in lieu of this intervention, every step possible must be taken to make sure that the dosage regimen within the study has been followed and that the proposed dosage regimen is capable of being adhered to by the unsupervised elderly patient.

The obviousness and sustained-release aspects of the transdermal patches certainly lead to improved, if not absolute, compliance. Conceptually, an implantable device capable of providing reliable long-term therapy does resolve the issue of compliance, but until variable rate control has been perfected, appropriate applications are few and the risks of this approach for therapy outweigh the benefits. An invariant rate of drug delivery, not able to be easily increased to respond to a temporary exacerbation or decreased to alleviate an adverse response, is restrictive and possibly dangerous. As outlined previously, devices play a very useful role in the clinical elucidation of pharmacodynamic information, but to design a potentially useful conventional dosage form without consideration of the ability of the patient to comply is legitimate only when the drug is strictly intended to be administered by a health professional.

The fact is that we will be administering drugs with "conventional" dosage forms for a number of years to come, and several factors that have a negative influence on patient compliance can be mitigated through rational dosage form design. The major concern with compliance in elderly patients is related to the fact that each individual is usually taking a number of prescription and nonprescription products simultaneously. Federal statistics indicate that, on the average, an elderly patient will receive 11 drugs yearly (1.4 per visit to the physician).[1] The Pennsylvania Blue Cross reports this number to be 18; in the Veterans Administration system, the average number dispensed per clinic visit is three.[5] Considering the permutations of multiple products given b.i.d., t.i.d., and q.i.d., a once-a-day dosage is a substantial advantage for the elderly patient. In cases where the drug disposition parameters are not appropriate (e.g., short half-life), this attribute may be achieved through the use of dosage forms ranging from coated tablets or beads to transdermal delivery systems.

No new drug has ever gone to the market initially, intentionally, as a sustained-release dosage form, and it would be obfuscatory to introduce the unknowns associated with sustained release in the early stages of clinical testing. As sustained-release technology develops, particularly with low-dose drugs, product development will be easier and release profiles more reproducible, and one can envision sustained-release dosage forms of new as well as existing therapeutic agents presented as dosage forms for the elderly. Obviously, these will have to be tested in the elderly leading to more clinical studies based on pharmaceutics rather than pharmacology.

As alluded to previously, difficulty in swallowing is often cited by elderly patients as a problem associated with tablets and capsules. It is not unusual to find a patient actually skipping doses because of this problem. It is also not unusual to find patients and caregivers distributing crushed tablets or the contents of a capsule in juice or applesauce as an ad hoc drug delivery system. To the extent that this practice improves compliance, it is a benefit, but there are cases where such an approach can be a detriment to therapy and may even be dangerous. If all the food in which the drug is distributed is not consumed, not all of the drug is administered. A tablet that is enteric coated or coated for sustained release should not be crushed; the protection of the coat is lost in the first case, and in the second the large quantity of drug intended to be released over a long period of time is presented immediately for absorption if the integrity of the coating is destroyed. A granular product coated for sustained release may offer an advantage in this case, because it can readily be distributed in a liquid or other foodstuff to facilitate administration without destroying the release-controlling mechanism. TheoDur Sprinkle℠ has recently been promoted as a dosage form for the elderly on this basis.

Few studies have examined the potential deleterious effects that administration of a product in food may have. Food substances are not necessarily inert media for drug distribution. There are a number of natural and synthetic polymers present that can bind the drug, and juices may have quite an acidic pH, presenting unexpected conditions for release of drug prior to administration (see Section 4).

Unfavorable organoleptic properties can also lead to compliance problems. Solid dosage forms can be coated to eliminate a bad taste, but liquid dosage forms present a greater challenge. Not only are generalizations about the flavor preferences of elderly patients dangerous, but often a dislike of a product may develop with chronic use. It is certainly wise to have alternate flavors available if possible, and to consult with the patient on a regular basis to make sure he or she has not discontinued a medication because it "tastes bad."

The dosage form chosen for a clinical study should be selected and designed with these considerations in mind. It is better to anticipate a problem and avoid it rather than to try to educate elderly patients with regard to the appropriate administration technique for each of the several drug products they will encounter.

4. DOSAGE FORM PERFORMANCE IN ELDERLY PATIENTS

Once a decision has been made regarding the dosage form to be used, the product should be formulated to avoid potential problems. Elderly patients present increased drug response variability. No real effort has been made to partition that variability into drug- and dosage-form-related components. Consequently, the following is speculation and there is a risk of perceiving a problem that is not really a problem.

Perhaps it should be emphasized that this section is not an exposition of factors that adversely effect bioavailability in elderly patients. The absorption, distribution, metabolism, and excretion processes may be

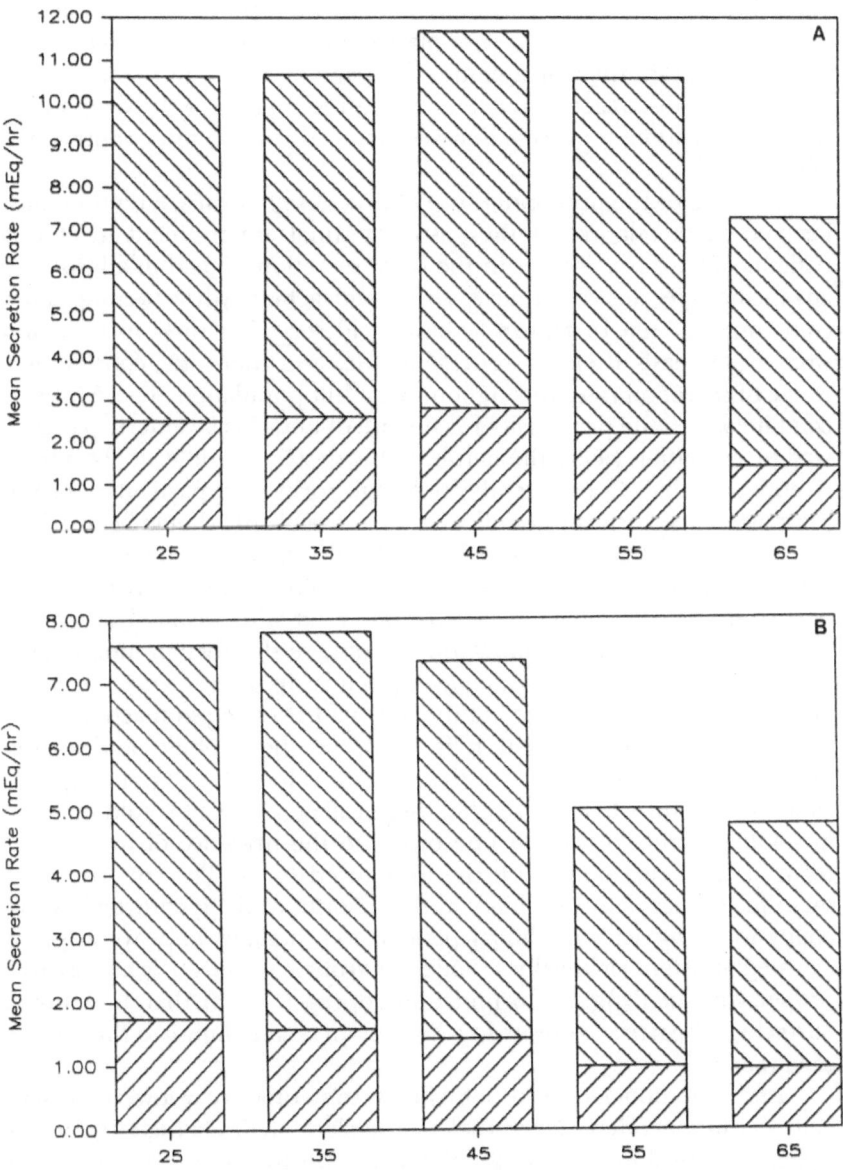

FIGURE 1   Basal acid secretion in (A) men and (B) women as a function of age. The value plotted on the abscissa at the interval midpoint of 65 is actually for 60 and greater. ▨ , mean rate; ▧ , mean + 2S.D. (Data taken from Ref. 10.)

altered by disease, aging, concomitant drug therapy, or all these factors. Such changes definitely influence plasma level versus time profiles, often requiring significant adjustments of dosage and dosage regimen. The focus of this discussion is related to the presentation of the drug to the site of absorption, a process that clearly needs to be under control in a clinical study.

Because of the preponderance of oral dosage forms, attention must be paid to the nature of the gastrointestinal environment in elderly patients. There is a change in the basal secretion of acid in adults with advancing age, and atrophic gastritis and achlorhydria become more common. As seen in Figure 1, basal acid secretion rates in men and women decrease after the fourth decade, and there is a considerable range of secretion rates throughout the adult population. These factors, combined with the propensity for the use of antacids by elderly patients, lead to a situation where the nature of the gastric and intestinal contents is unpredictable. Consequently, a dosage form or a particular formulation whose performance is influenced by variable pH should be avoided.

Generally speaking, most bioavailability problems have been encountered with nonelectrolytes and weakly acidic drugs. There have been few, if any, bioavailability problems observed with weakly basic drugs because they are quite soluble at low pH and are absorbed predominantly in the intestine. In solution, a weakly acidic drug with a pK_a of 3.0 is nearly completely un-ionized at the normal pH of the stomach; absorption of the drug in this state is favored, but the solubility and, consequently, dissolution rate are at their lowest values. If the pH of gastric contents is 3, for whatever reason, there will be about a twofold increase in solubility and dissolution rate owing to the presence of the ionized form of the drug along with the un-ionized form. The confluence of these factors leads to variability in the amount of drug absorbed; availability for weak acids in elderly patients may actually improve as pH is elevated above normal values. Since electrolytic character is an intrinsic property of the drug, no formulation approach, other than intentional buffering of gastric contents to a specific pH, can eliminate this source of variability.

Although nonelectrolyte solubility is sometimes very low, it is essentially pH independent. Nevertheless, the electrolytic character of an excipient may prohibit the design of a dosage form whose performance is completely independent of pH variation. Disintegration, drug dissolution, and the eventual concentration of drug available from a tablet designed to "dump" the drug into the gastric contents can be intentionally or unintentionally altered because of dosage form components other than the drug. The choice of excipients, particularly in the case of low-

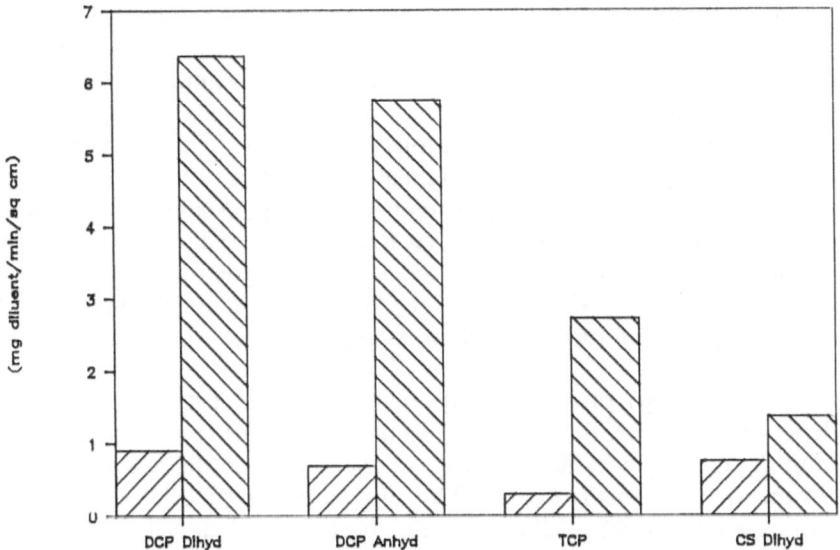

FIGURE 2 Effect of pH on the intrinsic dissolution rate of selected calcium salts used as tablet excipients. DCP Dihyd, dicalcium phosphate dihydrate; DCP Anhyd, anhydrous dicalcium phosphate; TCP, tricalcium phosphate; CS Dihyd, calcium sulfate dihydrate; ▨, pH2; ▩, pH1. (Taken from Koparkar.[11])

dose drugs, may determine the degree of pH variability in drug release. For example, dicalcium phosphate, tricalcium phosphate, and calcium sulfate are salts used frequently as tablet fillers. As shown in Fig. 2, the dissolution of these calcium salts is dramatically decreased as pH is elevated from a value of 1 to 2. If drug release from a tablet containing one of these calcium salts is dependent to any extent on the dissolution of the excipient, drug availability will be profoundly diminished by elevated gastric pH. Tablets containing one or more of these calcium salts, intended to be used extensively by elderly patients, should be formulated to ensure complete disintegration to primary drug particles. If any of these salts is used as a source of calcium, it can be expected that inhibited and incomplete dissolution will be encountered.

It would seem wise in situations such as this, where insoluble excipients are used, to employ one of the newer, extremely effective disintegrants currently available: croscarmellose sodium, crospovidone, or sodium starch glycolate. These agents, although showing some variation in physical properties in dispersion at different pHs,[6] virtually ensure complete disintegration of the tablet often in less than a minute.

The significance of an *in vitro* result is always legitimately open to question until an actual *in vivo* assessment has been completed. Since one of the unstated objectives of this discussion is to stimulate continued research into the degree to which dosage form variability affects therapy in the elderly, it is important to emphasize this warning. At the least, as much care as possible should be taken to assure that the *in vitro* test is as reasonable as possible. The pH-dependent drug–excipient interaction reported in the literature involving croscarmellose and also sodium starch glycolate[7] serves as an example of an apparent problem that apparently is not a problem. When studied at reasonable ionic strengths, the drug–excipient interaction appears negligible across the physiological pH range.[8]

The solubility or permeability of many tablet coatings is also pH dependent. In fact, the function of enteric coatings is to delay release until the tablet passes through the stomach and into the less acidic environment of the small intestine. The film coatings used to achieve enteric coating are essentially "all-or-nothing" barriers to drug release. Figure 3 presents the results of a dissolution study of enteric-coated aspirin

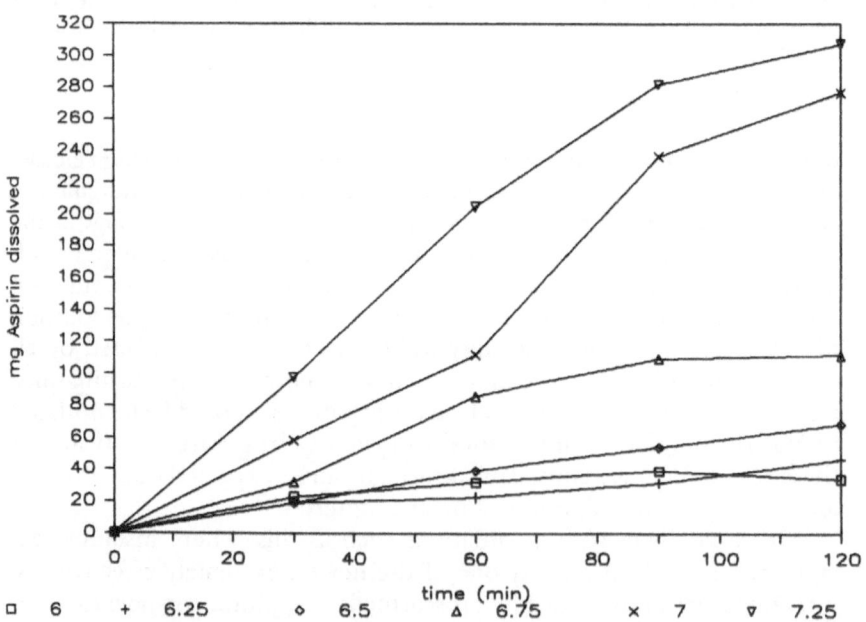

FIGURE 3 Extent of dissolution of aspirin from enteric-coated tablets (Ecotrin®) as a function of pH. Refers to pH of a 0.1 M phosphate buffer used as the dissolution medium, USP dissolution method (50 rpm). (Taken from R. Gary Hollenbeck, unpublished data.)

(Ecotrin℗) at various pHs ranging from 6 to 7.25. Although it is obvious that the enteric coating does its job quite well, in terms of preventing release of the drug in the stomach, it is reasonable to wonder whether a patient whose intestinal pH happens to fall below 6 is going to receive a complete dose of the drug.

Studies of the pH variability in performance of sustained-release dosage forms are practically nonexistent. With some exceptions, systems based on film coatings whose solubility or permeability is pH dependent would be expected to present greater variability than systems based on matrix erosion or osmotic pressure. Formulation of a sustained-release product is further complicated when the drug itself is a weak electrolyte,[9] and its solubility is pH dependent. And finally, delayed gastric emptying, often encountered in elderly patients, can perturb the plasma level versus time profile for any product given orally.

Drug–excipient, drug–drug, and drug–food interactions have the potential to reduce availability in any patient population. These interactions are of more concern in elderly patients, if only for the reason that they are generally taking more drug products than a younger, healthier population. As mentioned before, there is also the possibility that the drug product may actually be given with food. Mealtime is generally a scheduled daily event for elderly patients and consequently a convenient time for drug product administration. A dosing regimen based around mealtime probably improves compliance, at the expense of increased probability of food interaction.

If it were not for the overwhelming convenience and ease of oral administration, one would be tempted to recommend that this route be totally avoided in elderly patients; it is unreliable enough in the so-called normal, healthy population. Many of the concerns expressed here for this route of administration extend to others. Performance of a sustained-release topical drug delivery system, for example, may vary because of decreased hydration and histological changes in the skin as it ages.

5. Conclusions

Undoubtedly, unforeseen technological advances will completely alter the manner in which therapeutic agents are utilized by patients in the future. Use of infusion devices similar to those currently employed in hospitalized patients and constant-rate oral drug delivery systems similar to those used now in animal studies will be widespread in the near future. Prototype "artificial organs," which are really implantable devices whose operation is controlled through sensors and feedback loops, are already available. These advances will have more of an impact on the elderly population than on any other group.

In the meantime, it is essential to pursue research that identifies cases where compromised drug delivery can be indicted as a contributor to therapeutic failure in elderly patients, and that explains the failure on a physico-chemical basis. A database needs to be established to guide the formulator of products used exclusively or extensively by elderly patients.

Based on the information articulated in this chapter, it is reasonable to advocate the following with regard to the consideration of dosage forms to be used in clinical studies involving elderly patients: Basic pharmacological and pharmacokinetic information should be collected on these patients with a method of drug delivery that is absolutely controlled, and the prospective dosage forms to be utilized by unsupervised patients should be designed to minimize the frequency of administration and to avoid variability in performance due to the natural and pathophysiological changes in the body that are frequently encountered in the elderly.

REFERENCES

1. Baum C, Kennedy DL, Forbes MB, Jones JK: Drug use in the United States in 1981. *JAMA* 1984; 251:1293–1297.
2. Fara J: Recent advances in parenteral drug delivery systems. *J Parenteral Sci Technol* 1983; 37:20–25.
3. Franetzki M: Drug delivery by program or sensor controlled infusion devices. *Pharm Res* 1984; 6:237–244.
4. Kimminall MD: Spoons provide potential for dosing errors. *Am Pharmacy* 1979; NS19:25–27.
5. Lamy PP: Patterns of prescribing and drug use, in Butler RN, Bearn AG (eds): *The Aging Process: Therapeutic Implications.* New York, Raven Press, 1984, pp 53–82.
6. Shangraw R, Mitrevej A, Shah M: A new era of tablet disintegrants. *Pharm Tech* 1980; Oct:49–57.
7. Chien YW, Van Nostrand P, Hurwitz, AR, Shami, EG: Drug–disintegrant interactions: Binding of oxymorphone derivatives. *J Pharm Sci* 1981; 70:709–710.
8. Fan A: A basis for identification, characterization and assessment of the implications of drug interactions with selected cellulosic excipients *in vitro.* Ph.D. Thesis, Pharmaceutics Department, University of Maryland School of Pharmacy, 1985.
9. Lee VH, Robinson J: Drug properties influencing the design of sustained or controlled release drug delivery systems, in Robinson J: *Sustained and Controlled Release Drug Delivery Systems.* New York, Marcel Dekker, 1978; pp 76–82.
10. *Documentia Geigy Scientific Tables,* 7th ed, Ciba-Geigy, Ltd, Basel, Switzerland, 1970.

11. Koparkar A: The modification of an automated dissolution test apparatus for the rotating disk method of intrinsic dissolution rate measurement, its validation and use in evaluating tablet diluents. Ph.D. thesis, Pharmaceutics Department, University of Maryland School of Pharmacy.

CHAPTER 18

Clinical Trial Design— Industry Perspective

William B. Abrams and Bruce E. Rodda

1. Introduction

It is now well recognized that the elderly, defined as individuals over 65 years of age, are growing in numbers and as a proportion of the total population. It is also recognized that this group disproportionately uses medicines. At the present time, for example, the elderly represent about 11% of the U.S. population and are estimated to consume 20–25% of all medications.[1] Based on data from the U.S. Census Bureau,[2] it is predicted that by 2030 the elderly will increase to about 21% of the U.S. population and consume more than 40% of all drugs. Equally important, the percentage growth of the "old elderly," those 75 and over, will be even greater.[3] The high prevalence of disabilities in the very old increases the chances for therapeutic misadventures. Indeed, it is widely perceived that adverse drug reactions and drug interactions are more common in the elderly than in younger patients. A recent report of the Royal College of Physicians on Adverse Drug Reactions in the Elderly[4] identified inappropriate practices by treating physicians as central to this problem.

WILLIAM B. ABRAMS • Scientific Development, Merck Sharp & Dohme Research Laboratories, West Point, Pennsylvania 19485; and Jefferson Medical College of Thomas Jefferson University, Philadelphia, Pennsylvania 19107. BRUCE E. RODDA • Clinical Biostatistics and Research Data Systems, Merck Sharp & Dohme Research Laboratories, Rahway, New Jersey 07065.

On the other hand, a review by Klein *et al.*[5] suggests that the incidence of adverse drug reactions may not be greater in elderly patients but the opportunity for such events certainly is. Despite the extensive use of drugs by the elderly, guidance to physicians relative to this population is not widely available; i.e., dosage and precautions specific to the elderly are not commonly found in drug package inserts, therapeutic texts, or related documents at present. This lack of information is a result of the relative infrequency, until very recently, of clinical drug trials designed to measure age-related differences in response to drugs.

Part of the problem of drug use in the elderly lies in changes, mostly reductions, in many physiological functions associated with the aging process. The principal changes and their potential consequences for the disposition and actions of drugs are listed in Table I. The last item on the list, large interindividual variation, applies not only to drug responses but also to drug entities within classes. Therefore, general statements about the use of drugs in older patients are of limited value. In order to provide the specific information needed by physicians to properly treat the sick elderly, clinical investigations must be conducted with specific drugs in defined populations.

This chapter is concerned with the design of clinical trials of drugs in geriatric populations from the perspective of the pharmaceutical industry.

TABLE I   CHARACTERISTICS OF THE ELDERLY RELATIVE TO DRUG DISPOSITION AND RESPONSES[a]

Reduced renal function: accumulation of renally cleared drugs
Reduced serum albumin levels: increased free drug
Relative increase in body fat: increased distribution volume of fat-soluble drugs
Reduction in body size: decreased distribution volume of water-soluble drugs
Reduction in liver metabolizing capacity: accumulation of metabolized drugs
Increased sensitivity to CNS drugs: adverse effects
Decreased cardiac reserve: potential for heart failure
Decreased baroreceptor sensitivity: tendency to orthostatic hypotension
Decreased vital capacity and maximum breathing capacity: increased risk from beta blockade and CNS depression
Oversecretion of antidiuretic hormone: potential for water intoxication
Concurrent illnesses: disease interactions
Multiple drugs: drug interactions
Large interindividual variation: wide dose range

[a]Adapted from Vestal and Dawson,[1] Shock,[6] and Fleg *et al.*[7]

2. INDUSTRY-SPONSORED DRUG STUDIES

Clinical trials initiated by pharmaceutical firms are addressed to two related audiences, practicing physicians and regulatory agencies. Scientific excellence remains, of course, the prevailing standard. In the United States, the clinical investigation and marketing of new drugs are governed by the 1938 Federal Food, Drug and Cosmetic Act expanded by the Kefauver–Harris Amendments of 1962.[8] These statutes have been implemented by a series of regulations published by the Food and Drug Administration (FDA) in the *Federal Register.*[9] These laws and regulations require that prior to marketing, substantial proof of safety and efficacy for the claimed indications be provided.[10] Substantial proof is that gathered by "adequate and well-controlled" clinical trials. Information generated by these studies reaches practicing physicians indirectly through the medical literature and directly through drug package inserts published in the *Physician's Desk Reference.* The package insert is the core vehicle because, after approval by the FDA, it also serves as a standard for educational and promotional efforts by the industry and essentially represents "official labeling."[9]

To assist sponsors in designing and conducting clinical trials that would qualify as "adequate and well controlled," the FDA has prepared a series of clinical guidelines in collaboration with industry and academia. The guidelines recognize the common practice of designating drug trials by "phases."[8] Phase I studies represent the initial administration of a new drug to human beings, usually healthy volunteers. The objectives include human tolerance, absorption, metabolism, elimination, and pharmacodynamics. Similar studies conducted later in the development process are still considered phase I investigations. In phase II, the test drug is first used in the prevention or treatment of the disease for which it is intended. The goals are determination of efficacy, dosage ranges, and further proof of safety. Phase III studies comprise the broad clinical trials necessary to determine whether the drug is a candidate for marketing. The studies involved may range from multicenter, controlled, double-blind, efficacy, and safety trials conducted by practicing physicians to complex investigations of specific pharmacodynamic actions. Phases IV and V refer to programs conducted after marketing. Clearly, trial design will vary with phase, objective(s), and population.

The first FDA Guideline, *General Considerations for the Clinical Evaluation of Drugs,*[11] emphasizes the principles of good clinical design (Table II). Not surprisingly, the principles listed are consistent with those addressed in textbooks on the subject.[12,13] Other guidelines deal with bioa-

TABLE II   PRINCIPLES OF ADEQUATE AND WELL-CONTROLLED
CLINICAL INVESTIGATIONS[a]

Clear statement of objectives
Definition of selection criteria; comparability to target population
Documentation and verification of randomization procedure
Determination of sample size, including response criteria and risks of false-positive or
   false-negative results
Inclusion of appropriate comparison group(s)
Double-blind design whenever feasible
Objective methods of observation where possible
Defined response variables, including methods of observation and quantification
Specify limitations on protocol deviations

[a] Modified from Anello.[10]

vailability studies and specific drug classes. The principles and recommendations contained in these guidelines apply, of course, to geriatric drug studies, but they may require modification because of the special circumstances of the elderly.[14]

When a new drug is entered into clinical investigation, a comprehensive plan is prepared by the sponsor to chart the course of the drug candidate from the laboratory to the submission of a New Drug Application and beyond. Among the items addressed are the overall objectives (including claims to be pursued), the scope of the clinical studies to be conducted, and the design and statistical considerations that apply to these studies. In the context of this chapter, the approach to collecting appropriate information relative to the elderly should also be included in this plan. As described below, the main considerations are pharmacokinetics, dose ranging, drug interactions, and adequate efficacy and safety experience.

3. ELDERLY ISSUES

In recognition of the demographic and other issues noted previously, there have been considerable efforts by interested groups in academia, industry, and government to amend current clinical drug trial practices in order to provide the information needed to use drugs appropriately in the elderly. In 1983, for example, the FDA issued a *Discussion Paper on the Testing of New Drugs in the Elderly*.[15] The responses to this paper culminated in a workshop cosponsored by the American Society for Clinical Pharmacology and Therapeutics and the FDA.[16] The recommendations of the workshop were forwarded to the FDA as proposed guidelines and form the basis of many of the remarks that follow. The characteristics of drugs that may require special consideration for use in the elderly are listed in Table III and are similar to those proposed by Crooks.[17] Some of the factors that must be considered in planning drug studies in this population are listed in Table IV.

There is wide agreement that subjects should not be excluded from drug studies on the basis of (older) age per se. Positively stated, subjects and patients over 65 years of age should be recruited to participate in drug trials in numbers appropriate to the test drug's expected use in this population—if they otherwise qualify with the inclusion and exclusion criteria of specific protocols. These criteria, including the definition of "normal," will need careful consideration in the design of individual studies. For example, how much osteoarthritis, arteriosclerosis, systolic hypertension, glucose intolerance, or obstructive pulmonary disease will be "allowed" in addition to the target disease? Protocol interdiction of all drugs other than those being investigated could make patient enrollment difficult and possibly inappropriate. If the outcome variable involves the cardiovascular system, the cardiac enlargement, reduced heart rate responses, and diminished functional reserve characteristic of the elderly[7] must be taken into account.

4. PHASE I: PHARMACOKINETICS

Phase I and II studies are intended to establish the dosage recommendations for the large-scale phase III studies that follow. Part of the information needed is acquired in the baseline pharmacokinetics development program, which describes the absorption, distribution, metab-

TABLE III   DRUGS REQUIRING SPECIAL CONSIDERATION FOR USE
IN THE ELDERLY

Drugs expected to be widely used by the elderly (e.g., over 25% projected usage)
Drugs that affect or are affected by physiological homeostatic mechanisms, which may
  be deficient in the elderly
Drugs that act on the central nervous system
Drugs with a low therapeutic-to-safety ratio and
  are excreted largely by the kidney
  are subject to large first-pass effect
  are metabolized by oxidative mechanisms
  generate significant metabolites

TABLE IV   DESIGN PROBLEMS IN GERIATRIC DRUG STUDIES

Subjects: Elderly very heterogeneous
  Multiple disorders
  Multiple drugs
  Comprehension and verbal limitations
  Reduced response potential
Drugs
  Pharmacokinetic differences
  Pharmacodynamic differences
  Altered tissue responses
  Drug interaction potential
Trial conditions
  Subject availability
  Informed consent
  Validity of measurements
  Compliance and follow-up
Statistical issues
  Sample size determination
  Distribution of responses
  Independence of variables
  Assessment of adverse events

---

olism, and elimination of drugs.[18] Such programs generally include, in sequence, healthy volunteers, subjects with kidney and/or liver disease, and patients for whom the drug is intended. To prepare for the inclusion of the elderly in phase III trials, pharmacokinetic and dosing information should be obtained during the early studies. Of the various characteristics of the elderly listed in Table I, reduction in renal function is the most consistent and predictable. It has been proposed, therefore,[16] that when studying drugs that are excreted significantly through renal mechanisms, a single-dose study employing healthy elderly subjects be included in the baseline pharmacokinetic program. "Healthy elderly subjects" could be defined as independently living individuals who are within 20% of ideal weight and capable of being off all medication for 1–2 weeks prior to the study. Exclusion criteria should include clinically significant hematological or biochemical abnormalities and known hepatic or renal disease. It is to be emphasized that renal function cannot be estimated in elderly subjects by serum creatinine values alone because production of this metabolite is reduced in this population.[19] For clinical purposes, however, formulas and nomograms that include an age term are available for estimating renal function in older individuals.[19,20]

The analysis and interpretation of the single-dose study should take into account the expected decrease in renal elimination from reduced kidney function. If the results differ substantially from those obtained in a concurrent young control group or those predicted by models developed in younger subjects, additional studies may be required. For example, a study involving administration of radiolabeled drug to young and elderly subjects may identify not only gross, age-related differences, but also the basis of the differences. Multiple-dose investigations to steady state might be necessary with drugs whose cumulation potential is not readily predicted by single-dose information. Since information from younger subjects is available, statistical methods that use this information (such as a Bayesian approach[21]) may assist in identifying age-related differences.

The pharmacokinetic data obtained from these studies, plus quantitative efficacy and safety measurements acquired in phase II, should permit a rational judgment about whether dosage adjustments may be required in elderly people. Drug interaction studies may be conducted as in younger subjects, taking into account the medications, both prescription and over-the-counter, commonly used in the elderly.

5. PHASE II

The designing of phase II and III studies that are addressed to or incorporate geriatric patients can be a real challenge to clinical trialists. Some of the concerns are listed in Table IV.

As noted earlier, dose-ranging and safety information, in addition to pharmacokinetic data, are needed to prepare for phase III. Objectives relative to the elderly, such as dosage recommendations and claims to be pursued, must be clearly identified. If the claim involves an objective measure, e.g., reduction in blood pressure or elimination of pathogenic bacteria, including the elderly does not pose novel problems. Clearly, noninvasive measurement techniques are preferable. However, when the target disease is a chronic disorder or frequently associated with other ailments, drug efficacy may have to be measured in terms of functional improvement rather than cure.[14] A number of functional assessment techniques have been developed[22] that could be applicable for such purposes, but their validity and reliability would need to be established for each specific drug study.[14] Similarly, a variety of rating scales and other mental status instruments[22,23] are available for use in geriatric psychopharmacology, but their utility as drug response measures also requires validation. Comprehension capacity, hearing and visual impairment, verbal limitations, attention span, and even general energy levels must be considered in the application of these tools to specific elderly subgroups.

Phase II studies provide an opportunity to investigate possible altered tissue responses in elderly individuals. For example, older subjects exhibit decreased responses to both β-adrenergic agonists and antagonists.[24] Studies with propranolol demonstrate the complexities involved and the need for comprehensive programs. Although tissue responsiveness may be reduced as noted, the bioavailability of this drug is increased in the elderly.[25] Thus, dosage recommendations require good-quality clinical information, as well as the results of the special studies referred to previously. The possibility of altered tissue responses, as well as disposition differences, may also be suspected if the relationship among dose, drug blood level, and biological actions differs in older versus younger subjects. Blood level/effect correlations may also be useful in screening for drug interactions.

6. SUBJECT SELECTION

Subject selection, although critical in all clinical trials, is especially problematic in elderly studies. In phase I and II, elderly subjects must be healthy enough to define the effects of age per se without the confounding effects of interfering drugs and disease, whereas in phase III the study participants should be as representative of the ultimate target population as possible. When age limits are removed from protocols, some adjustment in admission criteria for older volunteers may be necessary for both enrollment and relevance purposes. For example, blood pressure limits may be raised to 160/100 and certain chronic diseases permitted. Acceptance of common chronic disorders was discussed earlier. On the other hand, when addressing a specific age-related question, inclusion and exclusion criteria must be sufficiently precise to focus on the target population. For example, a current multicenter trial in elderly hypertensives employs two principal inclusion criteria and 20 exclusion criteria to accomplish this objective (J. F. Walker and L. Nugent, personal communication, 1985). Finding a sufficient number of eligible older patient volunteers to complete a drug testing program and to obtain informed consent are well-known difficulties in geriatric research and represent urgent challenges to both sponsors and investigators.

7. Phase III

The broad objectives of phase III trials relative to the elderly encompass obtaining adequate efficacy, safety, and dosage information, at least for the drugs described in Table III. As noted previously, it is recommended that older subjects not be excluded from phase III trials. Design issues include sample size, subject inclusion and exclusion criteria, pharmacokinetic differences, if any, and the choice of appropriate outcome measures. Among the factors impacting on sample size determination are the variability of and quantitative reduction in response to therapeutic interventions characteristic of this population. The presence of such factors often calls for large numbers of subjects and limited analyses.[26] The need for a subgroup analysis by age also bears on sample size determination. In phase III, study participants should be generally comparable to the ultimate target population. Because older patients often have multiple disabilities, averaging almost three per person,[27] intrinsic population variability must be factored into experimental design and sample size estimations.

Experimental design and sample size estimations must also consider adverse events and the possibility of a high dropout rate. Needless to say, adverse events, drug related or not, are common in this group. Thus, a priori rules for identifying and recording such occurrences are essential. An adequate control group is also essential for interpreting the incidence and severity of undesirable reactions in geriatric studies. Indeed, with the confounding issues noted previously, controls are essential for efficacy determination. The study plan that best accommodates these concerns usually includes a randomized, double-blind, parallel design. Efficacy measurements will, of course, vary with the drug, but in this population, as in others, the measures should be as objective, quantitative, and appropriate as possible. "Appropriate" includes scientific relevance, feasibility, and subject understanding and comfort.

Another very important issue is the number of elderly patients that must be studied when considerable information is available from studies conducted in a younger population. If we assume that the variability of an observed response is about the same in the elderly as in younger patients, which is often not a good assumption, a rule of thumb suggests that if we are willing to estimate the effect in the elderly with one-half the precision as in the younger patients (twice as wide a confidence interval), then one-quarter as many subjects would need to be examined.

For example, suppose that a study in which the effect of a new antihypertensive agent on supine diastolic blood pressure was estimated has been completed in 100 younger patients. The results indicated a blood pressure reduction of 15 mm Hg and a 95% confidence interval, (13.5, 16.5). Given that degree of precision in younger patients (about 10% of the mean), the effect in older patients may be adequately determined if the precision is roughly 20% of the mean response. Assuming the effect in the elderly approximates that in younger patients, a study of elderly patients would require only about 25 patients and would yield a confidence interval 6 mm Hg wide (12.0, 18.0), rather than 3 mm Hg as in the younger patients. Given the information already obtained from the younger patients, this broader confirmation may be completely adequate. Although this example discusses a specific reduction in precision (50%), the acceptability of confirmatory, rather than definitive, results in the elderly will require fewer patients than in studies of a younger population.

This parsimony can also be realized if it is necessary to compare older and younger patients in the same study. Usually, the definitive studies in younger patients will require a high degree of statistical power so there will be a clear distinction between the new therapy and the control agent, active or placebo. In a study that directly compares younger and older patients with respect to the effect of the new agent, such a precise distinction is often unnecessary. The objective of these studies is to determine whether there are any clinically important differences between the responses of older and younger patients. Differences that would indicate the need for dosage adjustment or more intensive observation are usually quite large and would usually require relatively few patients for detection.

If, however, the response in the elderly is substantially more variable than that in younger patients, studies of a limited size would not be adequate for identification of an unusually high or low response. If there are pharmacological or pharmacokinetic reasons for anticipating a more variable response in the elderly that might be clinically important, larger studies, more limited objectives, or both may be necessary.

An additional consideration is utilization of Bayesian statistical procedures, which provide two advantages over conventional statistical procedures in the analysis of this type of study.[28] First, employing Bayesian analysis enables study designs for the elderly to take full advantage of the information gained from the early studies. Classical statistical procedures make inferences based on the results of a particular study, but only on that study. The Bayesian approach, although often mathematically complex and open to philosophical discussion, utilizes information relative to the magnitude and variability of response from early studies, and thus can reduce the number of patients necessary in an elderly program. These analytic techniques can also be formally incorporated in the analysis to substantially improve the precision of the estimates.

The second advantage of the Bayesian approach is that it gives a clear statement of the likelihood of a clinically important difference, given the results of the current study and information gleaned from prior studies. Following a classical analysis, two potentially troublesome issues may develop. First, there can be a substantial difference between the new and control treatments, yet the analysis may not indicate statistical significance. This outcome can result in difficulties in interpretation. The second issue is that the observed difference is small, and clinically unimportant, but *is* statistically significant. These potentially inconclusive results occur because the classical approach does not address the critical question, "Given the results of the study, what is the likelihood that a clinically important difference exits?" A Bayesian analysis can answer this question and may use the limited information from the elderly population more effectively.

For example, in studying an antihypertensive drug, assume that a clinically important change in blood pressure is 7 mm Hg. Using a Bayesian approach, the conclusion would take the form "Given the results of the study, the probability that the true treatment effect is at least 7 mm Hg is *P*." *P* is a function of the data based on which a complete probability distribution relating any clinically important effect could be derived. The technique thus permits a positive statement concerning clinically important effects in each study.

As discussed previously, variability of response is often greater in older patients that in younger individuals. Since this variability can be a critical consideration in patient care, the need to estimate it can be more important in this population. The issue is not whether this variability may be somewhat greater in the elderly, but whether it is of sufficient magnitude to be clinically important. Two types of variability are important, the degree to which one patient responds differently than another, and the consistency for a specific patient over time. It is certainly important to identify those patients who are hyper- or hyporesponders to therapy, and also to characterize those patients who respond in an unpredictable manner.

Unfortunately, studies designed to estimate the variability of treatment effects demand a much greater number of patients than studies that estimate actual treatment effect. For this reason, studies of the latter type would be employed only if there was a priori information, suggesting that large inter- or intrapatient variation could be anticipated, and that this variability was clinically important to patient care. Estimation of variation of a smaller magnitude might best be obtained in a postmarketing surveillance study, if necessary.

The reliance on postmarketing surveillance studies for developing the main profile of drug effect in elderly patients creates the potential for collecting misleading information. Unless drug effects in the elderly are formally studied, the data obtained will be observational and thus will have the potential for raising false impressions. If these were then pursued, patient care could be compromised, resources spent, and patients exposed unnecessarily. Acquisition of the correct amount of accurate and valid data in the elderly requires the prospective inclusion of the elderly in development programs for drugs that will be used in this population.

8. New Therapeutic Objectives

With the aging of Western societies, increased attention is being given to interventions that could improve the quality of life[14] and prolong the duration of personal independence.[29] Drugs that beneficially influence disorders common in the elderly, such as Parkinsonism, dementia, glaucoma, and osteoporosis, would, of course, accomplish such ends. The same may be true, however, for drugs that treat arthritis, hypertension, congestive heart failure, obstructive pulmonary disease, benign prostatic hypertrophy, and other chronic disorders. These types of objectives, i.e., personal independence and quality of life, broaden the scope and design of studies designed to prove the value of a drug for such purposes. Functional tests become "activities of daily living."[29,30] Assessments are made by all components of the elderly care community—e.g., geriatricians, nurses, social workers—and are analyzed by epidemiological techniques. Indeed, large-scale drug-related intervention trials have established benefits relative to mortality and morbidity.[31,33] Quality of life and independence are equally relevant outcomes for geriatric patients. Experimental design planning would include all the issues applicable to drug development, intervention trials, and geriatric research.[14,21,27]

9. Conclusions

Clinical drug studies conducted to provide information on the proper use of drugs in the elderly employ the same basic design considerations as in other populations, but require attention to the special characteristics of this group. The main considerations, all related, are possible pharmacokinetic differences, the need for dosage adjustment, drug interactions, disease interactions, and the gathering of adequate experience prior to marketing. Recommendations for changes in current clinical drug trial practices in order to generate this information are discussed.

References

1. Vestal RE, Dawson GW: Pharmacology and aging, in Finch CE, Schneider EL (eds): *Handbook of the Biology of Aging*, ed 2., New York, Van Nostrand, 1985, p 744.
2. US Department of Commerce, Bureau of the Census, 1970 Census of the Population.
3. Coe RM: Comprehensive care of the elderly, in Cape RDT, Coe RM, Rossman I (eds): *Fundamentals of Geriatric Medicine*. New York, Raven Press, 1983, p 3.
4. Report of the Royal College of Physicians: Medication for the elderly. *J Roy Coll Physicians London* 1984; 18:7–17.
5. Klein LE, German PS, Levine DM: Adverse drug reactions among the elderly: A reassessment. *J Am Geriatr Soc* 1981; 29:525–530.
6. Shock NW: Aging of regulatory mechanisms, in Cape RDT, Coe RM, Rossman I (eds): *Fundamentals of Geriatric Medicine*. New York, Raven Press, 1983, p 51.
7. Fleg JL, Gerstenblith G, Lakatta, EG: Pathophysiology of the aging heart and circulation, in Messerlie FH (ed): *Cardiovascular Disease in the Elderly*. Boston, Martinus Nijhoff, 1984, p 11.
8. Abrams WB: Introducing a new drug into clinical practice. *Anesthesiology* 1971; 35:176–192.
9. Leber P: Establishing the efficacy of drugs with psychogeriatric indications, in Crook T, Ferris S, Bartus, R (eds): *Assessment in Geriatric Psychopharmacology*. New Canaan, CT, Mark Powley Associates, 1983, p 1.
10. Anello C: FDA principles on clinical investigations. *FDA Papers* 1970; 4:14–15; 23–24.
11. FDA Guidelines: *General considerations for the clinical evaluation of drugs*. US Dept. of Health, Education, and Welfare, September 1977.
12. Friedman LM, Furberg C, DeMets DL (eds): *Fundamentals of Clinical Trials*. Boston, John Wright, PSG Inc, 1982.
13. Abrams WB: The development of clinical protocols, in Finkel MJ (ed): *Factors Influencing Clinical Research Success*. Mount Kisco, NY, Futura Publishing Co., 1976, p 7.
14. Williams ME, Retchin SM: Clinical geriatric research: Still in adolescence. *J Am Geriatr Soc* 1984; 32:851–857.
15. FDA Discussion Paper on Testing of Drugs in the Elderly, Department of Health and Human Services, Public Health Service, Food and Drug Administration. Memorandum from Temple R, September 30, 1983.
16. Workshop on Proposed FDA Guidelines for Clinical Evaluation of Drugs Being Developed for Use in the Elderly, sponsored by the American Society for Clinical Pharmacology and Therapeutics and Food and Drug Administration, Rockville, MD, Sept. 13–14, 1984, cochairmen: Reidenberg MJ, Abrams WB.
17. Crooks, J: Report prepared for the Committee on Safety of Medicines, England.
18. Rotenberg KS (Food and Drug Administration): IND/NDA Guidelines for Bioavailability Submission. Presented at 20th Annual International Industrial Pharmacy Conf., Austin, Texas, Feb. 24, 1981.

19. Lindeman RD, Klingler EL, Jr: Fluid, electrolyte, trace mineral and renal problems in the elderly: With emphasis directed at the effects of renal impairment on drug pharmacology, in Vestal RE (ed): *Drug Treatment in the Elderly*. Australia, ADIS Health Sc. Press, 1984, p 116.

20. Bjornsson TD: Use of serum creatinine concentrations to determine renal function. *Clin Pharmacokin* 1979; 4:220–222.

21. Rodda RE, Davis RL: Determining the probability of an important difference in bioavailability. *Clin Pharmacol Ther* 1980; 28:247–252.

22. Kane RA, Kane RL: *Assessing the Elderly: A Practical Guide to Measurement*. Lexington, MA, Lexington Books, 1981.

23. Crook T, Ferris S, Bartus R (eds): *Assessment in Geriatric Psychopharmacology*. New Canaan, CT, Mark Powley Associates, 1983.

24. Vestal RE, Wood AJJ, Shand DG: Reduced beta-adrenoceptor sensitivity in the elderly. *Clin Pharmacol Ther* 1979; 26:181–186.

25. Castleden CM, George, CF: The effect of ageing on the hepatic clearance of propranolol. *Br J Clin Pharmacol* 1979; 7:49–54.

26. Friedman JA, Chalmers TC, Smith H, et al.: The importance of beta, the type II error and sample size in the design and interpretation of the randomized control trial. *N Engl J Med* 1978; 299:690–694.

27. National Center for Health Statistics: Chronic Conditions and Limitations of Activity and Mobility: United States, July 1967 Vital and Health Statistics, PHS Pub. No. 1000-Series 10-No. 61. Washington, DC, Public Health Service, US Government Printing Office, 1971.

28. Box GED, Tiao GC: *Bayesian Inference in Statistical Analysis*, Reading, MA, Addison Wesley Publishing Co, 1973.

29. Katz S, Branch LG, Branson MH, Papsidero JA, Beck JC, Greer DS: Active life expectancy. *N Engl J Med* 1983; 309:1218–1224.

30. Katz S, Akpom CA: A measure of primary sociobiological functions. *Int J Health Serv* 1976; 6:493–508.

31. The Norwegian Multicentre Study Group: Timolol-induced reduction in mortality and reinfarction in patients surviving acute myocardial infarction. *N Engl J Med* 1981; 304:801–807.

32. Beta-Blocker Heart Attack Trial (BHAT) Research Group Report: A randomized trial of propranolol in patients with acute myocardial infarction: I. Mortality results. *JAMA* 1982; 247:1707–1717.

33. Lipid Research Clinics Program. The Lipid Research Clinics Coronary Primary Prevention Trial results: 1. Reduction in incidence of coronary heart disease, and 2. The relationship of reduction in incidence of coronary heart disease to cholesterol lowering. *JAMA* 1984; 251:351–364, 365–374.

CHAPTER 19

METHODOLOGICAL ISSUES
A REGULATORY PERSPECTIVE

PAUL LEBER

1. INTRODUCTION

Any discussion of methodology must consider that methods are no more than servants of purpose. If the reader is to understand, from a regulatory perspective, how and why one particular method or design feature is judged superior to another, some degree of familiarity with the domestic drug regulatory process and its purposes is important.

2. DOMESTIC DRUG REGULATION: ITS AIMS AND REQUIREMENTS

The Food and Drug Administration (FDA or agency) is the organization within the executive branch of the federal government that is responsible for enforcement of the Federal Food, Drug and Cosmetic Act. The agency interprets the Act, promulgates rules and regulations to ensure that its aims are achieved, and enforces these requirements.

Under the provisions of the Federal Food, Drug and Cosmetic Act, a new drug product may NOT be legally distributed within the United

PAUL LEBER • Division of Neuropharmacological Drug Products, Food and Drug Administration, Rockville, Maryland 20857.

States unless it has been shown to be effective, safe in use, and adequately labeled.

As directed by the Act, permission to market a drug is sought in the following manner. The individual or corporation seeking marketing approval (the sponsor) submits a New Drug Application (NDA) to the agency. The application is required to present comprehensive documentation of the new drug product's quality (i.e., identity, strength, composition, stability, freedom from contamination, and bioavailability). Most critically, the sponsor must provide "substantial" evidence obtained from "adequate and well-controlled" clinical trials that the product is effective, as well as evidence that the product is safe for use for the purposes recommended in its labeling (i.e., package insert for prescription drugs). In addition, the sponsor must propose labeling that provides adequate directions for the use of the product.

The FDA is required to evaluate all the information submitted and to approve the application, thus allowing distribution and sale of the product, *unless* it determines that the evidence and documentation submitted are inadequate to support the sponsor's claims that the product satisfies the requirements of the Act. Although the agency is directed to approve an application unless it finds it deficient, it is important to note that approval requires an affirmative act by the agency. Applications do *not* automatically become effective, as they once did (prior to passage of the 1962 amendments) simply because the agency fails to act on them within a specified interval.

It is critical to note, too, that any decision by the agency in regard to an application must be based on the requirements of the Federal Food, Drug and Cosmetic Act. The agency has no authority to approve or disapprove applications merely because it seems sensible to do so in the instant case. All agency actions must be referenced to the authority granted by the Act. This has several consequences that are not widely known. First, although it is a sponsor's burden to establish that a product is safe and effective, a sponsor need only present the minimum quantity and quality of evidence required to meet the standards of the Act. Second, the agency cannot properly refuse to approve an application once it has met these "minimum" requirements. Third, because the Act does not address the question of relative value of drug products, the agency must approve a product so long as it is not unsafe for use for the claims made in its labeling. As a consequence, several drug products of varying relative efficacy may be simultaneously marketed for the same use. However, because the Act does demand that drug products be accurately labeled, the agency is able to use a drug product's labeling to convey important information about each product's relative risks and benefits.

3. Drug Development Program Design

The importance of a product's ultimate indication and its labeling in the planning of a drug development program should now be clear. If a sponsor has an interest in developing a drug for some general use, it is critical, at the earliest possible time, to frame the general objective of the program in terms of a concrete, meaningful, and understandable labeling claim. The nature of the claim will determine essentially all other aspects of the development program.

Specifying the "right" regulatory claim can be difficult. Legally, a claim can be made for virtually any action "intended to affect [beneficially] the structure or any function of the body of man. . . ." However, for most practical purposes, the object of treatment should correspond to a symptom, condition, or disease that is recognizable and diagnosable. A product cannot be approved as a treatment for a patient population that cannot be described and/or diagnosed in a meaningful way. For example, the agency objected to a sponsor's claim for a product intended to treat "apathetic and withdrawn senile behavior" on the grounds that the states of apathy and withdrawal were not clinically defined or meaningful and that neither was linked to senility in a unique way (see discussion of pseudospecificity below). The point, of course, is not that anyone denies the existence of older apathetic individuals; rather, the concern is that the usage employed is so imprecise that the labeling claim could lead to the drug's administration to almost anyone on any given day (i.e., it provides inadequate directions for the product's use). Furthermore, simply from a clinical trial design perspective, how would one draft a protocol to select appropriate patients for participation in clinical trials for this indication? In short, the more closely linked a claim is to an established disease, syndrome, or phenomenon, the more readily it can be investigated without ambiguity.

It is also critical in specifying the claimed use that will guide a drug product's development to avoid the trap of "pseudospecificity." Indeed, this trap is especially troublesome to those who seek to develop drugs for the geriatric drug market.

Pseudospecificity arises when a sponsor seeks to tie a labeling claim to some patient characteristic by selectively studying the drug product in patients with that characteristic although neither the treatment nor the disease bears any essential relationship to the characteristic. Consider the example of the case where a sponsor selectively studied an antipsychotic drug in older psychotic patients. The sponsor was warned that documenting that the drug worked in such patients would not lead to a claim of age-related antipsychotic action. Admittedly, the sponsor might have developed evidence that the drug was effective in older psychotic patients, but there would be no basis to conclude that the product's effectiveness was in any way linked to the age of the patients studied. If one wishes to document that age influences a drug effect, one must study the drug in individuals of various ages.

Of course, the pseudospecificity of claimed indications is not always so clear-cut. If, for example, a drug were studied exclusively in older patients with cognitive impairment, a condition for which there is no established standard treatment in any age group, concerns about pseudospecificity would not be primary. However, the entire problem could be circumvented if the claim were linked to a disease rather than a phenomenon or symptom complex. In the example given, were all the patients recruited because they had senile dementia of the Alzheimer's type (SDAT), a claim for the "treatment of cognitive impairment in SDAT" would not be pseudospecific. Of course, the labeling claim would be even better if it specified precisely what cognitive function was improved (i.e., attention, short-term memory, recall, reason).

Whatever the nature of the claim decided upon, the sponsor cannot sensibly hope to plan the entire course of a drug development program before the product has been through some clinical testing. Drug development programs must have a grand strategy, but their design must permit flexible implementation so that information acquired during clinical testing can be used to modify the overall program as required.

Rigid implementation of a drug-testing scheme according to a timetable developed by a sponsor's management may appear efficient, but if the timetable becomes the overriding element controlling what is done, the potential for waste is great. For example, the dose at which a drug will be recommended for use is the dose at which (or around) large-scale efficacy and safety testing must be done. If early clinical tests do not properly identify the dose–response characteristics of a drug product, there is considerable risk that trials intended to provide "definitive" evidence of efficacy will fail. Indeed, dose response is not the only issue of importance. It seems sensible to examine systematically each trial at the time it is completed to determine whether its results suggest a reason to modify plans for the overall drug development program. If such a "real-time" sequential analysis is not carried out, important questions may not be answered, let alone raised, until the agency reviews the NDA for the product. Should the neglected question or questions be substantive, additional clinical trials may be required, delaying any chance for approval for months, even years.

Therefore, it is worth determining as soon as possible whether age influences the response to a drug. Admittedly, current regulations do not make evaluation of drugs in the elderly a sine qua non of drug approval. Nonetheless, although the Federal Food, Drug and Cosmetic Act does *not* specifically demand that drugs be evaluated in any particular set of circumstances, the law does require that drug products be evaluated by "all methods reasonably applicable" to ensure their safety. Consequently, the agency has the authority to impose requirements for testing a drug product in the elderly to ensure its safety in use. Of course, a similar argument could be made to require testing of drugs in virtually any identifiable group of patients (i.e., blacks, whites, the obese, the very young).

Given the costs involved, however, it makes little sense to impose routine testing requirements if none are needed in most cases. Thus, flexible and sequential planning of a drug's development allows the agency and the sponsor to determine on an ad hoc basis whether certain tests are needed.

Consider how this strategy can be applied to the evaluation of a drug product in the elderly. Because of the positive correlation between increasing age and illness prevalence, most drugs will, if arbitrary age-related exclusion criteria are avoided, be tested in patient samples that contain substantial numbers of older patients. The performance of the older patients in these studies can be monitored to determine whether age is a variable of importance in the use of the drug. For example, Temple has proposed that an examination of blood levels of drugs in older patients can provide a potentially valuable guide to the extent to which age, per se, may influence the metabolism and elimination of a drug.[1] Similarly, a higher frequency of side effects in the elderly may suggest a special sensitivity to the agent. In any case, if drug development programs include in their planning "budget" resources for exploring, on an ad hoc basis, questions raised during the course of the program, it may be possible to avoid routine requirements for specific studies.

4. SAFETY, EFFICACY, AND DIRECTIONS FOR USE

A clinical drug development program has three major goals. First, it must provide "substantial" evidence of efficacy from more than one investigational source. Second, it must provide sufficient experience with the drug to permit a reasonable conclusion that the drug product is "safe" within the meaning of the law. Finally, it must provide data on how the drug ought best be used (i.e., information that can serve as a basis to recommend how the drug product is to be administered to gain its therapeutic effect at minimal risk).

Experience suggests that the goals of a drug development program cannot be achieved in one or two massive clinical trials of omnibus design. A study seeking to establish the efficacy of a drug may incorporate features that make it a poor source of information about the risks of the drug under conditions of probable use. For example, early studies seeking to determine whether or not a drug is effective may reasonably attempt to enroll a homogeneous sample of patients selected for characteristics likely to increase the probability of showing a drug effect. Thus, placebo responders may be excluded with a placebo runin period, patients with very severe or very mild illness excluded, and some effort made, depending on the indication, to find subjects known to be drug responsive.

In contrast, a large-scale safety study, carried out after the efficacy of the product has been established, should attempt to recruit a heterogeneous patient sample. The aim of the large-scale safety study is *not* to demonstrate the efficacy of the drug (there is no penalty if it does, of course), but to collect enough clinical experience to identify relatively common problems that may be associated with the use of the drug.

4.1. EFFICACY ISSUES

Presuming that the phase I clinical testing has identified the doses of a drug that are reasonably tolerated, it is advisable to establish, as soon as possible, the doses of drug that cause a therapeutic effect. As noted earlier, it makes little sense to embark on either definitive efficacy or safety testing without this information.

Actually, dose–response and efficacy data can theoretically be generated simultaneously in fixed-dose, dose–response studies. The requirement for fixed dose, however, is critical. Although controlled investigations employing flexible dose titration can permit conclusions to be reached about drug efficacy, they are an imprecise source of data about dosage. Flexible dose designs tend to distort the dose–response relationship because the population of patients participating in a positive efficacy trial will necessarily include among those assigned to drug those who improved spontaneously as well as those who received high doses without any substantive improvement. In fact, if this flaw is not recognized, such a pattern of response can lead to a false conclusion that the product possesses an inverted U-shaped dose–response curve.

4.2. OUTCOME ASSESSMENT MEASURES

The measure or set of measures selected to follow a drug's effect in a clinical trial is determined, in part, by whether or not established treatments for the therapeutic indication already exist. For simply practical purposes, it is most appropriate to use standard outcome assessments if these exist, especially if there is a history of repeated use and acceptance by the agency. When outcome measures are well established, as they are in anxiety and depression, for example, agency guidelines are usually available that identify appropriate types of clinical trial assessment procedures.

Unfortunately, in the important area of drugs intended to treat the cognitively and emotionally impaired elderly, drugs with so-called psychogeriatric indications, controversy exists about the proper choice of outcome assessment measures.[2] In part, this is a reflection of the lack of any "benchmark" drug for the indication. Consequently, the FDA has not been able to endorse any particular approach, and it would seem prudent to seek the opinion of as many experts as possible about the best current choices for rating cognitive and emotional impairment. Moreover, because the field of geriatrics is continuing to evolve rapidly, frequent consultation with the agency about the currently preferred approach to specific questions seems advisable.

Another problem not unique to, but extremely common in, geriatric psychopharmacology is "multiplicity" of outcome assessment measures.[3] By evaluating a study on more than a single outcome measure, one increases the overall chance of making a type 1 error about the efficacy of a drug. Although there are mathematical approaches to the problem of multiplicity, it is much more satisfying if a drug development program can identify prospectively two or three outcome measures upon which the efficacy of a drug will be decided. Again, if a drug development program has been conducted in a sequential manner, the selection of appropriate outcome measures for definitive trials can be based on results in earlier exploratory studies.

Replication of positive results is another important regulatory issue. The Act demands that substantial evidence of efficacy be obtained from "investigations." The plural form has been held to refer to two independent sources of evidence of efficacy. This leads immediately to the question of how the agency interprets a "multicenter clinical trial," that is, a trial that is designed as a single study but carried out at several different sites. Policy has evolved on this point over time. It is now generally held that if a study is designed as a unit, it shall be analyzed as a single unit and counted as a single unit.

In part, this policy is intended to prevent the intellectually questionable and logically flawed practice of picking, after the fact, one or two winners from among many losers. Such retrospective selection causes the same sort of problem that was discussed earlier under multiplicity; it substantially increases the chance that an ineffective drug will be declared effective.

Post hoc pooling of several studies to obtain a single statistically significant result is also usually considered unacceptable. It, too, is a procedure that increases the risk of the agency falsely declaring an ineffective drug effective.

In summary, definitive clinical efficacy trials should be designed at a time when the dose, the treatment size effect, and the variation in response to treatment are well characterized. This information, of course, permits one to design studies with adequate power to detect the drug effect of interest.

4.3. SAFETY

The concept of a completely safe drug is, of course, illusory; nothing is absolutely safe, and safety, unlike efficacy, cannot be proved absolutely. From a regulatory perspective, safety is invariably considered in light of the uses of a particular product, a risk-versus-benefit assessment. Actually, the Act, itself, is rather vague, saying only that a sponsor must provide "adequate tests by all methods reasonably applicable to show . . . the drug . . . is safe [also not unsafe] for use under the conditions prescribed, recommended, or suggested in the proposed labeling. . . ."

In general, animal toxicity testing and phase I human pharmacology provide some general information about the potential risks of using a drug product. Ultimately, however, all assurances about a drug's safety depend on the accumulation of experience with the drug under conditions of actual use. It is from repeated experience of "safe passage" that one gains the impression that a drug is "safe."[4] Obviously, however, even after several hundred patients, or even thousands, have used a drug, many risks will remain unappreciated. Indeed, if a unique patient characteristic or trait affects the likelihood of suffering an adverse event upon drug exposure, there is a substantial chance that routine premarketing assessment procedures may not detect the risk simply because insufficient numbers of individuals with the trait have been exposed.

Considerations of this sort apply to the evaluation of risks of drug use that may be increased with age. The ability of a drug development program to detect age-associated risks will directly correlate with the number of older patients enrolled in the program. Of course, because the elderly, particularly the sick elderly, may be less able than healthy young adults to deal with untoward effects of a drug, it seems sensible to delay their exposure until the common pharmacological effects of the drug have been assessed in normal volunteers. In general, this approach seems reasonable even if a drug is being developed for use only in the elderly.

Admittedly, a case can be made to exclude older patients from definitive efficacy studies on the grounds that they are more likely than younger patients to increase experimental variation, thus reducing the efficiency of the study design. It seems probable, however, that if the demonstration of a drug's efficacy depends so heavily on the reduction of variance, it cannot have a very potent effect. Consequently, once the clinical pharmacology of a drug is known, there appears to be every reason *not* to exclude older patients from clinical trials.

In short, prior to any contemplated action, it is useful to know that large numbers of older patients used the drug successfully, without ill effect. In most instances, no special effort should be necessary to obtain this experience because patients suffering from most common diseases tend to be older. However, special recruitment efforts may be necessary when drugs are developed for treatment of symptoms that are common in all age groups (e.g., anxiety, depression, insomnia, pain). Certainly, once efficacy is established, the routine exclusion of older subjects, once a common feature of protocols, is counterproductive.

5. SPECIAL STUDIES

As noted earlier, a decision to conduct special tests, in any setting or age group, should be made only if the evidence suggests there is a reason to do so. As a rule, there seems to be no reason to conduct trials that will do little more than confirm that age affects human physiology in a predictable way. For example, if pharmacokinetic tests reveal that a drug is eliminated primarily by glomerular filtration and urinary excretion, and if the dosage adjustments needed for any given degree of renal impairment are defined, there is little reason to do studies in the elderly to document that the elimination of the drug is retarded. However, if phase III tests reveal marked variation in therapeutic response, unexpected or unusual adverse event profiles, and marked variation in drug blood levels in older subjects not accounted for by the factors known to affect pharmacokinetics, it may be necessary to conduct investigations in older subjects to answer specific questions. Depending on the nature of the problem and the urgency of society's need for the drug product, testing in the elderly may be deferred to phase IV, thus avoiding any needless delay in the approval of an effective drug judged safe in younger individuals.

6. CONCLUSION

The assessment of drug product efficacy and safety is an empirical undertaking. The requirements of the process will vary with the nature of the drug, the disease, the patient population studied, and the sponsor's claim.

Successful and efficient product development programs depend on early identification of pharmacological effects and clear specification of the drug's intended therapeutic claim. Drug development programs must be able to link the results of clinical investigations without ambiguity or questionable validity to the labeling indication claimed.

The size of the program needed to document a drug product's efficacy will vary inversely with the robustness of the drug's therapeutic effects. For powerful drugs effective in a substantial proportion of patients, efficacy may be established in a relatively small clinical testing program. For drugs of more modest therapeutic efficacy, very large programs may be required.

In contrast to the absolute demonstration of efficacy, the demonstration of safety is invariably a relative one and depends on extensive experience with the product under the probable conditions of its use. Consequently, the study of substantial numbers of older patients is essential if the common potential risks of a drug are to be adequately described prior to its approval.

It seems likely that most drug products can be evaluated without specific geriatric testing programs as long as care is taken *not* to exclude older patients from late phase II and phase III testing. In some instances, especially when drugs are intended to serve as symptomatic treatments, special efforts may be needed to ensure that adequate numbers of older subjects are recruited.

In every case, however, the real-time analysis of trial results and frequent consultations between the agency and the sponsor over the entire course of a drug's development are recommended.

REFERENCES

1. Temple, R: FDA guidelines for clinical testing of drugs in the elderly. Presented at the DIA Workshop on Geriatric Drug Testing and Development—Practical Applications, April 2, 1985, Bethesda, MD.
2. Leber, P: Establishing the efficacy of drugs with psychogeriatric indications, in Crook T, Ferris S, Bartus R (eds): *Assessment in Geriatric Psychopharmacology*. New Canaan, CT, Mark Powley Associates, 1983, Chap I.
3. Tukey, JW: Some thoughts on clinical trials, especially problems of multiplicity. *Science* 1977; 198:679–684.
4. Leber P: Safe passage (how good a guarantee?). *Psychopharmacol Bull* 1982; 18:6–10.

CHAPTER 20

STATISTICAL ANALYSIS OF DRUG DISPOSITION DATA

DAVID G. COVELL AND PREM K. NARANG

1. INTRODUCTION

Pharmacokinetics has become a rigorous science that attempts to relate the interaction of a drug with the biological environment into which it is introduced. Such a relationship is often described by proposing a mathematical model of the system and then using the data to define values for the unknown model parameters. Aris[1] points out that "being derived from 'modus' (a measure) the word 'model' implies a change in representation." That is, a model attempts to organize observations or measurements of the system under investigation into a form useful for hypothesis testing. The model parameters, which frequently must be determined from the observed data, function as measures for comparison of results within and between experiments. Based on this criterion, at least three tasks confront the analysts when faced with a data set: (1) to define a parametric model that best describes the system under investigation, (2) to estimate values for unknown model parameters, and (3) to determine whether the proposed model is a good or bad prototype of the system under investigation.

DAVID G. COVELL • Laboratory of Mathematical Biology, National Cancer Institute, National Institutes of Health, Bethesda, Maryland 20892.   PREM K. NAR-ANG • Clinical Pharmacokinetics Research Laboratory, Pharmacy Department, Clinical Center, National Institutes of Health, Bethesda, Maryland 20892.

A variety of methods are available to pharmacokinetists for achieving these goals. These methods depend in large part on the type of model proposed to describe the data. The less complex the model, the less difficult it is to estimate model parameters. In fact, the parameters of simple models can often be obtained by graphical means without use of a digital computer. More frequently, however, digital computers are used to perform sophisticated data analyses using very complex models. The latter approach uses a variety of computer software routines to obtain the solution to the proposed model and then to fit this solution to the measured data in order to define values for the unknown model parameters and finally to evaluate the appropriateness of the model.[2,3]

When the proposed model of the system takes the form of ordinary differential equations, the method is often referred to as compartmental analysis (see Jacquez[4]). The common approach is to specify a mass balance model, which describes the kinetics of drug input, absorption, distribution, metabolism, and excretion in a mathematical form. The time rate of change of the mass of material in each compartment is expressed as the difference between the rates of entry and exit of material from the compartment. The model parameters are the transfer rate constants, which define the fraction of material moving from one compartment to another connected compartment over time. This approach yields a system of differential equations that must be solved. The solutions are generally fit to the experimental data by modifying the unknown model parameters. An acceptable fit of the model solution to the experimental data reflects consistency between model and data. Additional conclusions depend very strongly on the particular system under investigation.

Examples of noncompartmental analysis are becoming more common in the scientific literature.[5] The approach is to specify a model, usually in the form of a polynomial function or a sum of exponential terms, and to fit the model to the data by adjusting the unknown model parameters. The parameters of a noncompartmental model are the coefficients and exponents of each term of the summation. By far the most common usage of a noncompartmental model is that of estimating the area under a kinetic curve.

The distinction between compartmental and noncompartmental analysis is for the most part semantic. In most systems the transfer rate constants of the compartmental model are algebraic functions of the exponents and coefficients of a sum of exponentials obtained from the noncompartmental approach. That is, the sum of exponentials solution and the solution of the compartmental model are often interchangeable.[6] The apparent drawback of the compartmental approach is that the analyst is forced to define the connectivity of the system. The advantage of the compartmental approach is that the analyst can begin to investigate which regions of the model are responsible for differences in the measured data. The selection of a method depends on the goals of the analyst. In general, however, it is safe to say that most pharmacokinetic data obtained from drug disposition studies are routinely subjected to compartmental analysis using a variety of computer software programs.

Customarily, estimation of unknown model parameters is performed by employing standard nonlinear least-squares regression analysis on a digital computer.[7] The results of such an analysis typically include estimated or fitted model parameters along with time course plots of the observed and model predicted data.[2,3] Frequently, little or no attention is given to the statistical significance of the parameter estimates or to the statistical validity of the proposed model. The tasks of statistically estimating model parameters and of examining model validity have become major issues in evaluating the results of compartmental analyses.[8] Statistical estimates of parameter means and standard deviations are two important features for risk assessment and safety evaluation related to the therapeutic modalities where dose extrapolations from few experimental frames are often necessary. Therefore, application of statistical principles must be regarded as a vital and integral part of any drug disposition study, executed with an aim to estimate pharmacokinetic model parameters.

Many different statistical approaches to data analysis have emerged in recent times. Most approaches differ in their treatment of how and where randomness enters the system under investigation. The simplest approach assumes that randomness enters the system strictly in the measurement of data. This approach is often referred to as the deterministic approach because the proposed model of the system under study is assumed to have no random component. More complicated approaches assume that randomness enters the system via many other routes, most notably via the proposed model. These latter approaches are often referred to as nondeterministic or probabilistic approaches and frequently require the treatment of stochastic processes.[9] An example of ways by which randomness can enter a system is useful.

Assume a simple experiment where we measure the width of a doorway. A standard 1-ft plastic ruler could be used in an end-to-end fashion, and the average of many measurements could be reported as the width. This approach assumes a known (deterministic) model for the measure of distance (the ruler) and that each measurement consists of a true value plus some random error. Here the randomness has nothing to do with the ruler itself but could be due to errors in end-to-end placement of the ruler or to errors by the observer recording the measurement. The random error is frequently assumed to have known statistical characteristics (i.e., mean and standard deviation) and can be used to estimate the mean and variance of the unknown model parameters. The key feature of the deterministic approach is that the model is assumed to be *known*. From a biological perspective, however, we know that all rulers are not identical. And in fact, a better approach may have been to measure the width of the doorway with many rulers because each ruler probably has some slight variation in its measure of length. This approach extends the previous approach by introducing randomness in the *model* as well as the measurement. Additional probabilistic features can be introduced if one chooses to measure the doorway with a 1-ft ruler, a yardstick, and a fabric tape. Again, each method adds more complication in terms of where and how randomness enters the system. Extensions of this theme are countless; you could ask whether more than one of the same type of doorway should be measured and these measurements averaged, or reversing the question, one could ask if the most recent measurement can be used to predict future measurements.

Suffice it to say that the latter approaches extend the deterministic description to include randomness in additional parts of the proposed model and then apply a probablistic approach to data analysis. The deterministic approach, however, is more prevalent when data is gathered from drug disposition studies evaluating therapeutic modalities in the preclinical or early clinical phases of drug trials. The intention of this work is to cover the deterministic approach in detail and to briefly expose the reader to the available techniques for nondeterministic analysis.

2. Deterministic Approach

The underlying premise describes the response of a system to a known input as predictable and reproducible. The physical laws that govern this response can usually be represented by mathematical equations. The observation or measurement of this response is subject to uncertainty such that no measurement is exactly predictable. When proceeding with an analysis of drug disposition data, such uncertainties are usually neglected initially, but statistical theory is employed to calculate the effects of this uncertainty.

2.1. Model Development

The goal of the analysis is that of data reduction. We seek to characterize the data set in terms of a model with a few representative model parameters. The compartmental approach proposes a system of n ordinary linear differential equations to describe the time course of material in the n compartments of the model:

$$\frac{dx_k(t)}{dt} = g(x,\lambda,t) \quad , k = 1,2,3 \ldots . n \quad \text{(Equation 1)}$$

$$x(0) = x_o$$

where

x_k = response of the kth compartment (e.g., drug in plasma, urine, tissue, etc.) over time;

x = an $n \times 1$ vector whose elements are the contents of the compartments of the model, with units of mass, fraction of dose, etc.;

λ = an $n \times n$ matrix whose elements are the compartmental rate constants with units of 1/time: these terms are the model parameters, some or all of which must be determined from data;

t = independent variable, usually time;

x_o = a vector of initial conditions.

The ith experimental observation y_i corresponds to the measurement taken at $t = t_i$ and is defined as follows:

$$y_i = f(x(t_i),\lambda) + \varepsilon_i \quad \text{(Equation 2)}$$

where $f(x(t_i),\lambda)$ is either a numerical or analytical solution for Equation (1) evaluated at t_i, and ε_i is the error in measurement associated with observation y_i. Additional details about compartmental analysis can be found in Jacquez,[4] Lambrecht and Resigno,[10] and Boston et al.[11]

If one chooses a noncompartmental model such as a sum of exponential terms, then the observations are modeled as follows:

$$y_i = \sum_{j=1}^{m} a_j \, \overline{e}^{\theta_j \, t_i} \quad \text{(Equation 3)}$$

where the coefficients a_j represent the eigenvectors of the system and the exponents θ_j represent the eigenvalues of the system. The number of exponential terms, m, can be difficult to determine accurately. Excellent details of this method can be found in the work of Bard.[12] It is important to reiterate that whenever we have an analytical solution of the form given by Equation (3), we are dealing with the solution of a system of linear ordinary differential equations with constant coefficients [i.e., Equation (1)]. The solution of this system is, however, a nonlinear function of the model parameters θ or λ.

2.2. DATA ANALYSIS

Now that we have proposed a mathematical model whose solution might describe the experimental measurements, we begin the analysis by outlining guidelines for relating this model to our data. Daniel and Wood[13] suggest that a good method of fitting model solutions to data should, in addition to possessing certain properties of robustness, (1) use all relevant data, (2) have reasonable parsimony in the number of unknown model parameters, (3) take into account the error in the data, (4) provide some measure for the precision of estimates for unknown model parameters, (5) be able to locate systematic deviations in data from the chosen model equations, and (6) provide some measure of how well the model will predict over future experimental frames. This approach therefore assumes (1) the correct form of the model equations is chosen, (2) the data include a representative sample of the system, (3) the measurement errors are independent, and (4) the measurement error distribution has a zero mean and known variance, σ^2. These assumptions are probably rather simplistic. However, it is known that the regression problem is quite robust to minor violations of these assumptions, at least in the linear case. When experimental data do not have the same variance, weighting of observations is usually resorted to, which will be discussed later. Needless to say at this juncture, the support given to a model by the experimentally observed results can be no better than the quality of data used in defining the system.

2.3. PARAMETER ESTIMATION

Once the data have been acquired from drug disposition studies, preliminary graphical techniques are usually employed to formulate a model. A widely used technique is to (1) graph the data on semilogarithmic paper, (2) determine the minimum number of exponential terms suggested by the data, and (3) use this information to propose a compartmental model.[8] The number of exponential terms defines the *minimum* number of compartments for a model of the data. This model may then be used to derive equations describing the system [e.g., Equation (1)].

The estimation of unknown model parameters is usually achieved by employing the method of least-squares.[7] In this approach, an expression of the form given by Equation (4) is generally minimized to find values for unknown parameters for which the model best describes the data.

$$J = \sum_{i=1}^{k} w_i \{y_i - f(x(t_i),\lambda)\}^2 \qquad \text{(Equation 4)}$$

where k is the total number of data points and w_i is the weight assigned to the ith datum.

Because of the form of Equation (4), quadratic minimization numerical software is frequently used to estimate the unknown parameters of the model. Many different approaches can be used to obtain the minimum of J. Some commonly used techniques[14,15] are (1) gradient methods (often referred to as steepest-descent methods), (2) simplex searches, (3) Gauss–Newton methods, (4) Levenberg–Marquardt methods, and (5) conjugate gradient methods. It is instructive to examine one of these approaches to give the reader a feel for how J is minimized and thus to demonstrate how the estimates for each parameter are obtained.

Gradient methods of optimization are based on Taylor's expansion of the function J about the current value of the unknown parameters, λ (neglecting higher-order terms):

$$J(\lambda + \Delta\lambda) \simeq J(\lambda) + \partial J(\lambda)/\partial\lambda \; \Delta\lambda \qquad \text{(Equation 5)}$$

where $\Delta\lambda = \lambda^{j+1 \text{ iterate}} - \lambda^{j\text{th iterate}}$. The last term on the right-hand side of Equation (5) is the scalar correction to the function J for the current values of λ, which yields an approximation to the function J evaluated at $\lambda + \Delta\lambda$. This correction term will be referred to as ΔE to give

$$J(\lambda + \Delta\lambda) = E + \Delta E \qquad \text{(Equation 6)}$$

First-order optimization methods use only the Jacobian gradient term, $\partial J(\lambda)/\partial\lambda$, to calculate ΔE. If second derivative terms are used, then the method is referred to as second order. A suitable change, ΔE, that *always* results in a decrease in J is found in the case of the gradient methods by evaluation of the Jacobian, $\partial J(\lambda)/\partial\lambda$:

$$\Delta E = \partial J(\lambda)/\partial\lambda \; \Delta\lambda = -\alpha \, | \, \partial J(\lambda)/\partial\lambda \, | \qquad \text{(Equation 7)}$$

where α is a scalar constant ($\alpha > 0$) that defines the size of the change in λ. That is, we select $\Delta\lambda = \lambda^{j+1 \text{ iterate}} - \lambda^{j\text{th iterate}} = -\alpha \, \partial J(\lambda)/\partial\lambda \, / \, |\partial J(\lambda)/\partial\lambda|$. Iterations on this procedure are continued until the change in J becomes exceedingly small, in which case J is said to have converged to a minimum and the values of the model parameters are obtained.

2.4. WEIGHTING OF DATA

The selection of an appropriate weight for each datum will strongly affect the final estimates for unknown model parameters.[8] Under most circumstances it can be shown[16] that the best linear unbiased parameter estimates are obtained when each datum is weighted by the inverse of the variance of the observation, $\sigma_{y_i}^2$:

$$w_i = 1/\mathrm{var}_i = 1/\sigma_{y_i}^2 \qquad \text{(Equation 8)}$$

Unfortunately, in most cases the analyst never has enough information to completely specify the true variance associated with a datum. If, however, enough independent replicates have been taken at each time point, an estimate of the variance can be made, $s_{y_i}^2$. Such an experimental design is both costly and difficult to implement.

An acceptable alternative is to obtain replicates at a few time points and then construct a plot of the mean versus variance for all replicates.[17,18] This information can be used to develop a model of the variance of the measured data. Examples of variance models are

$$\mathrm{var}_i = a + b\,y_i + c\,y_i^d \qquad \text{(Equation 9)}$$

The selection of values for the scalar terms in the variance model is problematic. Use of $a \neq 0$ and $b = c = 0$, can be appropriate if the analyst believes that all measurements are known with equal certainty. However, this is rarely true. Data from drug disposition studies with highly effective and specific target-oriented drugs come from sensitive analytical methodologies with error variances that are nonconstant over the domain of the standard calibration curves employed. In such cases, alternatives have been suggested based on statistical distributions of random variables. For example, when y_i is obtained from counting radioactivity, the variance of the observation has been shown to vary with magnitude and thus to approximate a Poisson distribution. Therefore, $s_{y_i}^2 = E(y_i)$, and one can use a weight equal to $1/E(y_i)$. This corresponds to values of $a = c = 0$ and $b = 1$ in Equation (9). When y_i displays a constant coefficient of variation in the domain of the measured values, resulting, perhaps, from an analytical procedure that entails an aliquoting process, the variance may be taken as proportional to $E[(y_i)]^2$.

The appropriate weight would then be $1/E[(y_i)]^2$, reflecting an integer "d" value of 2. Even though the integer values of 1 or 2 attempt to approximate $\sigma_{y_i}^2$, in many cases noninteger values of "d" may be a better approximation of the true variance.[18]

The scalar terms a, b, c, and d must be obtained by fitting the variance model [Equation (9)] to the standard deviations calculated from the replicates. After the parameters of a variance model have been defined, the weight for each datum can be calculated. The advantage of this approach is that all of the replicated data can be used to estimate a global variance model for the measured data versus using only a few replicates at each measurement time. Jacquez and Norusis[19] suggest that for replicate sizes of less than 10, the variance model approach should be used. For replicate sizes equal to or greater than 10, the standard deviation of the replicates is a good estimate of the variance of the ith datum.

An alternative variance model has been proposed by Nichols and Peck,[20] which attempts to estimate the parameters of the mathematical model of the system (λ or θ) concurrently with the parameters of the variance model (a,b,c,d). They propose the following variance model:

$$var_i = a + b\ f(x(t_i),\lambda) + c\ f(x(t_i),\lambda)^d \qquad \text{(Equation 10)}$$

At each step of the parameter estimation process, the weights are recalculated based on the updated values for parameters of the system model and parameters of the variance model.[21,22] The updating process is frequently referred to as iterative reweighting.

Note that in Equation (10) the observed measurement y_i has been replaced by the model calculated value $f(x(t_i),\lambda)$. In principle this is similar to selecting the expected value of the observation to weight the data. In this case, however, the expected value is obtained from the model calculated value and not the mean of the replicates for each time point. This approach can prove beneficial as it guards against possible outliers in the measured data.[23]

When the proposed model fits the data quite well, the observed data and the model calculated data will agree and the distinction between Equations (9) and (10) will be unimportant. When this situation does not exist, as is often the case when different models are being tested or good parameter estimates as initial guesses are not available, the task of obtaining model parameter estimates *and* variance parameter estimates can exceed the limits of many numerical optimization techniques. Regardless of which weighting scheme is selected, the individual weights used in the calculation of the final parameter estimates should be inspected to determine whether a subset of data may have been given relatively more weight, and if so, an explanation should be provided.

Other variance models have also been proposed by investigators to obtain weights for parameter estimation. Kramer *et al.*,[24] while analyzing their data on pharmacokinetics of dioxin, employed a nonstandard weighting algorithm ($w_i = 351e^{-0.29 \, y_i}$) based on a separate study of the variance of their digoxin assay. Berman and Weiss[2] used $1/y_i + 1/0.5y_i^2$ as the weighting in their analysis. MINIQUE (a minimum quadriatic unbiased estimate) has been developed by Rao[25] for linear equations for determining appropriate weights. One can also model the variance as a function of time. We suggest that the reader review the work from several reports[18,19,23–25] on the influence of various weighting algorithms on the accuracy and precision of model parameters estimates before selecting an appropriate weight for analysis of pharmacokinetic data. It is important at this juncture to note that the observation variance is often contaminated by multiple sources of random variation, and therefore, the residual variance following regression analysis of drug disposition data is not only due to error from analytical variance but also from deviations due to a priori model misspecification.

2.5. Model Validity

When a mathematical model of a system under investigation is proposed, one often is not certain about its appropriateness for that application. Judging the appropriateness of a model depends on a clear understanding of the concept of a model and its relationship to experimental data. In the context of this chapter a model has been defined as a prototype of a biological system. As such, a model refers to any set of equations that under certain conditions and for certain purposes provides an adequate description of the physical system. Models are frequently challenged by the notion that any model is an oversimplification of the real system. This challenge is easily dismissed by noting that simplification is the hallmark of most models. In other words, a model is valid insofar as it extracts some important feature of the system into a form that can be analyzed more easily.

Most commonly, with data collected from clinical trials, the aim of the analysis is to obtain model parameters that adequately describe the data and then to use this information to predict the behavior of the system over future experimental frames. In such cases significant effort should be directed first at examining whether the model is sensible and second at obtaining statistically valid estimates for unknown model parameters.

The most common approach to judging the validity of a model is to examine the unexplained or residual sum of squares. The residual, ε_i, is the difference between the ith observation and the corresponding model predicted value.

$$\varepsilon_i = y_i - f(x(t_i),\lambda) \qquad \text{(Equation 11)}$$

As noted earlier, the underlying assumptions for regression analysis assume ε_i to be independent random variables from a known statistical distribution. Consequently, if the chosen model is appropriate for the data, the observed residuals should then reflect the properties assumed for the distribution of ε_i. This forms the basis for residual analysis, a highly useful means of examining the appropriateness of a model.

Analysis of drug disposition data by fitting models to the data demands examination of residuals before declaring model adequacy. Residual analysis can be used to (1) evaluate departures between model and data, (2) reveal deficiencies of experimental design, (3) indicate whether the basic assumptions about data weighting are consistent, (4) detect outliers in the regression problem, and (5) detect correlation of model parameters. Excellent reviews on this subject can be found in the works of Anascombe and Tukey[26] and Mosteller and Tukey.[27]

Figure 1 shows an example of a plot of residuals against the independent variable, usually time, in a pharmacokinetic study. The purpose of this analysis was to examine the validity of the assumption of constancy in the variance of error terms. It can easily be seen that as the magnitude of the independent variable increases, so does the spread of residuals. One can also encounter error variances decreasing with increasing levels of the independent variable. Nonconstancy in error variance can also be effectively studied by plotting ε_i as a function of the model predicted value $f(x,\lambda)$.[21,22] Therefore, it is important that simple residual plots be examined routinely by plotting weighted residuals or standardized residuals as a function of time following initial model fitting to the pharmacokinetic data. In general, an overall horizontal band of residuals around the value zero, displaying no systematic tendencies to be positive or negative, would suggest random error in data and an appropriate model. If the horizontal nature of the residual band is seriously violated, heterogeneity of variance is suggested. Proper weighting, selecting alternative models, and transforming the observed variable are remedial measures for dealing with such heterogeneity of error variance.

Major departures from normality of error terms can be examined by plotting cumulative distribution of weights or standardized residuals.[22] The cumulative frequency distribution plots can be constructed

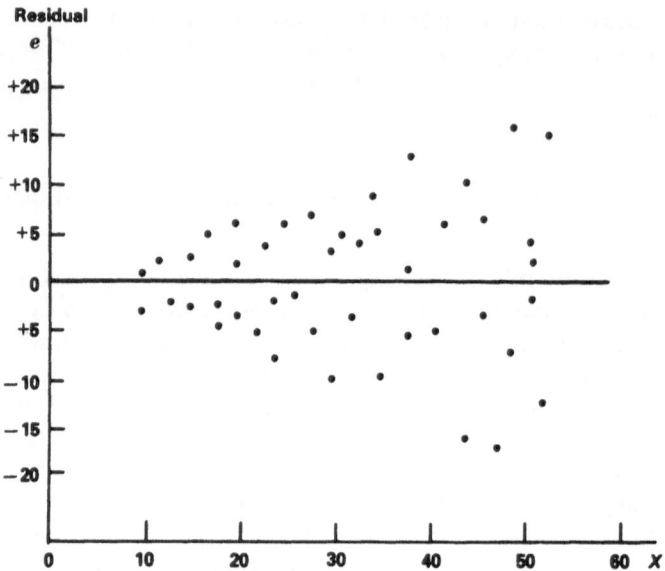

FIGURE 1   A plot of residuals, obtained following model fitting, against an independent variable.

on normal probability paper. Figure 2 shows such a plot, which should be linear if the distribution is normal. Substantial departures from a straight line are grounds for suspecting that distribution is not normal. One must, however, be careful in evaluating this plot, as random variation can be misinterpreted if the sample size is too small. Other statistical tests include the chi-square and Kolmogorov-Smirnov tests.[28] Such tests can help an analyst detect outliers in drug disposition data. To test whether the residuals show any systematic tendencies, nonparametric tests such as a Wilcoxin rank-sum test or a "run test" are used.[29] These tests do not have any distributional assumptions and hence are applicable to any model, linear or nonlinear in parameters. Details on other statistical tests on residuals can be obtained from the reports of Weber and Monarchi.[29]

When a fit of the model to drug concentration versus time data provides a satisfactory randomness of scatter and increasing the number of parameters does not significantly reduce the sum of weighted squared deviations, the model may then be considered "consistent."[8] Although a model is "consistent," it may still be "nonunique" or ill conditioned. This is true when the information contained in the data is insufficient to define the chosen model and more than one model could describe the data equally well. This is usually true for empirical models employed in analysis of drug disposition data.

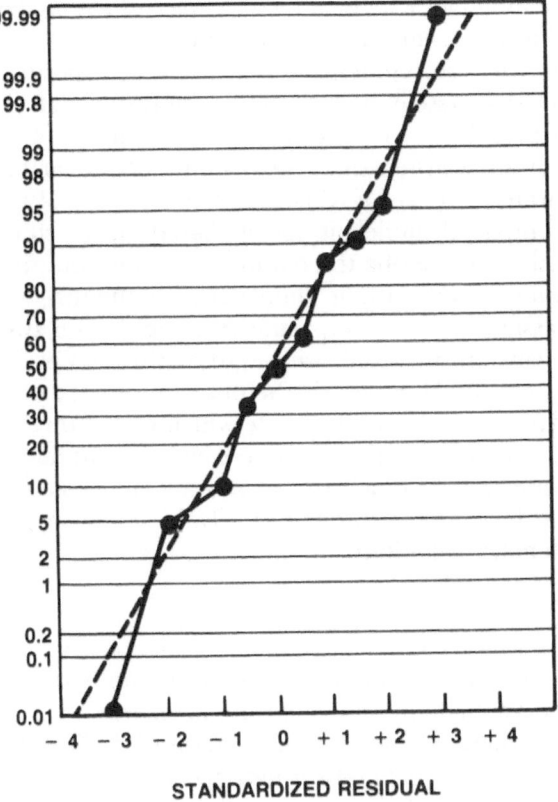

STANDARDIZED RESIDUAL

FIGURE 2   Cumulative frequency distribution of standardized residuals on normal probability paper.

2.6. MODEL DISCRIMINATION AND SELECTION

In the analysis of pharmacokinetic data obtained from drug disposition studies, equations with sums of exponentials are usually fitted to data. However, in any kind of modeling analysis, particularly in compartmental analysis, it is important that one should review the reasons for undertaking model building. The selection of appropriate equations for a pharmacokinetic model should depend predominantly on the use for which it is intended. If the sole purpose is to obtain a smooth curve to predict system behavior within some range of experimental conditions, as is usually the case in the analysis of pharmacokinetic data obtained from clinical trials, one usually employs empirical models. However, when it is desired to understand the underlying mechanisms and predict the system behavior beyond the experimental range, a mechanistic or compartmental model may be a better choice. Such models can also be very useful in the design of new experiments.

Wagner[30] has recommended that if drug disposition can be described by a linear system, it may not be necessary to determine the right model. However, it may be extremely important to obtain as complete a model as possible when the system shows nonlinear behavior, to be able to accurately forecast a response. The selection of a sum of exponentials or empirical models is usually based on the number of exponentials needed to describe the plasma drug concentration versus time curve following the drug input. If only one exponential term is involved, we usually assign the one-compartment model, and if two or three exponential terms are involved, we assign a two- or three-compartment model, respectively. However, if the observations are obtained in only one compartment, one cannot decide which of the three two-compartment models or which of the 13 three-compartment linear mammillary models actually applies to a given set of data. This dilemma is imposed on an analyst because the number of solvable rate constants, R, as suggested by Benet,[31] is limited.

$$R = 2n - 1 \qquad \text{(Equation 12)}$$

Where n represents the number of driving compartments, given the observations are made in only one compartment.

Although there are no absolute criteria regarding the selection of a model, two goals are generally sought when selection is for analysis of drug disposition data. It is necessary that the scatter of the observed data, in residual analysis, be randomly distributed around the fitted equation or model response, and second, the sums of weighted squared deviations should be minimal. To test whether the sums of weighted squared deviations have been sufficiently reduced to justify fitting with additional parameters, the standard F test[32] from multiple linear regression may be used[4]:

$$F' = \frac{(WSS_p - WSS_q)}{WSS_q} * \frac{df_q}{df_p - df_q} \qquad \text{(Equation 13)}$$

where WSS_p is the residual weighted sum of squared deviations obtained with the pth set of parameters and WSS_q is that obtained with q^{th} set of parameters; df is the degrees of freedom and is equal to the total number of observations used in curve fitting minus the number of parameters of the model ($df_p > df_v$). Under some general conditions, F' asymptotically has an F distribution with $df_p - df_v$ degrees of freedom. Then, if the F statistic is significant at 5% level (one-tailed), one accepts the model with the greater number of parameters. For discrimination among models, where precision in parameter estimates is of utmost importance, it is usually necessary to have a substantially larger number of observations than required for fitting model equations to data.

The second criterion in model selection can be evaluated by studying scatter by the analysis of residuals as described earlier.

2.7. VARIANCE ESTIMATES FOR MODEL PARAMETERS

Introducing errors while making measurements is almost unavoidable. Therefore, it is essential to ensure that they are relatively small in magnitude and random in nature. If errors are large, parameter estimates will be imprecise, though not necessarily inaccurate, and if they are nonrandom, the parameter estimates can be inaccurate and biased. For a parameter estimate to have any validity, uncertainties must be calculated and quoted. If drug disposition studies from different laboratories report different parameter estimates from similar experimental protocols without providing measures of their precision, it is impossible to know if their findings are challenged or confirmed. Several scientific issues related to drug disposition in the pharmacokinetic literature are confounded by problems where elaborate theories get built on dubious information.

The uncertainties associated with parameter estimates can be computed and expressed as either a standard deviation, coefficient of variation, and/or a 95% confidence interval. Accurate measures of parameter uncertainty can always be found by replicating experiments to produce many samples of data used to calculate sets of parameter estimates. These sets of parameter estimates are used to compute means and variances. Monte Carlo simulation has been used to investigate the statistical proporties of model equations and the role of experimental error or measurement uncertainties in parameter uncertainty. Figure 3 shows the results of a series of Monte Carlo simulations using a model with two exponentials, where the variation of uncertainty in parameter estimates (expressed as % coefficient of variation, %CV) is plotted as a function of experimental uncertainty (experimental standard deviation, SDEV). This technique, when appropriately applied to design of experiments, can lead to significantly better parameter estimation at the end of a clinical trial evaluating a pharmacotherapeutic agent.

Estimates of parameter variance can also be obtained from least-squares theory. We state without proof that an estimate of the covariance

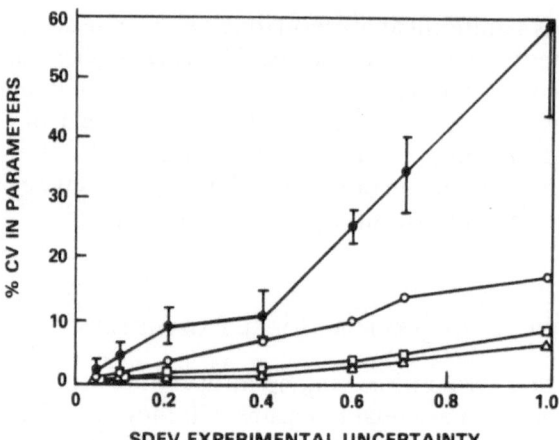

FIGURE 3   Results of a series of Monte Carlo simulations showing uncertainty in parameter estimates (% CV) as a function of experimental uncertainty (SDEV). Error bars equal one standard deviation and are shown for the estimate of parameter A2 only. ○, Parameter A1; ●, A2; △, A3; □, A4. Model: Y = A₁ · exp(− A₂ · +) + A₃exp(− A₂ · t). From McIntosh and McIntosh, 1980.[48]

matrix, V_λ, for unknown model parameters, λ, of a compartmental model is given by

$$V_\lambda = \frac{e'e}{m-n} (g'g)^{-1} \qquad \text{(Equation 14)}$$

where

$$e'e = \sum_{i=1}^{m} [y_i - f(x)(t_i),\lambda)]^2$$

and the prime stands for matrix transpose, m is the number of measurements, p is the number of unknown model parameters (i.e., the elements of the matrix λ), and g is the $n \times n$ matrix of partial derivatives:

$$g \equiv \{\partial f(x(t_i),\lambda)/\partial\lambda\} \quad i=1,m \qquad \text{(Equation 15)}$$

From these equations we can see that one way of getting a large variance estimate for a parameter is for any element in the $(g'g)$ matrix to have a small value [e.g., such that when the inverse matrix is calculated $(g'g)^{-1}$, a large value will occur]. This means that the change in model solution following a change in model parameter [i.e., calculation of the partial derivative $\partial f(x(t_i),\lambda)/\partial \lambda)$] will have a small value. In other words, the model solution will not be sensitive to that unknown parameter, and little or no information about the value of that parameter can be obtained from the experimental data. In such situations the parameter is commonly considered as nonidentifiable.[33] Such cases usually result from inappropriate data collection and/or model misspecification. If the parameter cannot be shown a priori to be identifiable, then its value must be obtained from other sources.

2.8. CORRELATION OF MODEL PARAMETERS

If during the analysis of drug disposition data an analyst finds the search for estimates of unknown model parameters to be slow to converge to a minima and the estimated error in one or more model parameters to be much larger than would be expected from experimental uncertainty in the data, the parameters are said to be "poorly determined." This situation can occur when the gradient calculation [Equation (15)] yields conflicting or redundant information, often owing to correlation among unknown model parameters.

The correlation of unknown model parameters is a common feature of models containing exponential terms, e.g., those most frequently employed in the analysis of pharmacokinetic data today. The problem is greatly exacerbated as one increases the number of exponentials used in the data analysis. Conditions easily arise when two exponential terms take on equal or nearly similar values for exponents and thus report redundant information. Consequently, analyses where a sum of many exponential terms is fitted to data *must* include estimates for the variance of the unknown model parameters before the credibility of those parameter estimates can be assessed. Parameter identifiability and/or parameter correlation must be regularly evaluated when any proposed model is employed in fitting pharmacokinetic data or forecasting the performance of drug-dosing regimens.

3. NONDETERMINISTIC APPROACH

Nondeterministic models deal with systems that are stochastic in nature. Predictions for such models can only be made in terms of a probability distribution. In the limit of many simulations of a stochastic system we recover the mean response or the deterministic response discussed earlier.

Stochastic variation may occur in numerous ways and as a result has led to much confusion in the literature. We shall briefly discuss three forms of such variation.

3.1. MARKOVIAN APPROACH

In the first example, we formulate the differential equations of the proposed compartmental model as a simple birth–death process.[34] Such a process can be viewed as systems with a short memory; behavior at time t depends only on the preceding instant. Hence, each particle entering any compartment at time t has a well-defined probability of leaving that compartment by the time $t' > t$. The mean time to exit usually has an exponential distribution function, and the rate constant describing the fractional rate of exit of material from a compartment per unit time is the *mean* time to exit. The actual time when a particle leaves a compartment is a random variable that must be selected from a known (exponential) probability distribution. Since the matrix of transport rate constants, λ, completely defines the kinetic characteristics of the system, it can determine the distribution of particles over time in a compartmental system. In fact it does; the product $\lambda_{ij}\Delta t$ represents, to first order in Δt, the probability that a particle will transfer from compartment j to compartment i in Δt units of time.[35–37] The quantity $(1 - \Delta t$ times the sum of rate constants leaving the ith compartment) represents the probability that a particle remains in compartment i in Δt. The fate of a particle in the ith compartment in the interval from t to $t + \Delta t$ is thus described as follows;

$$x_k(t + \Delta t) = x_k(t)(1 - \lambda_{kk}\Delta t) + \sum_{m \neq k} \lambda_{km}x_m(t)\Delta t \quad \text{(Equation 16)}$$

which, for Δt approaching 0, gives

$$\dot{x}_k(t) = \sum_{m \neq k} \lambda_{km}x_m(t) - \lambda_{kk} x_k(t) \quad \text{(Equation 17)}$$

where

$$\lambda_{kk} = \sum_{m \neq k} \lambda_{mk}$$

Equation (17) has the form of a system of linear ordinary differential equations with constant coefficients. In fact, Equations (1) and (17) are equivalent. The solution of Equation (17) can be used to obtain the mean behavior of the system. We state without proof that the mean and variance of $x_k(t)$ are

$$E[x_k(t)] = x_k(0) \, e^{-\lambda kk^t} + \sum_m \lambda_{mk} \, \lambda_{kk} \, E[x_m(t)]e^{-\lambda kk^t} \qquad \text{(Equation 18)}$$

$$\sigma[x_k(t)] = E[x_k^2(t)] - \{E[x_k(t)]\}^2$$

$$= x_k(0) \, e^{-\gamma kk^t} (1 - e^{-\lambda kk^t}) + \sum_m \lambda_{mk} \, \lambda_{kk} \, E[x_m(t)]e^{-\gamma kk^t}$$

$$\text{(Equation 19)}$$

Numerous authors have examined the covariance structure of stochastic compartmental models.[38–41] They suggest that a useful measure of the importance of introducing a nondeterministic model of the type described here is the coefficient of variation (CV = $\sigma[x_k(t)]/E[x_k(t)]$). Resigno and Matis[39] show that for a simple one-compartment system

$$\frac{\sigma[x_k(t)]}{E[x_k(t)]} = \frac{\text{sqrt}[x_o c^{-\lambda t}(1 - e^{-\lambda t})]}{x_o e^{-\lambda t}} = \text{sqrt}\left[\frac{1 - e^{\lambda t}}{x_o e^{\lambda t}}\right] \qquad \text{(Equation 20)}$$

The CV will depend on x_o, the number of particles initially introduced into the compartment. Clearly for a large x_o the CV is very small. Large CVs depend on the relative values of x_o, λ, and t. In the latter case the analyst must determine whether the covariance of the model response is small enough to warrant use of the simpler deterministic model.

3.2. POPULATION STATISTICS

An alternative approach, which has been popularized by Sheiner et al.,[42] treats model parameters as random variables obtained from a global or population distribution. Each model parameter consists of a true value plus some uncertainty, η:

$$\theta_1 = \text{model parameter} = \text{mean value} + \eta \qquad \text{(Equation 21)}$$

The variance of the model parameter across the population, $\sigma^2(\theta_1)$, equals the variance of η. Experimental responses of subjects will be different, in part, because the model parameters that define biodistribution kinetics will be slightly different owing to random variation. The mean response of the model is defined in terms of an analytical function of the model parameters:

$$y(t)_{observed} = f(\theta_1,\theta_2, \ldots \theta_n,t) + \varepsilon_{observed} \qquad \text{(Equation 22)}$$

where ε is the measurement error and θ_i are *estimates* of the true model parameters. The expected value of the response, $E[y(t)]$, is $f(\theta_1,\theta_2, \ldots \theta_n,t)$. The covariance matrix for the model parameters is

$$V_\theta = \begin{vmatrix} \sigma^2(\theta_1) & \text{cov}(\theta_1,\theta_2)\ldots \\ \text{cov}(\theta_2,\theta_1) & \sigma^2(\theta_2) \quad \ldots \\ & \ldots \ldots \\ & \ldots \ldots \\ \text{cov}(\theta_n,\theta_2) & \text{cov}(\theta_n,\theta_2)\ldots \end{vmatrix} \qquad \text{(Equation 23)}$$

The variance of the response $\sigma^2(y(t))$ equals the variance of a small perturbation about the solution $\delta y(t)$ and is denoted $\sigma^2[\delta y(t)]$. To estimate the variance, $\sigma^2[y(t)]$, we proceed first by calculating $\delta y(t)$ and then using this to estimate the variance of $\delta y(t)$ (dropping the notation of time):

$$\delta y = \partial y/\partial\theta_1\, \delta\theta_1 + \partial y/\partial\theta_2\, \delta\theta_2 + \ldots + \partial y/\partial\theta_n\, \delta\theta_n \qquad \text{(Equation 24)}$$

Since δy is a linear function of the $\delta\theta_i$, we can apply rules for the propagation of errors in linear relations:

$$\sigma^2(\delta y) = = \begin{vmatrix} \partial y/\partial\theta_1 & \partial y/\partial\theta_1 \ldots \partial y/\partial\theta_1 \end{vmatrix} \begin{vmatrix} V_{\delta\theta} \end{vmatrix} \begin{vmatrix} \partial y/\partial\theta_1 \\ \partial y/\partial\theta_2 \\ . \\ . \\ . \\ \partial y/\partial\theta_n \end{vmatrix} \qquad \text{(Equation 25)}$$

Since $\sigma^2(\delta y) = \sigma^2(y)$ and $V_{\delta\theta} = V_\theta$, we can easily calculate the variance of the response, $\sigma^2(y)$, as a function of the population variance of the model parameters, V_θ, and the partial derivatives of the response with respect to the model parameters, $\partial y/\partial\theta_i$.

For example, if we propose a one-compartment model of a system with unknown model parameters θ_1 and θ_2,

$$y(t) = [x_o/\theta_1]\, e^{-\theta_2\, t} \qquad \text{(Equation 26)}$$

and defining $a_1 = \partial y/\partial\theta_1 = -[x_o/\theta_1^2]e^{-\theta_2 t}$ and $a_2 = \partial y/\partial\theta_2 = -t[x_o/\theta_1]e^{-\theta_2 t}$, we can use Equation (25) to obtain

$$\sigma^2(y) = a_1[a_1\sigma^2_{\theta_1} + a_2\text{cov}(\theta_1,\theta_2)]$$
$$+ a_2[a_1\text{cov}(\theta_1,\theta_2) + a_2\sigma^2_{\theta_2}] \quad \text{(Equation 27)}$$

From this example we can see that the introduction of random variation in model parameters can be handled in a rather straightforward manner. However, the inclusion of random variation in a model forces the user to have some knowledge of the variance *and* covariance of model parameters.

The process of estimating unknown model parameters using this approach is simply a matter of minimizing the sum of squared weighted residuals [Equation (4)]. The weight for each observation is determined from the variance estimated for each observation. The sum of weighted squares will thus be an implicit function of the unknown system parameters and the unknown variance model parameters.

From these results we can conclude that the Markovian approach discussed earlier and the current approach differ primarily in the estimation of the *variance* of the model response. This can be seen easily by comparing the one-compartment variance calculation [Equation (19)] with the above result [assuming $\sigma^2_{\theta_1}$ and $\text{cov}(\theta_1,\theta_2)$ are zero]. The mean response for the one-compartmental model will be identical.

3.3. KALMAN FILTERING

Both Markovian and population approaches are useful advances to the more common deterministic analyses. They attempt to introduce into the analysis the fact that no biological system is exactly predictable. The mode by which the uncertainty enters the system is, however, quite different for the two cases. The population approach accepts as a model the exact solution of a differential equation. The parameters of this solution are treated as random variables that are specified with reference to a population distribution. Having selected the true parameters for a given sample of that population, the model then loses that component of its stochastic behavior and for most practical purposes becomes equivalent to a deterministic model.

The Markovian approach attacks the nondeterministic issue from a much different perspective. As in the case of the population approach, randomness enters the system via model parameters. In the Markovian approach their effect, however, is manifest *during* the dynamic evolution of the system response.[34] As such, the differential equations describing the system response include as an input a forcing function, which can neither be controlled nor be modeled deterministically. Special methods for solving these differential equations have been developed; the two most widely used are the Ito and the Stratonovich calculus.[43,44]

An approach to treating mean and variance estimations in systems modeled by differential equations was proposed by Kalman[45] and has been adopted by others.[46,47] This method brings together the static elements of the population approach with the dynamics of stochastic differential equations and is referred to as Kalman filtering. The most important difference between the population approach and Kalman filtering is that the latter takes into account the variation in the model's response over time. Therefore, Kalman filtering can be used to estimate the unknown parameters of a pharmacokinetic model, to monitor any changes in these parameters that may occur during treatment, and finally to predict the mean and variance of the patient's response to continued therapy.

To better understand the basic ideas behind the Kalman filter, we will consider monitoring a drug response. Suppose the measurement of a drug level yields a value z_1. The best estimate of the true drug level (denoted x) is then

$$x_1 = z_1 \qquad \text{(Equation 28)}$$

and the variance of the error of this estimate is

$$\sigma^2_{x_1} = \sigma^2_{z_1} \qquad \text{(Equation 29)}$$

If another measurement is made at the same time, z_2, with a higher or lower precision instrument, the best estimate of drug level can be obtained by weighting each observation by an averaged variance:

$$\mu = [\sigma^2_{z_1} /(\sigma^2_{z_1} + \sigma^2_{z_2})]z_1 + [\sigma^2_{z_2} /(\sigma^2_{z_1} + \sigma^2_{z_2})]z_2 \qquad \text{(Equation 30)}$$

$$= z_1 + [\sigma^2_{z_1}/(\sigma^2_{1} + \sigma^2_{z_2})][z_2 - z_1]$$

or if we define a Kalman gain as $K = \sigma^2_{z_1}/(\sigma^2_{z_1} + \sigma^2_{z_2})$, we get a solution of the form

$$\mu = z_1 + K [z_2 - z_1] \qquad \text{(Equation 31)}$$

Similarly, the variance of this mean can be calculated as

$$1/\sigma_x^2 = 1/\sigma^2{}_{z_1} + 1/\sigma^2{}_{z_2} \qquad \text{(Equation 32)}$$

or

$$\sigma_x^2 = 1/\sigma^2{}_{z_1} - K/\sigma^2{}_{z_1}$$

The careful reader will note that this approach is nothing more than weighted least-squares. In this form, the Kalman filter just recasts the variance-weighted least-squares problem into an iterative format. With such an approach the current estimates for mean and variance of the system response can be continuously updated as new data are collected.

The previous discussion considers only the static problem. Now we consider the dynamic problem. Suppose that the time course of the administered drug can be described by a one-compartment model. In this model the rate constant for excretion is changing over time because of improvements in the patient's health. We might propose the following model for this sytem:

$$dx(t)/dt = - a\, x(t) \qquad \text{(Equation 33)}$$
$$da(t)/dt = k \qquad + v(t)$$

where a is the excretion rate constant that is changing at the rate k. The term v(t) represents the nondeterministic noise in the system. Without going into the details, systems of this nature can be readily evaluated using a Kalman filter to estimate the mean and variance of the rate constant a and the drug level x over time. Such a tool can be useful for anticipating a patient's response and adjusting therapy accordingly.

We have examined some stochastic approaches to the mathematical analysis of biological data using compartmental techniques. All these approaches are consistent in the mean with their deterministic counterparts. In most cases the introduction of stochastic elements into the analysis permits a greater understanding of the uncertainty of the system. In many cases the covariance structure of a system can yield information that will assist pharmacokineticists in designing better treatment protocols.

ACKNOWLEDGMENT.   The authors would like to thank Ms. Karen Marconi for her helpful assistance in typing this manuscript.

REFERENCES

1. Aris R: *Mathematical Modeling Techniques.* London, Pitman Publishing Limited, 1979.
2. Berman M, Weiss MF: *SAAM-27: User's Manual.* NIH Publication No. 78-180. Bethesda, MD, National Institutes of Health, 1978.
3. Knott GD: MLAB—A mathematical modeling tool. *Comp Programs Biomed* 1979; 10:271.
4. Jacquez JA: *Compartmental Analysis in Biology and Medicine.* Amsterdam, Elsevier, 1972.
5. Yamaoka Y, Nakagawa T, Uno T: Statistical moments in pharmacokinetics. *J Pharmacokin Biopharm* 1978; 6:547–558.
6. Covell DG, Berman M, DeLisi C: Mean residence time—theoretical development, experimental determination and practical usage in tracer analysis. *Math Biosci* 1984; 72:213–244.
7. Lawson CL, Hanson RJ: *Solving Least Squares Problems.* Englewood Cliffs, NJ, Prentice-Hall, 1974.
8. Berman M: The formulation and testing of models. *Ann NY Acad Sci* 1963; 108:182–194.
9. Doob JL: *Stochastic Processes.* New York, John Wiley & Sons, Inc, 1953.
10. Lambrecht RM, Resigno A: *Tracer Kinetics and Physiologic Modeling.* Berlin, Springer-Verlag, 1983.
11. Boston RC, Greif PC, Berman M: Conversational SAAM—An interactive program for kinetic analysis of biological systems. *Computer Programs Biomed* 1981; 13:111–119.
12. Bard, Y: *Nonlinear Parameter Estimation.* New York, Academic Press, Inc, 1974.
13. Daniel C, Wood FS: *Fitting Equations to Data.* New York, Wiley-Interscience, 1971.
14. Wilde DJ, Beightler CS: *Foundations of Optimization.* Englewood Cliffs, NJ, Prentice-Hall, 1967.
15. Marquardt DW: An algorithm for least-squares estimation of nonlinear parameters. *Society of Industrial and Applied Mathematics Journal* 1963; 11:431.
16. Scheffe H: *The Analysis of Variance.* New York, John Wiley & Sons, Inc, 1959.
17. Rodbard D, Lenox RH, Wray HL, Ramseth D: Statistical characterization of the random errors in the radioimmunoassay dose–response variable. *Clin Chem* 1976; 22(3):350–358.
18. Finny DJ, Phillips, P: The form and estimation of a variance function with particular reference to radioimmunoassay. *Appl Statist* 1977; 26(3):312–320.
19. Jacquez JA, Norusis, M: Sampling experiments on the estimation of parameters in heteroscedastic linear regression. *Biometrics* 1973; 29:771–779.
20. Nichols AI, Peck, CC: *ELSNLR-Users Manual,* Technical Report No 5, Division of Clinical Pharmacology. Bethesda, MD, USUHS, 1981.
21. Draper NR, Smith H: *Applied Regression Analysis.* New York, John Wiley & Sons, Inc, 1963.
22. Netter J, Wasserman W: *Applied Linear Statistical Models.* Homewood, IL, RD Irwin, 1974.
23. Berman M: Information content of data with respect to models. *Am J Physiol* 1983; 245:620–623.
24. Kramer WG, Lewis RP, Cobb TC, et al: Pharmacokinetics of digoxin: Comparison of a two and a three compartment model in man. *J Pharmacokin Biopharm* 1974; 2:123.
25. Rao CR: Estimation of heteroscedastic variances in linear models. *J Am Stat Assoc* 1970; 65:161–172.
26. Anscombe FJ, Tukey JW: The examination and analysis of residuals. *Technometrics* 1963; 5:141.
27. Mosteller F, Tukey JW: *Data Analysis and Regression.* Reading, MA, Addison-Wesley, 1977.

28. Siegel S: *Nonparametric Statistics for the Behavioral Sciences.* New York, McGraw-Hill, 1956.
29. Weber JE, Monarchi DE: Performance of the Durbin-Watson test and WLS estimation when the disturbance term includes serial dependence in addition to first-order autocorrelation. *J Am Stat Assoc* 1982; 17:117.
30. Wagner JG: Do you need a pharmacokinetic model, and if so, which one? *J Pharmacok Biopharm* 1975; 3:457–478.
31. Benet LZ: General treatment of linear mammillary models with elimination from any compartmental as used in pharmacokinetics. *J Pharm Sci* 1972; 61:536–541.
32. Mandel J: The statistical analysis of experimental data. New York, Interscience, 1964.
33. Cobelli C, DiStefano JJ: Parameter and structural identifiability concepts and ambiguities: A critical review. *Am J Physiol* 1980; 239:R7–R24.
34. Bharucha-Reid AT: *Elements of the Theory of Markov Processes and Their Applications.* New York, McGraw-Hill, 1960.
35. Feller W: *An Introduction to Probability Theory and Its Applications.* New York, John Wiley & Sons, Inc, vol I, 1968.
36. Bailey NTJ: *The Elements of Stochastic Processes with Applications to the Natural Sciences.* New York, John Wiley & Sons, Inc, 1964.
37. Matis JH: On the stochastic theory of compartments: Solution for n-compartment systems with irreversible time-dependent transition probabilities. *Bull Math Biol* 1974; 36(5/6):489–504.
38. Purdue P: Variability in a single compartmental system: a note on SR Bernard's model. *Bull Math Biol* 1981; 43:111–114.
39. Rescigno A, Matis JH: On the relevance of stochastic compartmental models to pharmacokinetic systems. *Bull Math Biol* 1981; 43:245–255.
40. Matis JH, Wehrly TE: Stochastic models of compartmental systems. *Biometrics* 1979; 35:199–207.
41. Robertson JS: *Compartmental Distribution of Radiotracers.* Boca Raton, FL, CRC Press Inc, 1983.
42. Sheiner LB, Rosenberg B, Marthae VV: Estimation of population characteristics of pharmacokinetic parameters from routine clinical data. *J Pharm Biopharm* 1977; 5(5):445–479.
43. Stratonovich RL: A new representation for stochastic integrals and equations. *Society of Industrial and Applied Mathematics Journal Control* 1966; 4:362–371.
44. Mortensen RE: Mathematical problems of modeling stochastic nonlinear dynamic systems. *J Stat Physics* 1969; 1(2):271–296.
45. Kalman RE: A new approach to linear filtering and prediction problems. *Trans ASME (J Basic Eng)* 1960; 82D:35–45.
46. Gelb A: *Applied Optimal Estimation.* Cambridge, MA, MIT Press, 1974.
47. Maybeck PS: *Stochastic Models, Estimation and Control.* New York, Academic Press, Inc, vol 1, 1979.
48. McIntosh JA, McIntosh RP: *Monographs on Endocrinology, vol 16: Mathematical Modelling and Computers in Endocrinology.* Heidelberg, Springer-Verlag, 1980, p. 95.

CHAPTER 21

ANALYTICAL METHODS

LAWRENCE J. LESKO

1. INTRODUCTION

It is becoming increasingly obvious in clinical studies of the elderly that the quantitation of drugs and/or their metabolites in biological fluids should be an integral part of trial designs.[1] Serum drug concentrations often provide unique answers to important questions pertaining to the pharmacokinetics and pharmacodynamics of therapeutic agents. Good drug assay data may be helpful not only in elucidating the disposition characteristics of drugs, but also in the discovery of metabolites as new therapeutic moieties. Serum drug concentrations may correlate with therapeutic or toxic effects of drugs and that may help define mechanisms of drug action. The technological revolution in analytical methodology occurring during the past decade has provided the clinical investigator with an exciting array of assay tools for identification and quantitation of drugs and metabolites in various body fluids. To utilize this technology optimally the clinical investigator needs to have a basic understanding of the principles of drug analysis and quality control. It is important to recognize that the overall inherent quality of a clinical study that includes drug concentration determinations can be no better than the quality of the assay methods used in the laboratory. Clearly,

LAWRENCE J. LESKO • School of Pharmacy, Clinical Pharmacokinetics Laboratory, University of Maryland at Baltimore, Baltimore, Maryland 21201.

one must pay close attention to the details of assay procedures and any special analytical concerns invoked by geriatric clinical studies.

The objectives of this chapter are to review some of the major general considerations pertinent to the selection of analytical methods for quantifying drug and metabolite levels in various biological fluids. The focus of the chapter is not necessarily on the principles of operation of analytical instrumentation or on the advantages/disadvantages of drug extraction schemes. Rather, the emphasis is on the aspects of drug assays that are most pertinent to drug studies in the elderly.

2. INSTRUMENTATION

It is helpful to have some knowledge of the analytical systems used for the measurement of drug concentrations because they do differ in their specificity, sensitivity, and cost, and these may be important considerations in geriatric study designs. The following overview is a brief characterization of the most frequently used assay systems for drugs.

High pressure liquid chromatography (HPLC) is the most broadly applicable assay system for drugs. It is versatile and permits the assay of drugs and their metabolites simultaneously. It usually requires small sample volumes, which may be important to the geriatric patient. HPLC is particularly useful for nonvolatile drugs that cannot be assayed by gas liquid chromatography (GLC). Equipment cost is moderate, but reagent costs are relatively low. HPLCs are easy to operate, trouble-shoot, and maintain. Their performance in drug assays is typically reliable over long periods of time.

GLC was the standard for drug assays for many years. It has recently been surpassed by HPLC but it still holds an important place in the menu of assay systems. GLC has many of the advantages of HPLC discussed previously, but it has many more disadvantages. Derivitization, as a pretreatment for specimens, is often required to detect the drug of interest. Assays on GLC are often more difficult to trouble-shoot, and down-time is greater than with HPLC. Typically, GLC with electron capture detection offers greater sensitivity for many drugs than do comparable assays by HPLC.

The use of gas chromatography–mass spectrometry (GC–MS) in geriatric drug studies is increasing. Mass selective detectors and selected ion monitoring provide for extremely sensitive assays when compared to classical GLC and/or HPLC. GC–MS procedures are also valuable because they often permit identification or confirmation of the nature of drug metabolites by "mass fingerprinting." GC–MS equipment is expensive, difficult to operate and maintain, and it has relatively small sample through-put. Recently, new developments in HPLC–MS have made mass detection possible for nonvolatile drugs following their separation and resolution by HPLC, but these detection techniques are still in the research stage.

Immunoassays are attractive assay systems for geriatric drug studies, and they have expanded considerably in the last few years. Radioimmunoassay (RIA), using either ^{125}I or ^{3}H as tracer, is a very sensitive assay tool which utilizes small sample volumes. Gamma counters for ^{125}I and liquid scintillation counters for ^{3}H are automated for good sample through-put and are adaptable to many different assays using labeled drugs. Consideration must be given to specificity of antibodies when using RIA since many drug metabolites cross-react with the antibodies to increase the apparent drug concentration. The disadvantages of RIA include limited shelf-life of antibodies, limited sources of reagents, and relatively high reagent costs. Disposal of radioactive material is also a problem in some laboratories. When both ^{125}I and ^{3}H tracers are available, the former is much easier to use in quantifying drug concentrations. Nonisotopic tracers, including some with a fluorescent label, have been useful in enzyme immunoassays (e.g., EMIT), and the latter procedure may be applicable to drug studies in the elderly. Enzyme immunoassays are attractive because they require no sample preparation and very small sample volumes. EMIT assays are automatable, easy to perform, and generally reliable. However, reagent sources are limited, reagent costs are high, and immunoassays are not as versatile as HPLC or GLC in quantifying metabolite and drug concentrations simultaneously.

3. Method Development

The process of assay method development for drug studies in the elderly is similar to that for any other target population. The process should be logical and systematic. It usually begins with a thorough literature search for published assays for the drug of interest or for structurally related drugs. One then normally gathers physicochemical data about the drug with respect to its solubility in water and organic solvents, its pKa, its ultraviolet absorbance and fluorescent characteristics, and any special properties regarding stability and lability under typical assay conditions. This background information is essential to develop appropriate extraction techniques, assay procedures, and detection methods. The selection of an internal standard, depending on the assay system, may be critical. The internal standard is often a structural analog of the drug of interest, and it should have physicochemical properties similar to those of the drug to be measured. Often, unpublished assay procedures for drugs may be obtained directly from the pharmaceutical manufacturer of the drug by contacting the appropriate person in a quality control division. After preliminary information on the drug assay is generated and a specific assay procedure is chosen, the most important part of the assay is the quality control and the assay validation. During the validation process it may be relevant to consider age-related factors that may affect analytical results, although there is little information in the literature pertaining to this issue. It is well known that the composition of body fluids may be influenced to varying degrees by the aging process.[2] For example, there may be inter- and intrapatient fluctuations in endogenous substances such as blood urea nitrogen or plasma lipids. As a consequence the range of quantitative values for these substances in the elderly may be quite different than in a younger adult population. These differences may introduce random errors into drug assay procedures by interfering with the extraction and/or measurement of drug analytes. In validating an assay it may be prudent to utilize control serum from an elderly population to account for these potential sources of error.

4. QUALITY CONTROL

When selecting an analytical method for a drug study in the elderly, or in any other population, it is imperative to consider the appropriate validation of the assay procedure prior to embarking on the drug study. The basic goal of assay validation is to assure the investigator that the analytical laboratory will produce reliable drug concentration data. An ideal assay method is accurate, precise, sensitive, specific, and reproducible.[3]

Accuracy is defined as how close a measured assay value is to the true assay value. Accuracy is expressed as a precentage of the true value, and it is generally checked closely at the drug concentrations nearest the assay's limit of sensitivity. *Precision* describes how close repeated measures of drug concentration in a single specimen agree with each other. Precision is defined by the relative standard deviation obtained by replicate analysis of specimens containing drug concentrations at the low, middle, and high parts of the range of concentrations expected in the study. Precision may be determined for either intraday or interday analysis with the precision of the former usually better than the latter. Interday precision may be especially important in geriatric drug studies that are designed as longitudinal trials to be conducted over months or years. *Sensitivity* is the lowest concentration of drug that can be assayed with acceptable accuracy and precision. Often, sensitivity is arbitrarily defined as that concentration of drug which produces a signal-to-noise ratio of 3:1 or 2:1. In drug studies the assay sensitivity should be such that serum concentrations may be measured for at least four elimination half-lives of the drug or that serum concentrations equal to one-tenth the peak concentration may be measured after a drug dose. *Specificity* in an assay is the ability to detect the drug of interest without interference from coadministered drugs, metabolites, and circulating endogenous substances. Specificity in drug assays for geriatric studies is particularly important because of the large number of different medications taken by geriatric patients and the unusual accumulation of metabolites and/or endogenous substances that may occur in elderly with age-related decreases in kidney function. *Reproducibility* of an assay is the ability to duplicate the results of an assay procedure when the assay is performed by different technicians in the same laboratory or in different laboratories. Since many geriatric clinical trials are multicenter, it would be critical to determine the reproducibility of a drug assay when performed in the laboratories at the respective study centers.

5. SPECIMEN COLLECTION AND STORAGE

Depending on the nature of the clinical study, there may be a need to measure drug concentrations in either blood, plasma, serum, urine, saliva, synovial fluid, and/or cerebrospinal fluid (CSF). Initially attention must be given to the proper collection, handling, and storage of these various specimens in order to assure the stability and integrity of the drug species of interest. In the early stages of assay development the device to be used for specimen collection should be evaluated as a possible source of assay interference. In some studies, indwelling catheters or heparin locks may be used to obtain multiple blood samples over time. The major concern with these devices is inadvertent dilution of the blood sample by the heparin solution used to keep the devices patent. In other studies, microsampling devices, rather than venipuncture devices, may be used to collect blood specimens and in some cases they have given lower assay values.[4] The most common collection devices are syringes or tubes made of glass or plastic. Drugs, such as antidepressants, that are weak bases have a tendency to adsorb to glass surfaces[5] while plastic materials may leach plasticizers, antioxidants, or other materials into the specimen. Several brands of collection tubes have stoppers or serum separators that are made of polymeric materials which may adsorb drugs or leach components into the specimens. Of course, adsorption of drugs to surfaces may lead to an underestimation of the actual concentration present in specimens. Leaching of tube or syringe components may directly interfere with analyte quantitation. In some cases plasticizers may displace protein-bound drug in plasma causing a redistribution of unbound drug into red blood cells and a reduction in the apparent concentration of total drug in plasma.[6,7] To minimize adsorption and leaching, the contact time between the specimen and the collection device should be kept as short as possible. In some cases silanization of glass surfaces will reduce loss of drug due to adsorption.

When *whole blood* is used, consideration should be given to the hematocrit of the patient as a source of assay variability. Since drugs may reequilibrate between plasma water and red blood cells upon standing after the blood sample is obtained from the drug recipient, storage times prior to actual drug concentration measurements should be standardized to minimize variability in extraction efficiency and subsequently assay results. *Plasma* is used in many drug studies in the elderly. When this fluid is used for drug assays, concern must be given to the effects the anticoagulant might have on measurement of drug concentrations. Heparin and EDTA are the most commonly used anticoagulants in blood specimens intended for drug analysis. Heparin has been known to increase the concentration of free fatty acids in plasma *in vitro*.[8] These lipids may displace weak acid drugs from their plasma protein binding sites and alter the normal whole blood/plasma concentration ratio. EDTA is preferred as an anticoagulant sometimes because it is a chelating agent which may stabilize unwanted oxidation of a labile analyte.[9] *Serum* is often the preferred fluid for drug studies because it is collected in the absence of any anticoagulant. However, to obtain neat serum it is necessary to allow the whole blood sample to stand for about 45 min at room temperature before the serum is harvested. *Urine* concentrations of drugs are useful to monitor compliance, to determine fractional recoveries of drug doses, or to calculate pharmacokinetic parameter values. For extended storage at freezer temperatures, urine must contain a preservative (e.g., toluene) to prevent bacterial overgrowth, which may limit assay sensitivity. After thawing and before drug analysis is begun, urine aliquots should be centrifuged and/or filtered to remove any debris. *Saliva* collection is often a part of geriatric drug studies because saliva is thought to be an ultrafiltrate of plasma. Saliva drug concentrations may be useful to monitor therapy or to estimate the extent of plasma protein binding. However, considerable care must be given to saliva collection because of the potential interference in drug quantitation due to the adsorption of drug to food particles in the collected saliva or to the presence of part of the orally administered drug dose in the collected saliva. In some cases parotid saliva, rather than mixed saliva, produces a cleaner specimen for drug analysis. In some geriatric drug studies there is an interest in quantifying drug concentrations in a fluid closer to the site of action than is blood or plasma. Thus, *synovial fluid* or *CSF* may be collected. Since these body fluids are basically protein-free and equivalent to an ultrafiltrate of plasma, they are relatively trouble-free when it comes to drug assays. However, the pH of these fluids may increase upon storage, and this may affect the recovery of drugs. To obviate this problem the pH should be checked and adjusted, if necessary, to provide uniform recoveries and more precise assay results.

Samples that are not to be assayed immediately should be frozen at $-20°C$ without delay to assure the stability of the analyte.[10] Serum or plasma should be separated from whole blood before the sample is frozen, to avoid hemolysis, which may interfere with many methods of analysis. Consideration should be given to the container in which the biological sample is stored. Again, many drugs bind to the surface of untreated glass or plastic, which reduces the recovery of the drug in the assay procedure. In some cases extraneous materials may leach out of plastic containers into the biological sample during storage and interfere with the assay. Since some compounds are light-sensitive, care should be given to protecting the sample at all stages of its handling to avoid marked photodegradation. Since serum/plasma samples may lose carbon dioxide upon standing, the pH of these samples may rise; it may be critical to readjust the pH of these samples to a designated value before proceeding with extraction and measurement of the drug. Some drugs are prone to oxidation when stored in serum or plasma.[11] It may be necessary to add stabilizers, such as antioxidants, to the specimen before storage in order to prevent chemical degradation. Many of these handling and storage processes may be controlled, or monitored, by preparing control specimens and handling/storing them in a manner identical to that of the patient specimens. Extended storage of drug-containing specimens may lead to altered concentrations of the drug because of desiccation. Similarly repeated freezing and thawing of specimens may result in dilution of the drug concentration because of condensation of water vapor on the storage container.

6. METABOLITE MEASUREMENT

The measurement and quantitation of drug metabolites are special concerns in geriatric drug studies. Some drug metabolites may be therapeutically active. Because of age-related decreases in renal function, metabolites normally excreted unchanged or as glucuronides by the kidney may accumulate in the plasma, especially in multiple-dose studies. Knowledge of these changes in metabolite disposition may help us understand altered sensitivity to drugs in the elderly. In many geriatric studies involving drugs that are extensively metabolized by the liver, one often finds an age-related reduction in the plasma clearance of the drug due to decreases in hepatic function. It is important in these situations to determine which metabolizing pathway is responsible for the overall reduction in drug clearance. Information such as this is necessary if the goal of the study is to ascertain the effects of aging on drug disposition. Metabolite quantitation also offers greater insight into other clinical problems in the elderly. For example, knowledge of metabolite excretion profiles would help determine if the influence of renal and/or hepatic diseases on drug disposition and effects is any different in the elderly versus a younger adult population. The significance and mechanisms of drug–drug interactions in the elderly may also be studied in detail through the measurement of metabolite levels.

Many assay procedures for parent compounds may be altered to quantitate the major metabolites. This is especially true if the analytical system is HPLC or GLC. Metabolites are generally more polar than parent compounds, and by varying the conditions of assay, one or more metabolites may be resolved from the parent drug. One limitation of metabolite quantitation is the availability of pure analytical standards of the metabolites. In some cases small amounts of metabolite standard may be obtained from a pharmaceutical company after submitting a "statement of investigator" form. In other cases it may be necessary to synthesize the metabolite and purify it for use as a standard. Some metabolites, such as glucuronides, are excreted in the urine as conjugates. In these instances it may be necessary to pretreat the sample with heat, glucuronidase, and a pH change in order to convert the conjugate to an unconjugated species prior to analysis. Many examples of assay approaches for metabolite quantitation may be found easily in the scientific literature.

Measurement of drug metabolites should be a priority in geriatric drug studies only under certain conditions. Metabolite quantitation becomes most important if the parent compound is not pharmacologically active but serves only as a prodrug, if the pharmacological activity or the serum concentrations of the metabolite are identical to or greater than the parent compound, and if the metabolism of the parent drug is subject to marked inter- and intrasubject variability. This is the case for drugs that undergo a significant first-pass effect.

7. Plasma Protein Binding

In the elderly, the pharmacokinetics and pharmacodynamics of unbound (free) drug are extremely important to determine. Free drug in circulating plasma is able to distribute into tissues and achieve an equilibrium with drug at the site of action. Therefore, plasma protein binding of drugs may play an important part in the disposition and effects of drugs in the elderly. In many studies the profiles of total (free plus bound) drug concentration may be similar in the elderly and in younger adults. Pharmacokinetic parameter values, such as those for plasma clearance, may be identical in the two populations suggesting the absence of any age-related changes in pharmacokinetics. However, it should be appreciated that the apparent lack of age effects may be due to offsetting changes in the elderly. Clearance, for example, may be decreased on one hand by reduced organ function in the elderly, but on the other hand, it may be increased because of reduced plasma protein binding. The net result is no apparent change in plasma clearance. By measurement of the unbound fraction of drug in plasma samples by ultrafiltration or by equilibrium dialysis, the pharmacokinetics of free drug may be delineated and the true effects of age on plasma clearance may be clarified.

The measurement of free drug in plasma samples following its separation is not much different in geriatric drug studies than in drug studies in younger adults. One has to determine the loss of unbound drug through adsorption to the apparatus used to separate bound from unbound drug in preliminary experiments. Since drugs may be greater than 90% bound in some cases, the concentrations of free drug may be considerably lower than total drug. Standards containing a range of drug concentrations should be carried through the entire procedure of separation and analysis of free drug in order to accurately determine the concentration of free drug in the unknown specimen. Analytically, one might also consider looking at the concentrations of free fatty acids and/or lipids in protein binding studies in the elderly since there are often age-related increases in these endogenous substances which may alter the unbound fraction of many drugs. If the intent of protein binding studies is to determine the binding dynamics of drugs in terms of the number of binding sites or the association constants, then quantitation of albumin, α-1-acid glycoprotein, and total protein in plasma should be seriously considered.

Priority should be given to determining the unbound concentration of drugs in geriatric drug studies when the drug is highly protein bound (>80%), when the protein binding is concentration-dependent, when protein binding is subject to significant intersubject variability, and when protein binding may be altered owing to diseases or to drug–drug interactions.

REFERENCES

1. *Report of the American Association of Clinical Pharmacology and Therapeutics: Drugs in the Elderly.* Norristown, PA, American Society of Clinical Pharmacology and Therapeutics, March 1984.
2. Keller H, Guder WG, Hansert E, Stamm D: Biological influence factors and inter-ference factors in clinical chemistry; general considerations. *J Clin Chem Clin Biochem* 1985; 23:3–6.
3. Smith RV, Stewart JT: *Textbook of Biopharmaceutic Analysis.* Philadephia, Lea & Febiger, 1981.
4. Perel JM, Stiller RL, Sallee FR, Lin FC, Narayanan S: Comparison between micro-sampling and venipuncture techniques for therapeutic drug monitoring of tricyclic antidepressants. *Clin Chem* 1985; 31:940.
5. Proelss HF, Lohmann JH, Miles DG: High performance liquid chromatographic si-multaneous determination of commonly used tricyclic antidepressants. *Clin Chem* 1978; 24:1948–1953.
6. Pile E, Shuterud B, Kierulf P, Fremstad D, Sayal SMA, Lunde PKM: Binding and displacement of basic, acidic and neutral drugs in normal and orosomucoid-deficient plasma. *Clin Pharmacokin* 1981; 6:367–374.
7. Devine JE: Drug-protein binding interferences caused by the plasticizer TBEP. *Clin Biochem* 1984; 17(6):345–347.
8. Kessler KM, Leech RC, Spann JF: Blood collection techniques, heparin and quinidine protein binding. *Clin Pharmacol Ther* 1979; 25:204–210.
9. Wong SH, Jain N, Jain P, Santiago C, Lin FC, Narayanan S: Effect of anticoagulants in blood collection system on analysis of tricyclic antidepressants by HPLC. *Clin Chem* 1982; 28:1644.
10. Chiou W: Pharmacokinetics of drugs in blood III: metabolism of procainamide and storage effect of blood samples. *J Pharm Sci* 1983; 72:572–574.
11. Kochak GM, Mason WD: Determination of free methyldopa in plasma by high pres-sure liquid chromatography and electrochemical detection. *J Pharm Sci* 1980; 69:897–900.

CHAPTER 22

Nursing Perspectives on Clinical Trials in Geriatrics

Barbara R. Heller, Maureen E. Power, and Georgeanne Cox Santolla

1. Introduction

The role of the nurse in clinical trials has long been overlooked. Traditionally seen as ancillary and supportive, the nurse is capable of a more significant contribution to research than previously expected. As an integral member of the research team, the nurse can do much to facilitate the research process and often plays a prominent part in coordination and implementation of a research protocol. The purpose of this chapter is to describe the nursing role in biomedical research with particular emphasis on clinical drug trials in geriatrics.

The nurse's involvement in clinical trials will vary depending on the specific protocol. Some of the common areas of nursing responsibility in clinical trials in geriatrics include: subject screening for eligibility; assisting with obtaining informed consent; preserving patient comfort and safety; patient and family counseling and teaching; monitoring and

BARBARA R. HELLER • Department of Nursing Education, Administration and Health Policy, University of Maryland School of Nursing, Baltimore, Maryland 21201.   MAUREEN E. POWER • Aging Research Nursing Service, Clinical Center, National Institutes of Health, Bethesda, Maryland 20205.   GEORGEANNE COX SANTOLLA • American Healthcare Institute, Silver Springs, Maryland 20910.

fostering protocol compliance; administration of investigational agents; controlling variables such as subject diet and activity levels; scheduling and performing study tests; collection and storage of laboratory data; documentation of expected and unexpected responses; management of untoward responses; and discharge planning and follow-up.

All these functions require that the nurse have a thorough understanding of the research protocol and work cooperatively and collaboratively with the principal investigator and other members of the research team. A team approach is most conducive to the accomplishment of research protocols and reduces the opportunities for error. Nurses are accustomed to working in a collaborative relationship with a variety of health care professionals. They are not only comfortable with this type of interaction but expect it, since they share responsibility for the well-being of patients under their care.

Settings in which clinical research is conducted offer additional, and sometimes unusual, challenges to professional nursing. As active participants in the research project, nurses may exercise clinical judgment yet must perform duties in a standardized manner, rigorously adhering to the protocol. This requires a dedicated staff who share a commitment to the goals of the trial. At the outset it is important to determine the level of commitment since nurses can unintentionally "hijack" a research study because of lack of understanding, ambivalence, or conflict over the value of the research. Involvement of nurses in the planning phases as well as an in-service education program for participating nurses may be the key to a successful research outcome.

2. Planning the Protocol

Creating a mechanism for nursing input during the design of the clinical trial can ensure that any special requirements for patient care are incorporated into the protocol. Participation of nurses in the planning process can and should occur at several stages. During initial stages, nurses can assist with the incorporation of specific data into the research protocol; for example, nurses can frequently identify exclusion criteria based on their experience with patients who have specific conditions or disabilities and their knowledge of the logistics of implementing protocols and controlling variables. Nurses can also provide insight into the specific nursing care that may be necessary to accomplish a given goal as well as type and extent of patient teaching that may be required.

Among the many other issues that need to be considered during subsequent stages of the planning process are: the division of responsibilities; development of step-by-step procedures in order to assure consistency; preparation of participating staff; scheduling of tests and procedures; and anticipating and planning for both usual and unexpected responses. Nurses, knowledgeable in the day-to-day functioning of the research unit and its staff, can provide useful information about these technical aspects of care, thus forestalling problems that may otherwise arise.

Nurses are also aware of constraints that may be imposed on the research by the organization in which it is being conducted. These organizational constraints may include: staffing patterns and availability; coordination of services required of other departments; availability of services during peak and off-hours; timely procurement of drugs or equipment; and the time required for the return of test results. Head nurses, in particular, must take into account the impact of the protocol on the workload and arrange staffing accordingly.

Identification and coordination of individuals and departments involved in the research effort and clarification of their specific roles and responsibilities prior to implementation may prevent potential conflict and omissions. Delineation of appropriate channels of communication during the planning of the protocol should avoid confusion and ensure that concerns and questions are dealt with as quickly as possible.

Regular planning meetings of the research team should be scheduled at predetermined intervals to review the logistical progress of the study and to modify procedures or arrangements as necessary. Hubbard and Devita[1] suggest that nursing staff be included in all rounds and conferences concerning the research project and individual subjects.

The nurse who has contributed to the development of the protocol, and is familiar with the technologies and equipment often used in drug trials, may also help to evaluate the feasibility of the proposed procedures and suggest more efficient or effective methods.

3. IN-SERVICE EDUCATION

Prior to implementation of a study, an in-service education program should be conducted by the principal investigator. The overall purpose and implications of the study should be discussed as well as the purpose of each test and procedure involved in the data collection process. Screening criteria and baseline studies should also be described.

Content of the in-service program should include a review of the pharmacological properties of the investigational agent, its known side effects, and adverse reactions. The nurse must be informed about the interactions and potential incompatibilities between agents and be able to recognize additive toxicity early. Procedures to be followed in the event of specific adverse reactions should be stipulated. A detailed description of the dosage, schedule, and route of administration contributes to consistency in administration. The nursing staff will also need to know whether a physician needs to be present during the infusion of an investigational drug.

Restrictions that need to be maintained, such as diet and activity limitations, as well as restrictions on the use of other medications, will be a major nursing responsibility and should be addressed completely during the in-service education programs. The purpose and duration of the restriction need to be identified and alternatives specified. Procedures to be followed in the event that a restriction is violated should also be indicated.

In many cases, nurses will be responsible for the collection of samples. The timing and sequence of these samples as well as the specifics of the collection technique should be outlined in the in-service education program including the use of preservatives and storage requirements, e.g., refrigeration of samples. Nurses also need to know the patient preparation required for all evaluation methods.

A thorough review of the consent form is also necessary for nurses participating in the consent procedure to enable them to respond appropriately to the questions of patients and their family members. While reviewing the form, nurses should be encouraged to examine it critically and anticipate questions that patients are likely to ask. Protocol flowsheets and all other forms used in the documentation of data, patient responses, and findings should also be reviewed at this time. The need for careful documentation of all facts pertaining to the conduct of research should be stressed.

Equipment and technical procedures should be explained in detail during the in-service program. Nurses should have opportunity for "hands-on" experience and practice with new equipment and procedures in order to master the technical skills necessary to implement the protocol safely and efficiently. Practice is particularly important with complex procedures such as rapid serial sampling, which requires that a large number of tasks be accomplished in sequence within a very short time frame.

When feasible, it is extremely helpful to have the in-service presentation videotaped. Nurses can then review the material when their schedule allows, thus solving the difficult problem of scheduling meetings. In addition to the verbal presentation, a written description of instructions is important to continuous referral. It is generally advisable to abstract pertinent information from the protocol to facilitate fast reference. A concise way to present this information is via standardized physician orders to protocol patients.

Protocol guidelines, which contain more detailed and explicit instructions such as common trouble spots, location of items, hints on problem solving, contact persons, and telephone numbers, can expedite the smooth implementation of the protocol. These documents should be available at the time of the in-service education session so nurses can review and comment on the information in order to minimize variations in interpretation.

Although these documents may emphasize "how to" perform various aspects of protocol implementation, the reasons for the protocol and method of implementation are equally important. Understanding the purpose and importance of each step in the research process is necessary when the nurse is required to make on-the-spot decisions during the conduct of the clinical trial.

4. Subject Screening

Nursing personnel can play a significant role in subject screening procedures and assessment of eligibility for inclusion in the research study. Nurses are routinely responsible for collecting baseline data such as vital signs, height, and weight, as well as historical information such as allergic responses, prior illnesses, and medication history. In addition, nurses can identify other essential information that can be of value in selecting patients and effectively managing their participation. For example, nurses can participate in gauging the competence of an individual subject to provide informed consent. The nurse can assess for level of understanding, presence of confusion, educational level, dependence on native language if other than English, ability to read and write English, and the existence and significance of auditory and visual problems common in the elderly.

5. Informed Consent

Although nurses usually do not assume primary responsibility for obtaining informed consent, they must ensure that patients understand what has been proposed, what the risks and alternatives are, and what the probability is that they will experience personal benefit from the investigational therapy. For the elderly subject in particular, this may require frequent repetition and clarification. It is essential for the nurse to routinely evaluate the patient's level of understanding of the proposed intervention and determine whether consent is truly informed and voluntary.

Informed consent is an ongoing process and a learning experience for the patient. The nurse who knows the patient well and understands the research protocol can spend time exploring the patient's and/or family's feelings about the proposed intervention serving as a conduit for information and communication between the patient and the principal investigator.

6. PRESERVATION OF PATIENT COMFORT AND SAFETY

An overriding concern when considering the ways and means of protocol implementation is to structure procedures so that patient inconvenience and discomfort are minimized as much as possible while attempting to maintain patient safety at the highest level.

The nurse must continuously monitor and interpret changes in pulse rate, temperature, blood pressure, respiratory response, skin manifestations, and pupillary response for signs of discomfort or pain. Nurses should be able to recognize signs of unexpected or unusual drug side effects and be capable of providing skilled emergency care if needed. Knowledge of specific age-related changes that can mask or mimic side effects is also important for nurses who are involved in geriatric drug trials.

7. ADMINISTRATION OF THE INVESTIGATIONAL AGENT

In all trials, the nurse is responsible for administration of a complex sequence of drugs in a safe, predictable, and consistent way. Prior to the initiation of the study, all equipment and supplies need to be on hand and evaluated to ascertain whether they are in optimum working condition. Nurses can assist in the evaluation of supplies and equipment to confirm that the specific needs of the research procedures are met: for example, quality control margins for monitoring equipment, intravenous solutions, and infusion pumps are set to meet clinical needs and may not be sufficiently accurate for the research procedure. Prefilled intravenous bottles may contain 10% more or less solution than the stated amount. This may not be significant for routine administration of intravenous infusions, but such variations can potentially affect research results. Manufacturers can provide information regarding the average percent deviation. The manufacturer, the pharmacy, or the hospital biomedical engineering department can perform the necessary modifications to enhance consistency.

Studies requiring the intravenous infusion of study agents face the difficulty of impaired venous access frequently found in older subjects. Nurses, through careful site selection, strict adherence to aseptic technique, and appropriate selection of needles, catheters, and methods of securing the line, can assure the patency of the intravenous infusion and minimize the risk of infection. Judicious use of restraints to maintain venous access routes should be considered if necessary. As with all elderly patients receiving parenteral therapies, nursing personnel should monitor for overhydration and underhydration.

A common occurrence in drug trials that may present a particular challenge to nursing staff is participation in double-blind studies. When routinely administering medications, the nurse always carefully identifies a medication prior to administering it to the patient. During a double-blind study, the nurse is unaware of the actual drug she is administering and may therefore feel an uneasiness about her ability to completely verify that the drug being administered is truly the drug the patient is to receive.

A major nursing responsibility in drug trials is monitoring for drug actions and side effects. Although actions and side effects may have been described in the in-service education program, nurses do not know which patient received which pharmacological agent in double-blind studies. This makes the task of monitoring drug effects much more difficult since signs and symptoms may be misinterpreted or may not be noticed as quickly.

Finally, when participating in double-blind studies, or any randomized trials, nurses may develop a bias toward or against an investigational agent. This can negatively impact on the results in two ways. The nurses' expectations may unconsciously prejudice her observations and documentation of patient response. Nurses' bias may also be unintentionally conveyed to subjects. This may have an even more immediate impact on patient outcomes. Avoiding these pitfalls is difficult, but an appreciation of the research process and goals of the specific protocol may lessen their occurrence as well as the occurrence of similar problems associated with any research protocol that requires administration of therapeutic agents on a randomized basis.

8. Scheduling and Collection of Data

Another nursing responsibility in clinical trials is ensuring that all study data are obtained and recorded accurately and that all tests are performed on schedule. Routines should be established that are consistent with the patient's life-style and preferences. Elderly patients in particular may have long-established routines that provide stability in their day-do-day activities. It is important to respect these routines to enhance patient comfort as well as compliance. As part of their initial and ongoing assessment, nurses can evaluate an individual patient's routine and contribute this information when tests are being scheduled.

Tests and procedures should be scheduled to allow for patient preparation and to permit the patient adequate rest between procedures. Elderly persons may become tired quickly and require longer rest time between tests. Overzealous scheduling can leave elderly subjects exhausted and discouraged. Nurses can contribute significantly to scheduling since they are in a position to gauge the elderly person's endurance and length of time needed for recovery between procedures. Nurses are also very familiar with the usual schedule on the research unit and how this may affect the scheduling of specific tests.

9. Documentation

Clinical observations made by the nurse may be helpful in aiding the principal investigator to interpret the clinical data. The importance of careful documentation of all facts and information pertaining to the conduct of the research cannot be overemphasized. Flowsheets that are clear and easily readable with sufficient room for additional comments concerning events that occur during the investigational period enhance the documentation process. Nurses must be familiar with any flowsheets used in the documentation of data, patient responses, and findings or valuable information may be lost.

10. Controlling Variables

Dietary and activity modifications, as well as restrictions on the use of other medications, must be explained to patients. Suggestions by the nursing staff of alternative food and drug choices as well as permitted exceptions may enhance participants' compliance. Dietary modifications or restrictions can become difficult to manage. Consideration needs to be given to the patient's usual nutritional status, eating habits, concurrent medical restrictions on diet, mechanical restrictions, and food preferences. Nurses can participate significantly in the management of these problems by assessing nutritional considerations and by encouraging patients to adhere to restrictions.

Activity restrictions may also pose a problem with elderly subjects. Elderly persons who participate in studies requiring intermittent or on-going bedrest are more prone to the hazards of immobility, such as atelectasis, constipation, and thrombophlebitis. Assessment and prevention of these complications is a vital nursing role during the implementation of the research study.

11. COMPLIANCE

Nurses can play a significant role in preparing the patient for carrying out their drug protocol as well as remaining within dietary and activity restrictions. Consideration of how to prepare the patients for discharge in order to enhance compliance has long been a nursing responsibility. Monitoring the patient's compliance after discharge is also within the scope of nursing.

Compliance with medication schedules is frequently a problem encountered with an elderly population. Nurses can perform pill counts, an effective and relatively easy method of determining compliance. Compliance can also be facilitated through the use of reminders, schedules, and dispensers. Identifying and involving a significant other such as a spouse or child in the administration of medications can often be helpful.

Compliance with dietary restrictions following discharge may be complicated by the limited selection at small neighborhood grocery stores frequented by elderly persons. For this reason it may also be necessary to involve others in shopping or cooking. Particular difficulties may arise if the patient participates in Meals-on-Wheels or community meals as part of his living arrangements or day care. Complicated or unusual dietary restrictions frequently require the services of a registered dietitian in order to ensure palatability and sound nutrition and to enhance compliance. A dietitian should be available to staff, patient, and family members in any study requiring dietary modification. Elderly patients often have concomitant problems such as dentures or other mechanical difficulties that make chewing difficult. These problems must be also taken into consideration.

12. DISCHARGE PLANNING AND FOLLOW-UP

The patient's living arrangements and social support system should be evaluated prior to discharge since strong supports may enhance compliance. Transportation to and from the research site for follow-up outpatient visits should be discussed prior to discharge. Special arrangements that often need to be made can be initiated by the participating nursing staff.

Nurses can assess not only the patient's medication-taking behavior with respect to prescribed medications, but also the use of other, self-prescribed, over-the-counter medications. The elderly tend to take many over-the-counter drugs, including laxatives, analgesics, sleeping medications, and vitamins. Because the use of such pharmacological agents can affect the results of the research study, patients should be instructed in the appropriate use of these agents or possible substitutions while participating in the study. Pharmacists are an invaluable resource when it is necessary to identify the active agents in over-the-counter drugs. The older patient's regular pharmacist may also be helpful in establishing exactly which prescription drugs the patient is taking since the elderly patient may see many physicians and no central medication record may exist other than at the local pharmacy. The pharmacist may have an awareness of his elderly patients' customary use of over-the-counter preparations as well.

13. PATIENT TEACHING AND COUNSELING

Patient teaching is vital to several areas of the research process, including informed consent, activity and diet restrictions, and patients' compliance. The nurse is often the focal person who keeps the patient and family informed about the progress of the drug trial.

Nurses are in an ideal position to initiate individualized patient teaching sessions using principles of adult learning. The content and instructional methods must be relevant to the needs of the older learner. Material should be presented at a pace that allows for assimilation with sufficient time allocated for the elderly patient to consider what has been said. Vision and hearing impairments also need to be considered when teaching elderly patients. Praise and repetition may assist in reinforcing learning.

Concern for the emotional well-being of the patient must also be considered and extended to the family who share the burden of uncertainty and fear of participating in an "experiment." Rapport established between the patient and nurse also may be a contributing factor in maintaining participation and compliance throughout the research study.

By encouraging patients and family members to become involved as much as possible and to ask any questions and air concerns, nurses can help to gain consent, trust, and confidence. A nurse may be seen as more approachable than other health care professionals and is often the person to whom feelings are expressed. Through simple listening and reassurance, the nurse may provide invaluable support for the patient and family. Nurses can also initiate referrals to other support services such as clergy and social workers if necessary.

14. CLINICAL NURSE SPECIALIST

Clinical trials can be a complicated undertaking that require the coordination of many disciplines and departments working in the institutional setting. A logical choice for coordinator of the many research components that must be organized and monitored is the masters-prepared clinical nurse specialist. Of the various levels of nursing that can be involved in a clinical trial, from staff nurse to head nurse to nursing administration, the clinical nurse specialist is the most adequately prepared for this coordinator role. Clinical nurse specialists' education prepares them to be expert practitioners, educators, consultants, change agents, and researchers. As an expert practitioner in gerontological nursing, a gerontological clinical nurse specialist may help both the primary investigator and the staff anticipate clinical problems that may arise as a result of the study. As an educator and consultant, the clinical nurse specialist can interpret the protocol to the nursing staff and assist in the planning of staff and scheduling needs. As change agent, the clinical nurse specialist can ease the disruption that may accompany the introduction of a new protocol into the daily nursing routine. Finally, through the knowledge and appreciation of research gained through masters-level preparation and personal practice, the clinical nurse specialist holds certain values and beliefs about the ultimate goal of clinical research that can prove valuable in facilitating the research process.[2]

15. THE NURSING ADMINISTRATOR

The nursing administrator can also facilitate clinical research and encourage willing and enlightened nursing participation in research initiative by serving as a role model and by motivating staff. The nursing administrator can influence research by developing institutional policies that eliminate organizational constraints that hinder research efforts. Staff development programs instituted by nursing administrators can promote the development of a positive research attitude among staff nurses by increasing knowledge about research goals and procedures and thereby reducing anxiety that may result from uncertainty about the appropriate nursing role in these activities. In addition, the appointment of nurses to institutional review boards can provide opportunities for nursing input regarding the care of research patients and would help to anticipate potential problems associated with the fulfillment of research objectives. Finally, nurses must be given sufficient incentives for participation in research activities by developing procedures that recognize and reward their important contribution to the acquisition of new knowledge through clinical research.

16. CONCLUSION

Increasingly, nurses are working with complex protocols and are administering a larger number of investigational drugs than ever before as initiatives in geriatric research and clinical drug trials develop. As an integral member of the research team, the nurse can make a significant contribution to the planning and implementation of a research protocol.

The role of the nurse in clinical trials offers unique opportunities as well as responsibilities. Whether or not nurses enhance certain aspects of a trial such as recruitment, compliance to prescribed regime, or overall quality and performance at the clinical unit level needs further investigation. The development of improved care to geriatric patients through clinical research is a common goal shared by all health professionals.

REFERENCES

1. Hubbard S, Devita V: Chemotherapy research nurse. *Am J Nursing* 1976; 76:561.
2. Hodgman E: The CNS as researcher, in Hamric A, Spross J: *The Clinical Nurse Specialist in Theory and Practice*. New York, Grune & Stratton, 1983, p 73.

INDEX